Tao and Kokotović / ADAPTIVE CONTROL OF SYSTEMS WITH ACTUATOR AND SENSOR NONLINEARITIES

Tsoukalas and Uhrig / FUZZY AND NEURAL APPROACHES IN ENGINEERING

Van Hulle / FAITHFUL REPRESENTATIONS AND TOPOGRAPHIC MAPS: From Distortion- to Information-Based Self-Organization

Vapnik / STATISTICAL LEARNING THEORY

Werbos / THE ROOTS OF BACKPROPAGATION: From Ordered Derivatives to Neural Networks and Political Forecasting

Yee and Haykin / REGULARIZED RADIAL BIAS FUNCTION NETWORKS: Theory and Applications

KNOWLEDGE-BASED RADAR DETECTION, TRACKING, AND CLASSIFICATION

KNOWLEDGE-BASED RADAR DETECTION, TRACKING, AND CLASSIFICATION

EDITED BY

Fulvio Gini and Muralidhar Rangaswamy

WILEY-INTERSCIENCE

A JOHN WILEY & SONS, INC., PUBLICATION

Published by John Wiley & Sons, Inc.
Published simultaneously in Canada

Library of Congress Cataloging-in-Publication Data:

Gini, Fulvio.
Knowledge based radar detection, tracking, and classification / Fulvio Gini, Muralidhar Rangaswamy
p. cm.
"A Wiley-Interscience publication."
ISBN 978-0-470-14930-0
1. Tracking radar. 2. Expert systems (Computer science). 3. Automatic tracking.
4. Target acquisition. 5. Adaptive signal processing. I. Rangaswamy, Muralidhar. II. Title.
TK6592.A9G56 2008
621.389'28—dc22

2007043610

Printed in the United States of America

10 9 8 7 6 5 4 3 2 1

CONTENTS

CONTRIBUTORS

Raviraj Adve, Department of Electrical and Computer Engineering, 10 King's College Road, University of Toronto, Toronto, ON, M5S 3G4 Canada, e-mail: rsadve@comm.utoronto.ca

Chris J. Baker, UCL, Department of Electronic and Electrical Engineering, University College, London, Room 1003, Roberts Building, Torrington Place, London WC1E 7JE, UK, e-mail: c.baker@ee.ucl.ac.uk

Edward J. Baranoski, Defense Advanced Research Projects Agency (DARPA), Strategic Technology Office, 3701 N Fairfax Drive, Arlington, VA 22203, USA, e-mail: Ed.Baranoski@darpa.mil

Alessio Benevoli, University of Florence, Dipartimento di Sistemi e Informatica, Via Santa Marta 3, 50139 Firenze, Italy, e-mail: benavoli@dsi.unifi.it

Igal Bilik, Department of Electrical and Computer Engineering, Hudson Hall Room 129, PO Box 90291, Duke University, Durham, NC 27708-0291, USA, e-mail: ib16@ee.duke.edu

Shannon D. Blunt, Department of Electrical Engineering and Computer Science, University of Kansas, 1520 W. 15th St. (Eaton Hall), Lawrence, KS 66045, USA, e-mail: sdblunt@eecs.ku.edu

Christopher T. Capraro, Capraro Technologies, Inc., 311 Turner Street — Suite 410, Utica, New York 13501, USA, e-mail: ccapraro@caprarotechnologies.com

Gerard T. Capraro, Capraro Technologies, Inc., 311 Turner Street — Suite 410, Utica, New York 13501, USA, e-mail: gcapraro@caprarotechnologies.com

Luigi Chisci, University of Florence, Dipartimento di Sistemi e Informatica, Via Santa Marta 3, 50139 Firenze, Italy, e-mail: chisci@dsi.unifi.it

Antonio De Maio, Università degli Studi di Napoli "Federico II", Dipartimento di Ingegneria Elettronica e Telecomunicazioni, Via Claudio 21, I-80125 Napoli, Italy, e-mail: ademaio@unina.it

Alfonso Farina, SELEX Sistemi Integrati, Via Tiburtina Km. 12.400, 00131 Rome, Italy, e-mail: afarina@selex-si.com

Karl Gerlach, Radar Division (Code 5341), US Naval Research Laboratory, 4555 Overlook Ave. SW, Washington, DC 22375, USA, e-mail: karl.gerlach@nrl.navy.mil

Fulvio Gini, University of Pisa, Dipartimento di Ingegneria dell'Informazione, via G.Caruso 16, 56122 Pisa, Italy, e-mail: f.gini@ing.unipi.it

Hugh D. Griffiths, Cranfield University, Defence College of Management and Technology, Shrivenham, Swindon SN6 8LA, UK, e-mail: hgriffiths.cu@defenceacademy.mod.uk

Joseph R. Guerci, 2509 N Utah St., Arlington, VA 22207, USA, e-mail: JRGuerci@IEEE.org

Todd B. Hale, School of Engineering and Management, Air Force Institute of Technology, 2950 Hobson Way, Wright-Patterson AFB, OH 45433-7765, USA, e-mail: Todd.Hale@verizon.net

Simon Haykin, McMaster University, Adaptive Systems Laboratory, CRL-103, 1280 Main Street West, Hamilton, ON L8S 4K1, Canada, e-mail: haykin@mcmaster.ca

Sandro Immediata, SELEX Sistemi Integrati, Via Tiburtina Km. 12.400, 00131 Rome, Italy, e-mail: simmediata@selex-si.com

Sergio Luis de Carvalho Miranda, Rua Visconde do Rio Branco, 767/603, São Domingos, Niterói, RJ, Brazil, e-mail: slc_miranda@yahoo.com.br

Muralidhar Rangaswamy, Air Force Research Laboratory Sensors Directorate, AFRL/SNHE 80 Scott Drive, Hanscom AFB, MA 01731-2909, USA, e-mail: Muralidhar.Rangaswamy@hanscom.af.mil

Aaron K. Shackelford, Radar Division (Code 5341), US Naval Research Laboratory, 4555 Overlook Ave. SW, Washington, DC 22375, USA, e-mail: aaron.shackelford@nrl.navy.mil

Joseph Tabrikian, Department of Electrical and Computer Engineering, Ben-Gurion University of the Negev, PO Box 653, Beer-Sheva 84105, Israel, e-mail: joseph@ee.bgu.ac.il

Luca Timmoneri, SELEX Sistemi Integrati, Via Tiburtina Km. 12.400, 00131 Rome, Italy, e-mail: ltimmoneri@selex-si.com

Michael C. Wicks, US Air Force Research Laboratory/Sensors Directorate, 26 Electronic Parkway, Rome, NY 13441-4514, USA, e-mail: Michael.Wicks@rl.af.mil

Karl Woodbridge, Department of Electronic and Electrical Engineering, University College London, Torrington Place, London WC1E 7JE, USA, e-mail: k.woodbridge@ee.ucl.ac.uk

1

INTRODUCTION[1]

Fulvio Gini and Muralidhar Rangaswamy

The use of surveillance for a variety of applications in the dynamically changing civilian and military environments has led to a great demand for innovative sensors and sensing configurations based on cutting-edge technologies, such as knowledge-based (KB) signal and data processing, waveform diversity, wireless networking, robotics, advanced computer architectures, and supporting software languages [1]. Improved sensor signal and data processing performance will be gained from KB and a priori information, multiple processing paradigms, and sensor fusion. A knowledge-based system (KBS) uses a priori information to improve the performance of deterministic and adaptive systems. Although the exact form of this prior knowledge is problem-dependent, a KBS consists of a knowledge base containing information specific to a problem domain and an inference engine that employs reasoning to yield decisions.

With maturing electronics and radar hardware, advanced radar systems will use KB techniques to perform signal and data processing cooperatively within and between platforms of sensors and communication systems while exercising waveform diversity, as well as reconnaissance, surveillance, imaging and communications within the same sensor system. In addition, these sensors will cooperate with other users and sensors, sharing information and data. Sensor system performance can be enhanced by changing a sensor's algorithms as the environment changes. This is the fundamental concept underlying KB or cognitive radar, known to the radar community since the pioneering papers of Vannicola and colleagues [2, 3], Haykin [4], and Baldygo et al. [5]. The operational radar environment is subject to rapid spatio-temporal variation. Hence, the key to efficient adaptation is real-time

Knowledge-Based Radar Detection, Tracking, and Classification. Edited by Fulvio Gini and Muralidhar Rangaswamy

exploitation of a priori knowledge pertaining to the operational environment. For example, if an airborne radar system is aware of certain features of the Earth and its surroundings, then it can significantly improve performance by exploiting degrees of freedom such as the transmit waveform, polarization, frequency, phase, power, modulation, and coding. The adaptive and optimal use of all available degrees of freedom is broadly termed "waveform diversity." Waveform diversity is the technology that will allow one or more sensors onboard a platform to automatically change operating parameters [e.g. frequency, gain pattern, pulse repetition frequency (PRF)] to meet the varying environments. Also, the system of sensors should operate with multiple goals managed by an intelligent platform network that can control the dynamics of each sensor to meet the common goals of the platform, rather than each sensor operate as an independent system. Intelligent software processing is required at all stages of signal, data, and system processing from the filtering, detection, tracking, imaging, and identification stages to the communications, command, and control (C3) stages. Examples of a priori knowledge are archival radar data, Geographic Information Systems (GISs), Digital Terrain Elevation Data (DTED), Land Cover Land Use (LCLU) data, information on the radar kinematical parameters, off-board sensor data, roadway maps, and background of air/surface traffic. Recent advances in environmental measurements, DTED, future information quality and accessibility, digital processing, mass and random-access memory technologies, have opened up many possibilities, unrealizable in the past, for radar systems to improve their on-line performance. New real-time processing techniques are required for [e.g. for the constant false alarm rate (CFAR) behavior of the radar system [6]] to take advantages of these advances to bring radar performance back to optimum under difficult operation conditions such as littorals that include mixed sea and variable terrain.

The great interest in the application of KB techniques to adaptive radar signal and data processing is evident from the following examples:

1. The Defence Advanced Research Projects Agency (DARPA) has been pioneering the development of the first ever real-time knowledge-aided adaptive radar architecture. In particular, the Knowledge Aided Sensor Signal Processing and Expert Reasoning (KASSPER) program has as its aim the development and application of a revolutionary new approach to demanding multidimensional adaptive sensor systems, with a near-term focus on military applications of Ground Moving Target Indicator (GMTI) radar and Synthetic Aperture Radar (SAR). Annual KASSPER workshops started in 2002 to allow the exchange of ideas across the spectrum of R&D activities, including knowledge-based space–time adaptive processing (KB-STAP), environmental knowledge-base generation and maintenance, and real-time KB embedded computing [7].
2. The US Air Force Research Laboratory's Sensors Directorate has been pursuing some of the most progressive work in employing KB techniques in the radar signal processing chain, specifically in the CFAR portion of the chain [5, 8].

3. The US Air Force (USAF) has an ongoing project called Autonomous Intelligent Radar System (AIRS) that is performing research in applying KB techniques to radar signal processing. The AIRS architecture design leverages advanced technologies developed by the World Wide Web Consortium (W3C) and the DARPA Agent Markup Language (DAML) program to define the next-generation Internet, also called the Semantic Web [9].
4. A series of lectures has been devoted to Knowledge-Based Radar Signal and Data Processing [10]. They were sponsored by the NATO Research and Technology Organization (RTO) with the following scope: promoting cooperative research and information exchange to support the development and effective use of national defense research and technology to meet the military needs of the alliance; maintaining a technological lead; and providing advice to NATO decision makers. This Lecture Series was held in Sweden, Hungary, and Italy in 2003; Poland and Spain in 2004; and in the Czech Republic, Belgium, and the UK in 2006.
5. A special section of the *IEEE Signal Processing Magazine* devoted to "Knowledge-Based Systems for Adaptive Radar: Detection, Tracking, and Classification," published in January 2006, edited by Fulvio Gini [11].
6. A special section of *IEEE Transactions on Aerospace and Electronic Systems* devoted to "Knowledge-Aided Sensor Signal and Data Processing," published in July 2006, co-edited by William Melvin and Joseph Guerci [12].

The aim of this book is to highlight recent advances in both knowledge-based systems and radar signal and data processing, in a common forum, in order to present a range of perspectives and innovative results with potential to enable practical adaptive radar systems design. The chapters of this book describe the current developments in the area and present examples of improved radar performance for augmented and upgraded systems, and project the impact of KB technology on future systems.

1.1 ORGANIZATION OF THE BOOK

The book is organized into ten chapters. This first chapter is the introduction to the concept of KB radar. The remaining nine chapters focus on the application of KB concepts to a specific radar function, that is, detection, tracking, or classification. Each of them is essentially self-contained, starting with introductory remarks, following with a discussion, and ending with a list of references. Their contribution is briefly summarized in the following.

Chapter 2, entitled "Cognitive Radar" (by Haykin), discusses the idea of cognitive radar. The radar environment is usually nonstationary, and adaptivity is the method implemented in modern radar systems for dealing with nonstationarity. In current designs of radar systems, adaptivity is usually confined to the receiver. In this chapter it is argued that for the radar to be cognitive, adaptivity has to be extended to the transmitter too. Three important conclusions are drawn:

1. Intelligence is a necessary requirement for the radar to be cognitive;
2. Feedback from the receiver to the transmitter is the facilitator of intelligent signal processing; and
3. The preservation of information in radar returns is of crucial importance to improved receiver performance.

Two potential applications of cognitive radars are finally presented, one dealing with multifunction radars and the other dealing with a network of noncoherent marine radars.

Chapter 3, entitled "Knowledge-Based Radar Signal and Data Processing: A Tutorial Overview" (by Capraro, Farina, Griffiths, and Wicks), describes the role of KB processing in exploiting available information such as positioning, waveform selection, and modes of operation to enhance radar performance. This chapter provides a brief overview of artificial intelligence (AI) and a rationale for knowledge bases and robotics, which are the two main areas of emphasis for bringing KB into fielded radar systems. Also, the role of Semantic Web technologies in KB radar systems is discussed. An end-to-end radar signal and data processing architecture for airborne surveillance radar and its over-arching KB processing and control are described in detail. The chapter ends with the authors' view of the future of KB radar research, including waveform diversity and intelligent sensor systems.

Chapter 4, entitled "An Overview of Knowledge-Aided Adaptive Radar at DARPA and Beyond" (by Guerci and Baranoski), provides a breezy tour of the KASSPER program, highlighting both the benefits of knowledge-aided (KA) adaptive radar, key algorithmic concepts, and a new "look-ahead" radar scheduling approach that is the cornerstone of High Performance Embedded Computing (HPEC) architectures. Methods in which prior knowledge can be incorporated into the space–time adaptive beamformer, which is the most demanding component of modern GMTI radar, are described in some detail. Finally, the chapter introduces the notion of extending KA processing to the adaptive MIMO (Multi-Input Multi-Output) radar problem. The methods described here are potentially applicable in many other adaptive sensor signal processing systems such as hyperspectral imaging, lidar, sonar, and other multidimensional sensor arrays where environmental disturbance is a dominant source of interference.

Chapter 5, entitled "Space–Time Adaptive Processing for Airborne Radar: A Knowledge-Based Perspective" (by Wicks, Rangaswamy, Adve, and Hale), provides an overview of radar STAP from its inception to state-of-the-art developments. The topic is treated with regard to both intuitive and theoretical aspects. A key requirement of space–time adaptive processing is knowledge of the spectral characteristics underlying the interference scenario of interest. However, these are seldom known in practice and must be estimated using training data. Two central problems arise in the application of STAP:

1. The homogeneity of the sample support needed to train the adaptive filter; and
2. The computational load of the algorithm. No algorithm is the best one and the only practical approach suggested in this article is to use a KB scheme that best matches the signal processing to the interference scenario at hand. The article illustrates the immense potential of KB approaches in solving these problems.

Chapter 6, entitled "CFAR Knowledge-Aided Detection and its Demonstration Using Measured Airborne Data" (by C. Capraro, G. Capraro, De Maio, Farina, and Wicks), addresses the design and analysis of a KA detector for airborne radar applications. The two building blocks of the proposed processor are the training data selector and the detector. The training data selector has the goal to choose the secondary cells that best represent the clutter statistics in the cell under test. It is a hybrid algorithm, which pre-screens training data through the use of terrain information from the United States Geological Survey (USGS). The second stage of processing is a data-driven selector, which attempts to eliminate residual training data heterogeneities. The performance of the proposed KA detector is analyzed using measured airborne radar data, obtained from the Multi-Channel Airborne Radar Measurements (MCARM) program, and is compared with alternative detectors proposed in the open literature.

Chapter 7, entitled "STAP via Knowledge-Aided Covariance Estimation and the FRACTA Meta-Algorithm" (by Blunt, Gerlach, Rangaswamy, and Shackelford), describes the development of a KB approach to airborne/space-based radar for GMTI in the presence of severely heterogeneous training data. In particular it addresses the benefit provided by model-based prior knowledge when used to supplement the FRACTA meta-algorithm, a multistage/multimetric approach that is robust to training data heterogeneity. The FRACTA meta-algorithm utilizes three stages of detection, which, individually, systematically identify potential targets while eliminating data contamination (censoring), detect targets within the clutter-suppressed environment (cell-averaging CFAR), and eliminate false alarms that may arise due to undernulled clutter and/or space–time filter sidelobes (Adaptive Coherence Estimator (ACE) detector). In the chapter it is demonstrated how approximate prior knowledge in the form knowledge-aided covariance estimation (KACE) further improves the robustness of the detector by supplementing interference covariance estimation in scenarios with insufficient sample support that would otherwise lead to "sample starvation" problems.

Chapter 8, entitled "Knowledge-Based Radar Tracking" (Benavoli, Chisci, Farina, Immediata, and Timmoneri), describes how to efficiently exploit a priori knowledge in the tracking of multiple radar targets. In many scenarios, heterogeneity of the surveillance region makes conventional tracking systems (not using the KB) very sensitive to false alarms and/or missed detections. In this chapter it is demonstrated that an effective use of a priori knowledge at various levels of the tracking algorithms significantly reduces the number of false alarms, missed detections, false tracks, and improves true target track life The main ingredients of the tracker are (1) Extended Kalman filtering to take into account nonlinearities; (2) Interacting Multiple Model for managing the target maneuvers; (3) Nearest Neighbour Cheap Joint Probabilistic Data Association for robust plot–track association; (4) *M* out of *N* logic for track initiation; (5) use of the Knowledge Base (geographical maps and targets characteristics) and of Amplitude Information; (6) use of fuzzy logic for classification of the surveillance region. The proposed algorithm is tested against simulated and live data pertaining from a SELEX-SI naval surveillance radar. The results demonstrate that the KB approach provides meaningful advantages, allowing for the reduction of false and tentative tracks while permitting the continuous track of useful targets.

Chapter 9, entitled "Knowledge-Based Radar Target Classification" (by Bilik and Tabrikian), addresses the problem of automatic target recognition by means of ground surveillance Doppler, in particular, the classification between a walking person, a pair of walking persons, and a slowly moving vehicle. The maximum likelihood (ML) and the "majority voting" decision rules were applied to the proposed classification problem. Two sources of knowledge were considered for target classification: statistical and physical. Statistical knowledge is obtained from a training database of recorded target echos, and physical knowledge is available by developing locomotion models for the different targets. The statistical classifier was applied to a seven-class problem of radar targets such as walking person, group of walking persons, tracked vehicle, wheeled vehicle, animals, and clutter. The human operator's performance has also been evaluated. In many cases, a training database may not be available, and in some cases, it may be insufficient to represent the different classes. On the other hand, the inaccuracy in the locomotion models results in limited classification performance. In the chapter it is shown that the best performance is achieved via a combined approach, which incorporates both the statistical and physical knowledge sources. The performances of the physical, statistical, and combined knowledge-based algorithms are tested using real data records from three classes: one person, two persons, and vehicle.

The final chapter, entitled "Knowledge-Based Resource Management for Multifunction Radar" (Miranda, Baker, Woodbridge, and Griffiths), focuses on the multifunction radar (MFR) resource management problem, that is, the allocation of finite resources in an optimal and intelligent way. The dynamic and interactive interplay between the setting of radar parameters to optimize the tasks to be carried out and perception of environment motivates the centrality of knowledge-based data processing in determining MFR performance. The chapter focuses on two related aspects of radar resource management: scheduling and task prioritization. Two different methods of scheduling are examined and compared, and their differences and similarities highlighted. The analysis indicates that prioritization is a key component to determining overall performance. A fuzzy logic approach for prioritizing radar tasks in changing environment conditions is described. By assessing the priorities of targets and sectors of surveillance according to a set of rules, an attempt is made to imitate the human decision-making process such that the resource manager can distribute the radar resources in a more effective way. Results suggest that the fuzzy approach is a valid means of evaluating the relative importance of the radar tasks; the resulting priorities are adapted by the fuzzy logic prioritization method, according to how the radar system perceives the surrounding environment.

We hope that this book will stimulate the interest of the scientific community in this new and exciting field of research, which offers a rich set of challenges and problems spanning a broad spectrum of basic and applied research.

ACKNOWLEDGMENTS

We thank all the authors for their hard work and outstanding contributions that made this book possible. Finally, we are extremely grateful to Prof. Simon Haykin, Editor-in-Chief of the Wiley Science Series on Remote Sensing, for encouraging us to publish this book as an outgrowth of the special section on knowledge-based systems for adaptive radar featured in the January 2006 issue of the *IEEE Signal Processing Magazine*. We are also thankful to Ms. Mercy Kowalczyk of the IEEE Signal Processing Society for her promptness in processing the requests from authors of articles in the *Signal Processing Magazine* for permission to re-use their work in this book.

REFERENCES

[1] M.C. Wicks. Radar the next generation — sensors as robots. *Proc. 2003 International Radar Conference* (3–5 September 2003). 2003;8–14.

[2] V.C. Vannicola, J.A. Mineo. Applications of knowledge-based systems to surveillance. *Proc. 1988 IEEE National Radar Conference* 1988;157–164.

[3] V.C. Vannicola, L.K. Slaski, G.J. Genello. Knowledge-based resource allocation for multifunction radars. *Proc. 1993 SPIE Conference on Signal and Data Processing of Small Targets* 1993;1954:410–425.

[4] S. Haykin. Radar vision. *Proc. Second International Specialist Seminar on the Design and Application of Parallel Digital Processors* 1991;75–78.

[5] W. Baldygo, M. Wicks, R. Brown, P. Antonik, G. Capraro, L. Hennington. Artificial intelligence applications to constant false alarm rate (CFAR) processing. *Proc. IEEE 1993 National Radar Conference*, Boston, MA, April 1993.

[6] E. Conte, A. De Maio, A. Farina, G. Foglia. Design and analysis of a knowledge-aided radar detector for doppler processing. *IEEE Transactions on Aerospace and Electronic Systems* (special section on "Knowledge-aided sensor signal and data processing") 2006;42(3):1058–1079.

[7] J.R. Guerci. DARPA KASSPER overview. *Proc. 2004 DARPA Workshop on Knowledge-Aided Sensor Signal Processing and Expert Reasoning (KASSPER)*, Clearwater, FL, 5–7 April 2004.

[8] G.T. Capraro, G.B. Berdan, R.A. Liuzzi, M.C. Wicks. Artificial intelligence and sensor fusion. *Proc. 2003 International Conference on Integration of Knowledge Intensive Multi-Agent Systems* (30 September–4 October 2003). 2003;591–595.

[9] G. Capraro, W. Baldygo, R. Day, J. Perretta, M. Wicks. Autonomous intelligent radar system (AIRS) for multi-sensor radars. *Proc. First International Workshop on Computational Advances in Multi-Sensor Adaptive Processing* 13–15 December 2005 Puerto Vallarta, Jalisco, Mexico. Invited to the special session on "Computational Methods for Multi-Sensor Radar."

[10] G. Capraro, A. Farina, H. Griffiths, M. Wicks. Knowledge-based radar signal and data processing. *NATO RTO Lecture Series 233*, November 2003.

[11] Special Section on "Knowledge-Based Systems for Adaptive Radar." *IEEE Signal Processing Magazine* 2006;23(1).

[12] Special Section on "Knowledge-Aided Sensor Signal and Data Processing." *IEEE Transactions on Aerospace and Electronic Systems* 2006;42(3).

2

COGNITIVE RADAR

Simon Haykin

In this chapter, we discuss "cognitive radar," the idea of which was first published in reference 1. Cognitive radar builds on three basic ingredients:

1. *Intelligent signal processing*, which itself builds on learning through interactions of the radar with the surrounding environment;
2. *Feedback from the receiver to the transmitter*, which is a facilitator of intelligence;
3. *Preservation of the information content of radar returns*, which is realized by the Bayesian approach to radar signal processing.

All three ingredients feature in the echo-location system of a bat, which may therefore be viewed as a physical realization (albeit in neurobiological terms) of what we mean by cognitive radar.

The chapter concludes with two potential applications of cognitive radar, one dealing with multifunction radars, and the other dealing with a network of noncoherent radars for homeland security.

2.1 INTRODUCTION

Radar is a remote-sensing system that is widely used for surveillance, tracking, and imaging applications, for both civilian and military needs. In this chapter, we focus attention on future possibilities of radar with particular emphasis on the notion of *cognition*. As an illustrative case study along the way, we consider the radar surveillance problem.

Knowledge-Based Radar Detection, Tracking, and Classification. Edited by Fulvio Gini and Muralidhar Rangaswamy

According to the *Oxford English Dictionary*, cognition is "knowing, perceiving, or conceiving as an act...". Given three distinct capabilities,

1. the inherent ability of radar to sense its environment on a continuous basis and thereby getting to *perceive* it,
2. the ability of phased-array antennas to electronically scan the environment in a fast manner, and
3. the ever-increasing power of computers to digitally process signals,

it is our conviction that it is not only feasible but also highly beneficial to build a cognitive radar system using today's technology. Indeed, if ever there was a remote-sensing system well suited for cognition, radar is it.

From the moment a surveillance radar system is switched on, the system becomes electromagnetically linked to its surrounding environment, in the sense that the environment has a strong and continuous influence on the radar returns (i.e. echoes). In so doing, the radar builds up its knowledge of the environment from one scan to the next, and makes decisions of interest on possible targets at unknown locations in the environment. The locations are not known before the radar is switched on, but they become determined by the radar receiver once the targets under surveillance are declared.

From signal-processing and control theory, we know that it is not necessary for the radar to keep the entire record of past data. Rather, by adopting a *state-space model* of the environment, and recursively updating the state vector representing an estimate of certain parameters pertaining to the environment, the need for storing the entire history of radar data on the environment is eliminated. The challenge is how to formulate the state-space model of the environment.

The requirement to update estimation of the environmental state is necessitated by the fact that the radar environment is *nonstationary*. Primary causes of nonstationarity include statistical variations in the weather, the presence of unknown targets at unknown locations, and the ever-present radar *clutter*, which refers to radar returns from unwanted objects. Recursive updating of a state is synonymous with *adaptivity*, which is the natural method for dealing with nonstationarity. In current designs of radar systems, however, adaptivity is usually confined to the receiver. For the radar to be cognitive, adaptivity has to be extended to the transmitter too, hence the need for a feedback channel from the receiver to the transmitter. Moreover, the radar has to learn from experience on how to deal with different targets, large and small, and at widely varying ranges, all in an effective and robust manner. We may therefore say that a cognitive radar implies adaptivity, but not the other way round.

2.2 COGNITIVE RADAR SIGNAL-PROCESSING CYCLE

The dictionary definition of cognition mentioned above also includes "conceiving," which might be taken to mean the following statement:

> The formulation of a hypothesis, and then testing that hypothesis for the likelihood of its correctness.

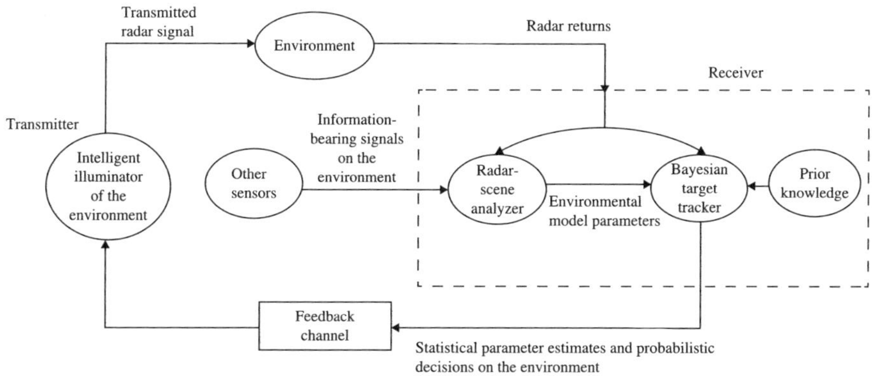

Figure 2.1 Block diagram of cognitive radar viewed as a closed-loop dynamic system.

This statement is in the spirit of the Bayesian approach to state estimation, with a probabilistic rating of alternatives. We are therefore emboldened to embrace the idea of Bayesian inference under the umbrella of cognitive radar.

This way of thinking leads us to the block diagram of Fig. 2.1, which depicts a picture of the cognitive radar signal-processing cycle. The cycle begins with the transmitter illuminating the environment. The radar returns produced by the environment are fed into two functional blocks: the radar-scene analyzer, and the Bayesian target-tracker. The tracker makes decisions on the possible presence of targets on a continuing time basis, in light of information on the environment provided to it by the radar-scene analyzer. The transmitter, in turn, illuminates the environment in light of the decisions made on possible targets, which initiates the next cycle of operation. The cycle is then repeated over and over again. Unlike a communication system, the feedback mechanism — a necessary requirement of a cognitive system — is easy to implement as the radar transmitter and receiver are usually co-located. Note also that although the process of target detection is not explicitly shown in the cognitive cycle of Fig. 2.1, it is part and parcel of the Bayesian target-tracker, which performs "detection through tracking" as explained later.

Based on the picture depicted in Fig. 2.1, a cognitive radar distinguishes itself from an adaptive radar in three important respects:

1. The radar continuously learns about the environment through experience gained from interactions of the receiver with the environment and, in a corresponding way, continually updates the receiver with relevant information on the environment.
2. The transmitter adjusts its illumination of the environment in an intelligent manner, taking into account such practical matters as the size of the target and its range, and consequently, making adjustments to the transmitted signal in an effective and robust manner.
3. The whole radar system constitutes a closed-loop dynamic system, encompassing the transmitter, the surrounding environment, the feedback channel, and the receiver. In other words, we have global feedback acting around the whole system.

It is well known that feedback is like a double-edged sword, in that it can become harmful if it is used improperly. Care must therefore be exercised in how the transmitter is designed in relation to the environment and receiver, so as to maintain a stable and reliable operation at all times.

One other important comment is in order. In reality, cognition is a two-way process, one being inside-out and the other being outside-in. These two parts of the cognitive process are so referred to, depending on whether the source of information leading to cognition resides inside or outside the receiver, respectively, as explained in the following[1]:

1. The "inside-out" part of cognition is represented by prior knowledge on the environment; it is an integral part of the receiver, as shown in Fig. 2.1. The form of prior knowledge is naturally application-dependent. For example, it may take the form of a geographic map, a clutter map of the environment, an elevation model, or kinematics of noncooperative targets. The Bayesian target-tracker retrieves information from the prior-knowledge base and utilizes it for improved radar performance on a need-be basis. Prior knowledge may therefore be viewed as the long-term memory of the receiver.
2. In contrast, the "outside-in" part of cognition may be viewed as short-term memory, which is developed by the receiver on the fly. It is initiated by the radar-scene analyzer in response to information-bearing signals gathered on the outside environment by the radar itself as well as other sensors working cooperatively with the radar.

2.3 RADAR-SCENE ANALYSIS

The function of the radar-scene analyzer is to provide the receiver with information on the environment on a continuous basis. This information is of critical importance to the decisions made by the receiver on possible targets of interest. This function builds on two sources of information-bearing signals:

1. radar returns, which are produced by the environment in response to the radar's own transmitted signal.
2. other relevant information on the environment (e.g. temperature, humidity, pressure, sea state), which is gathered on the fly by sensors other than the radar itself.

These two sources of inputs constitute the stimuli for the outside-in part of radar cognition.

1. The knowledge-based (KB) radar system described in subsequent chapters of the book may be viewed as a kind of inside-out cognitive system, embodying heuristics for determining how and when the signal-processing chain in the radar should be changed. The heuristics are developed through prior experimentation using a KB approach to target detection with human intervention; the human intervention is subsequently captured and then embedded into the receiver as a KB system.

In a surveillance scenario, radar performance is affected significantly by the unavoidable presence of interference. Typically, the interference is dominated by clutter (i.e. radar returns produced by undesired targets). Accordingly, to design a target tracker that embodies target detection, we need two kinds of information: one pertaining to the clutter acting alone, and the other pertaining to the target plus clutter.

2.3.1 Statistical Modeling of Statistical Representation of Clutter- and Target-Related Information

In order to describe how these two pieces of information can be addressed in specific terms, consider the case of a coherent radar dwelling on a particular patch of the ocean surface. With the radar being coherent, the radar returns contain amplitude as well as Doppler information on that patch. Correspondingly, the baseband version of the radar returns will be complex-valued. Now, the dwelling process can be of a long-term nature, in which case the nonstationary character of the radar returns becomes quite noticeable. In situations of this kind, we may be forced to avoid modeling the actual Doppler spectrum (i.e. plot of average power versus frequency) of the radar returns, and do so by exploiting the following intuitively satisfying observations:

> The Doppler spectrum of clutter by itself is relatively smooth, whereas the spectral content of the radar echo from a target appears essentially as a line component.

However, when the target cross-section is small and the target-to-clutter power ratio is therefore low, we need to enhance the line component due to the target. This enhancement may be achieved by performing the following transformation [2, 3]:

> Divide the average power in each Doppler bin of the spectrum (pertaining to the range-azimuth resolution cell of interest) by the mean of its neighboring bins, say k in number.

This transformation has the desired effect of accentuating the narrow peak of the line component due to the target and, at the same time, lowering the relatively wide peak of the clutter. Inspiration for the transformation, called a "peak filter," is traced to the "grouped periodogram test" described by Priestly [4], which was itself inspired by earlier work by Tukey in 1949. The statistics of the peak filter output, in the absence of a target, may now be evaluated under three assumptions [2, 3]:

1. None of the k neighboring Doppler bins in the power spectrum contains a target.
2. Inside a spectral window encompassing $(k + 1)$ Doppler bins, the continuous clutter power spectrum (that is always present) is approximately constant.
3. All $(k + 1)$ ordinates of the power spectrum are sampled independently.

Under these three assumptions, the individual ordinates of the actual power spectrum have a χ^2 distribution with two degrees of freedom [4]. Correspondingly, the peak-filter output, which divides each spectrum ordinate by k others, has a hypergeometric distribution, specifically an F-distribution with (2, 2k) degrees of freedom [2, 3]. On this basis, the clutter statistics are described by the distribution $F_{2,\ 2k}(z)$, where z is a random variable (i.e. average clutter power measurement). It is noteworthy that in reference 5, a similar observation is made using stochastic differential equation theory.

Turning next to the target, which is typically unknown, modeling its statistics is unfortunately not straightforward. For ease of implementation, and due to a lack of detailed knowledge about the target, it may be prudent to assume that the target has the same distribution that governs the clutter, but with a difference. (This assumption may hold in the case of a small target moving on an ocean surface, in which case the underlying dynamics of the clutter and the target are closely coupled.) Accordingly, if the clutter distribution is described by $F_{2,2k}(z)$, the target distribution is taken to be $\frac{1}{\gamma}F_{2,2k}\left(\frac{z}{\gamma}\right)$, where z is a power spectrum measurement and γ is the target-to-clutter power ratio [2, 3]; the scalar parameter z is not to be confused with the vector $\mathbf{z}$ introduced later.

In addition to the target statistics, the receiver needs to have a model that accounts for the motion of the target. To this end, we may assume that the target has a Gaussian-distributed acceleration with variance σ^2, which characterizes the agility of the target. For a low standard deviation σ, the target is seen by the radar when it is not accelerating. On the other hand, for a high σ, the task of target detection may become difficult due to possible confusion of the target with small clutter peaks, hence the likelihood of the radar making a decision error.

In summary, for an ocean environment under surveillance by a coherent radar, information on radar returns processed by the radar-scene analyzer for a particular range-azimuth cell may be modeled as follows:

1. *Clutter-statistics*, described by the F-distribution $F_{2,2k}(z)$, where z is a power spectrum measurement and k is the number of neighboring Doppler bins over which the measurement is averaged.
2. *Target-plus-clutter statistics*, described by the scaled F-distribution $\frac{1}{\gamma}F_{2,2k}\left(\frac{z}{\gamma}\right)$, where γ is the target-to-clutter power ratio.
3. *Target motion*, described by a Gaussian-distributed acceleration with a variance σ^2, which accounts for the target's agility.

It must be re-emphasized, however, that this model is appropriate for the specific case of a target moving on an ocean surface. For other environmental scenarios, the radar designer is challenged to develop appropriate statistical models to describe the information content of radar returns on clutter and targets.

2.4 BAYESIAN TARGET TRACKING

Previously, we mentioned that the Bayesian paradigm is a logical choice for coherent radar. We now describe a Bayesian strategy for the coherent radar detection of small targets in the presence of sea clutter. Unlike conventional tracking algorithms that perform intermediate detections (i.e. hard decisions) on the radar returns, the new algorithm processes the radar returns directly.[2] Specifically, the algorithm, referred to as a *direct tracking algorithm*, consists of three basic steps:

1. For a given search area, radar returns are collected over a certain period of time.
2. For each range-azimuth resolution cell in the search space, the probability that the cell contains a target is computed.
3. With the evolution of the target probability distribution resulting from the recursive computation of step 2 over time, target tracks are detected and corresponding hard decisions on possible targets are subsequently made.

In effect, the algorithm (formulated in probabilistic terms) may be viewed as a soft-decision procedure on target detection.

To set the stage for the Bayesian framework, let there be a total of R range-azimuth resolution cells in the search space S, and let $r \in S$ denote a resolution cell in question. Let ε_t^r denote the event of a single target occurring in resolution cell r at discrete time t. Let the vector $\mathbf{z}_t$ denote the frame that is made up of the spectral measurements for all R resolution cells at time t. The matrix

$$\begin{aligned}\mathbf{Z}_t &= [\mathbf{z}_t, \mathbf{z}_{t-1}, \ldots, \mathbf{z}_2, \mathbf{z}_1] \\ &= [\mathbf{z}_t, \mathbf{Z}_{t-1}]\end{aligned}$$

denotes the full set of all the available frames extending up to and including time t. Then, according to this notation, the vector $\mathbf{z}_t$ denotes the current frame and the remaining matrix $\mathbf{Z}_{t-1}$ denotes the combined set of all past frames. By the same token, $\mathbf{Z}_{t+1}$ denotes the combination of a future frame $\mathbf{z}_{t+1}$, the current frame $\mathbf{z}_t$, and all past frames $\mathbf{Z}_{t-1}$.

Following the traditional approach to state estimation, we may now identify three different forms of the Bayesian target-tracker:

1. *one-step predictor*, whose output is described by the conditional probability $\mathbf{P}(\varepsilon_t^r|\mathbf{Z}_{t-1})$;
2. *filter*, whose output is described by the conditional probability $\mathbf{P}(\varepsilon_t^r|\mathbf{Z}_t)$;
3. *smoother*, whose output is described by the expanded conditional probability $\mathbf{P}(\varepsilon_t^r|\mathbf{Z}_{t+1})$.

2. In reference 6, Bruno and Moura also describe a Bayesian approach to the tracking problem. Given a search space of R range-azimuth resolution cells and M possible targets, their algorithm is designed to track any of the targets. The algorithm does so by first computing the probability of each of the 2^M different target combinations. Specifically, the centroid of each target can be in any of the R resolution cells, or else be absent. The Bayesian tracking approach described in this chapter is however different, in that it is formulated in such a way that the algorithm can also operate in a smoothing mode, with the probability distribution of the smoothed output being conditional on both past and future observations.

Smoothing uses more information than both prediction and filtering, and may therefore be more accurate than both of them in a statistical sense. On the other hand, however, only prediction and filtering can be implemented in real time.

2.4.1 One-Step Tracking Prediction

Consider the *joint event* $(\varepsilon_t^r, \varepsilon_{t-1}^q)$, which describes a target occurring in resolution cell q at time $t-1$ and then moving into resolution cell r at time t. From probability theory, we may express the output of the tracking predictor at time t as

$$\begin{aligned}\mathbf{P}(\varepsilon_t^r|\mathbf{Z}_{t-1}) &= \sum_{q=1}^{R}\mathbf{P}(\varepsilon_t^r, \varepsilon_{t-1}^q|\mathbf{Z}_{t-1})\\ &= \sum_{q=1}^{R}\mathbf{P}(\varepsilon_t^r|\varepsilon_{t-1}^q, \mathbf{Z}_{t-1})\mathbf{P}(\varepsilon_{t-1}^q|\mathbf{Z}_{t-1}).\end{aligned} \tag{2.1}$$

However, given the fact that the event ε_{t-1}^q has occurred at time $t-1$, it makes the previous measurements matrix $\mathbf{Z}_{t-1}$ irrelevant. In other words, occurrence of the event ε_{t-1}^q conveys exactly the same amount of information as the joint event $(\varepsilon_{t-1}^q, \mathbf{Z}_{t-1})$. Accordingly, Equation 2.1 reduces to the simpler form

$$\mathbf{P}(\varepsilon_t^r|\mathbf{Z}_{t-1}) = \sum_{q=1}^{R}\mathbf{P}(\varepsilon_t^r|\varepsilon_{t-1}^q)\mathbf{P}(\varepsilon_{t-1}^q|\mathbf{Z}_{t-1}). \tag{2.2}$$

The conditional probability $\mathbf{P}(\varepsilon_{t-1}^q|\mathbf{Z}_{t-1})$ is the output of the tracking filter working on resolution cell q at time $t-1$. We also recognize $\{\mathbf{P}(\varepsilon_t^r|\varepsilon_{t-1}^q)\}_{q,r}$ as the set of probabilities that event ε_t^r follows event ε_{t-1}^q. This set of probabilities is referred to as the *transition matrix* of the tracker, the formulation of which exploits the statistical model of target motion as supplied by the radar-scene analyzer. It is noteworthy that the less agile the target is, the smaller the jumps the target is expected to make in the search space S, thereby causing the transition matrix to be more sparse. In any event, given the tracking filter output at time $t-1$ and the transition matrix, we may use Equation 2.2 to compute the output of the corresponding tracking predictor at time t.

2.4.2 Tracking Filter

Consider next the issue of computing the output of the tracking filter at time t, which is defined by the *posterior probability* $\mathbf{P}(\varepsilon_t^r|\mathbf{Z}_t)$. Applying *Bayes' theorem* to this probability yields

$$\mathbf{P}(\varepsilon_t^r|\mathbf{Z}_t) = \frac{p(\mathbf{Z}_t|\varepsilon_t^r)\mathbf{P}(\varepsilon_t^r)}{p(\mathbf{Z}_t)}, \tag{2.3}$$

where $p(\mathbf{Z}_t|\varepsilon_t^r)$ is the conditional probability density function of the current measurements matrix $\mathbf{Z}_t$ given the occurrence of event ε_t^r, and $\mathbf{P}(\varepsilon_t^r)$ is the *prior probability* of that event. The probability density function $p(\mathbf{Z}_t)$ in the denominator is the *evidence*, which acts merely as a normalizing function. As, by definition, $\mathbf{Z}_t = (\mathbf{z}_t, \mathbf{Z}_{t-1})$, we may rewrite Equation 2.3 by expanding the numerator, as shown by

$$\begin{aligned}\mathbf{P}(\varepsilon_t^r|\mathbf{Z}_t) &= \frac{p(\mathbf{z}_t, \mathbf{Z}_{t-1}|\varepsilon_t^r)\mathbf{P}(\varepsilon_t^r)}{p(\mathbf{Z}_t)} \\ &= \frac{p(\mathbf{z}_t|\varepsilon_t^r, \mathbf{Z}_{t-1})p(\mathbf{Z}_{t-1}|\varepsilon_t^r)\mathbf{P}(\varepsilon_t^r)}{p(\mathbf{Z}_t)}.\end{aligned} \tag{2.4}$$

Recognizing that the occurrence of event ε_t^r makes past measurements $\mathbf{Z}_{t-1}$ irrelevant, we may simplify Equation 2.4 by setting $p(\mathbf{z}_t|\varepsilon_t^r, \mathbf{Z}_{t-1}) = p(\mathbf{z}_t|\varepsilon_t^r)$. Thus,

$$\mathbf{P}(\varepsilon_t^r|\mathbf{Z}_t) = \frac{p(\mathbf{z}_t|\varepsilon_t^r)p(\mathbf{Z}_{t-1}|\varepsilon_t^r)\mathbf{P}(\varepsilon_t^r)}{p(\mathbf{Z}_t)}. \tag{2.5}$$

The first term $p(\mathbf{z}_t|\varepsilon_t^r)$ in the numerator of Equation 2.5 is the probability density function of measurement $\mathbf{z}_t$ given that there is a target in cell r at time t. The second term $p(\mathbf{Z}_{t-1}|\varepsilon_t^r)$ is computed by using the *recursive formula*

$$\begin{aligned}p(\mathbf{Z}_{t-1}|\varepsilon_t^r) &= p(\mathbf{z}_{t-1}, \mathbf{Z}_{t-2}|\varepsilon_t^r) \\ &= \sum_{q=1}^{R} p(\mathbf{z}_{t-1}|\varepsilon_{t-1}^q)p(\mathbf{Z}_{t-2}|\varepsilon_{t-1}^q)\mathbf{P}(\varepsilon_{t-1}^q|\varepsilon_t^r),\end{aligned} \tag{2.6}$$

where, as before, $p(\mathbf{z}_{t-1}|\varepsilon_{t-1}^q)$ is the input supplied to the receiver by the radar-scene analyzer, and $p(\mathbf{Z}_{t-2}|\varepsilon_{t-1}^q)$ is the one-step delayed version of $p(\mathbf{Z}_{t-1}|\varepsilon_t^r)$, hence the reference to Equation 2.6 as a recursive formula. The matrix of probabilities $\{\mathbf{P}(\varepsilon_{t-1}^q|\varepsilon_t^r)\}_{q,r}$ is the *inverse transition matrix*, which is defined by the probabilities that event ε_{t-1}^q preceded event ε_t^r. The term "inverse" is used here merely to imply the role reversal of these two events with respect to the transition matrix under Equation 2.2.

The following two points are noteworthy:

1. The recursive formula of Equation 2.6 is identical to the hidden Markov model (HMM) filter for a Markov chain $\{\varepsilon_t\}$ with transition probabilities $\{\mathbf{P}(\varepsilon_{t-1}^q|\varepsilon_t^r)\}$.
2. Given the posterior probability distribution of Equation 2.4, the conditional mean estimate (i.e. minimum mean-square estimate) of the event ε_t^r over the entire search space S can be computed as the summation $\sum_{r=0}^{R}\varepsilon_t^r\mathbf{P}(\varepsilon_t^r|\mathbf{Z}_t)$.

We may also compute the conditional probability density function $p(\mathbf{Z}_{t-1}|\varepsilon_t^r)$ in another way by recasting the recursive formula of Equation 2.6 as follows:

$$p(\mathbf{Z}_{t-1}|\varepsilon_t^r) = \frac{p(\mathbf{Z}_{t-2})}{\mathbf{P}(\varepsilon_t^r)}\sum_{q=1}^{R} p(\mathbf{z}_{t-1}|\varepsilon_{t-1}^q)\mathbf{P}(\varepsilon_{t-1}^q|\mathbf{Z}_{t-2})\mathbf{P}(\varepsilon_t^r|\varepsilon_{t-1}^q). \tag{2.7}$$

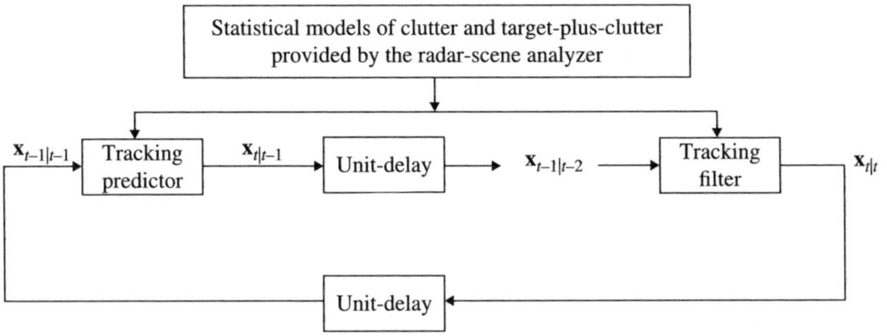

Figure 2.2 Block diagram of the Bayesian direct filtering system. Notation: t, discrete-time; $\mathbf{x}_{t|t}$, filtered state vector of probabilities of targets being present in the search space at t given spectral measurements up to and including time t. The other data vectors in the diagram are similarly defined.

Then, substituting Equation 2.7 into Equation 2.5, we obtain the new formula for computing the posterior probability at the output of the tracking filter:

$$\mathbf{P}(\varepsilon_t^r|\mathbf{Z}_t) = \frac{p(\mathbf{z}_t|\varepsilon_t^r)}{p(\mathbf{z}_1, \mathbf{z}_2|\mathbf{Z}_{t-2})}\sum_{q=1}^{R} p(\mathbf{z}_{t-1}|\varepsilon_{t-1}^q)\mathbf{P}(\varepsilon_{t-1}^q|\mathbf{Z}_{t-2})\mathbf{P}(\varepsilon_t^r|\varepsilon_{t-1}^q), \tag{2.8}$$

where the probability $\mathbf{P}(\varepsilon_{t-1}^q|\mathbf{Z}_{t-2})$ is a delayed version of the tracking predictor output, and the probabilities $\{\mathbf{P}(\varepsilon_t^r|\varepsilon_{t-1}^q)\}_{q,\ r}$ are elements of the transition matrix.

On the basis of Equations 2.2 and 2.8, we may now construct the block diagram of Fig. 2.2 for the Bayesian direct filtering system. The diagram is in the form of a closed-loop feedback system that operates by propagating a state vector of probabilities from one iteration to the next. Most importantly, the right relationship must be established between the radar parameters and statistical characteristics of clutter and target-plus-clutter for the tracker to maintain a stable operation.

2.4.3 Tracking Smoother

An attractive feature of the Bayesian tracker, described herein, is the fact that it is straightforward to make its operation conditional on both past and future spectral measurements. The result of this expansion is a *target tracking smoother*, for which the output is expressed as

$$\mathbf{P}(\varepsilon_t^r|\mathbf{z}_t, \mathbf{Z}_{t-1}, \mathbf{z}_{t+1}) = \frac{p(\mathbf{z}_t|\varepsilon_t^r)p(\mathbf{Z}_{t-1}|\varepsilon_t^r)p(\mathbf{z}_{t+1}|\varepsilon_t^r)\mathbf{P}(\varepsilon_t^r)}{p(\mathbf{z}_t, \mathbf{Z}_{t-1}, \mathbf{z}_{t+1})}. \tag{2.9}$$

The factorization of terms in the numerator of Equation 2.9 assumes that the radar is treated as a first-order Markov model, in which case the conditional dependence of the distribution of past measurements $\mathbf{Z}_{t-1}$ on the future measurements $\mathbf{z}_{t+1}$ may be ignored; that is, we may set $p(\mathbf{Z}_{t-1}|\varepsilon_t^r, \mathbf{z}_{t+1})$ equal to $p(\mathbf{Z}_{t-1}|\varepsilon_t^r)$.

The additional factor $p(\mathbf{z}_{t+1}|\varepsilon_t^r)$ in the numerator of Equation 2.9 is computed by running the right-hand side of the recursive Equation 2.6 backwards in time. Thus, whereas the target-tracking filter operates in the forward direction only, the target-tracking smoother operates in the forward as well as backward directions. Accordingly, decisions made on possible targets using the tracking smoother contain more information than the corresponding tracking filter, and may therefore be more reliable. However, this improvement in performance is gained at the expense of two factors: [1] increased computational complexity and [2] non-real-time operation. Simply put, for every gain in radar performance, there is a corresponding price to be paid.

2.4.4 Experimental Results: Case Study of Small Target in Sea Clutter

In references 2 and 3, the performance of the Bayesian target detector was evaluated using real-life radar data under varying conditions. The data were collected by means of the McMaster IPIX radar, which is a highly configurable coherent multi-function X-band radar built specifically for research purposes. For a subset of the database collected at a site in Dartmouth (Nova Scotia), the radar was operated in the dwell mode with a 1° pencil beam and fixed radio frequency of 9.39 GHz. The radar was mounted about 30 m above sea level, with the target of interest being located about 2.5 km offshore. The target was a sphere (1 m in diameter) made up of wire covered in foam. Radar range was sampled at 15 m intervals, obtained by using a 200 ns rectangular pulse. (The actual range resolution of the radar was 30 m.) The pulse-repetition frequency (PRF) was 2 kHz, but the pulse alternated between horizontal (H) and vertical (V) polarization, so that the effective single-polarization PRF was 1 kHz. For each pulse, both H and V polarizations were recorded simultaneously, resulting in a matrix of four possible transmit/receive polarizations: HH, HV, VH, and VV. For each combination in the matrix, the amplitude and phase of the radar returns were stored in the form of in-phase (I) and quadrature (Q) components.[3]

In reference 3, three data sets from the Dartmouth database were used to test the Bayesian target detector. The results pertaining to one of those data sets are reproduced in Fig. 2.3. Figure 2.3a displays the Doppler-time image of the raw radar data set, using a 64-sample sliding window. Figure 2.3b displays the resulting output of the Bayesian direct tracking smoother. Each pixel in the image represents the probability of a target being present in the corresponding resolution cell. The darker the pixel, the higher the probability of target occurrence. Note also that the dark traces included along the 500 Hz line indicate the points in time where the target was invisible to the radar or when the radar failed to detect the target.

3. To promote further research in the radar area, we have created a comprehensive website from which IPIX radar data sets are available; the data sets have been collected under different environmental conditions. For details see http://soma.mcmaster.ca/ipix.

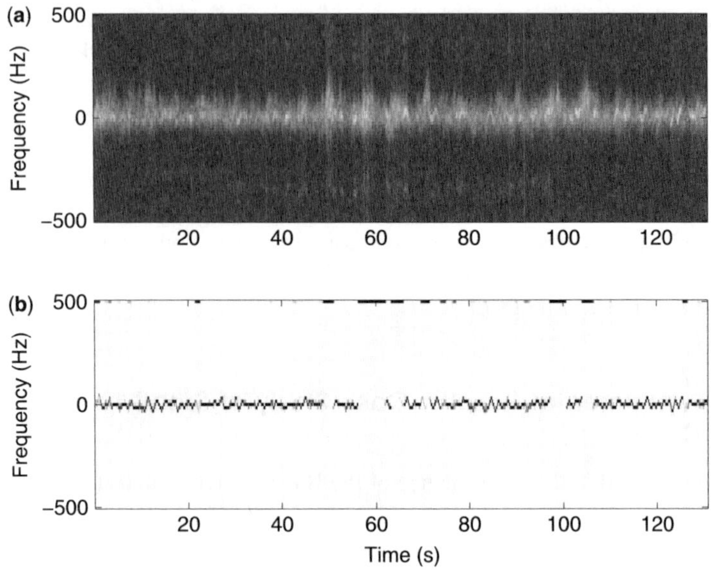

Figure 2.3 (**a**) A 64-sample sliding window Doppler-time image of raw radar data set 3 from reference 3. (**b**) Output of the Bayesian tracking smoother, where each pixel represents the probability of a target in the corresponding resolution cell.

Figure 2.3 and several other results reported in reference 3 attest to the effectiveness of the Bayesian direct tracker. In particular, even for a data set with an average target-to-clutter power ratio as low as -7 dB, Fig. 2.3 clearly demonstrates the visibility of the target most of the time.

2.4.5 Practical Implications of the Bayesian Target Tracker

To the best of the author's knowledge, the Bayesian target tracker described in detail in references 2 and 3 and highlighted herein is the first to be studied regarding the feasibility of direct target-tracking without intermediate detections. The use of a Bayesian approach to direct tracking, combined with complete reliance on soft decisions (i.e. avoiding hard decisions through intermediate detections), has some important practical implications:

1. Unlike hard decisions, the soft decisions made by the Bayesian target tracker preserve the information content of the radar returns; this approach follows the "principle of information preservation," inferred from Shannon's information theory [7].
2. The Doppler-time image produced by the Bayesian direct target tracker makes it possible for the radar to see the motion of the target in a manner comparable to the human eye. Indeed, we conjecture that an experienced human operator could not do a better job of following the target than the Bayesian tracker,

especially so when the tracker is operated in the smoothing mode; in this mode, the tracker exploits the combined benefit of forward and backward computations.

3. The basic idea behind the Bayesian approach is to view the information contained in the radar image as a probability distribution that characterizes the likelihood of a particular resolution cell containing a target. The distribution, in the form of a posterior probability density function, is determined in part by the statistical structure of the radar scene (i.e. the outside world), and in part by the way in which echoes from the world are actually encoded by the radar itself. Accordingly, the Bayesian approach distinguishes itself from other approaches by invoking an explicit statistical structure of the world that, in reality, is a fundamental necessity.
4. The book by Knill and Richards [8] presents a number of theoretical frameworks for studying visual perception that, in varying degrees, are all founded on Bayesian principles. In a way, this book lends further support to the Bayesian radar-target tracker, the theory of which is embodied in equations 2.1 through 2.9, depending on the mode of operation.

One last comment is in order. Using two different real-life radar data sets and computer-simulated data, a comparative evaluation of the Bayesian approach to target-detection-through-tracking has been made against a new detection strategy called the *correlation anomaly receiver*, which follows from the theory of stochastic differential equations [5]. The results of this evaluation, reported in reference 9, show that the Bayesian receiver's performance is superior to that of the correlation anomaly receiver.

2.5 ADAPTIVE RADAR ILLUMINATION

As it stands, there is no optimization being performed on the posterior probability distribution $\mathbf{P}(\varepsilon_t^r|\mathbf{Z}_t)$ computed by the Bayesian target tracker. This important matter, however, can be taken care of by making the transmitter, responsible for illuminating the environment, *adaptive*.[4] The practical issue with adaptive radar illumination (transmission) is how to observe past radar returns and extract useful information in order to decide or select the radar waveform for the next transmission in some optimal fashion.

In an implicit sense, the present spectral measurements at time t, denoted by $\mathbf{z}_t$ and the past measurements denoted by $\mathbf{Z}_{t-1}$, are all dependent on the transmitted signal. This dependence suggests that the whole radar system can be made adaptive by adjusting certain parameters in the transmitted signal in response to the probabilistic decisions made by the Bayesian tracker on the environment under surveillance. Note,

4. The book by Dag Gjessing in reference 10 is devoted to an experimental study of a form of radar illumination referred to as *target adaptive matched illumination*.

however, that by doing so, the radar system assumes the form of a *stochastic control system* involving a state-space model governed by the posterior distribution of Equation 2.4; the optimal solution to such "partially observable" stochastic-control problems is NP hard. Fortunately, there are suboptimal procedures such as reinforcement learning that can yield acceptable solutions; this issue is discussed later in Section 2.7.

There are many ways in which parameters of the transmitted signal can be adjusted.[5] One practical way is to use burst waveforms, with each burst made up of a sequence of uniformly spaced, nonoverlapping subpulses of fixed duration. The pulse amplitudes are held constant for two reasons:

- Unforeseen difficulties with dynamic range requirements are avoided.
- The target-to-clutter power ratio may not be sensitive enough to pulse-amplitude adjustments.

The logical strategy is then to adjust the phase of each transmitted radio-frequency (RF) pulse in accordance with feedback sent to the transmitter from the receiver. Here we have the choice of a phase response that varies with time according to a square law that results in linear frequency modulation (FM), or a cubic law that results in nonlinear FM. Both of these configurations are well known for their pulse-compression characteristics, with the nonlinear FM being more effective than the linear FM [12].

2.5.1 Simulation Experiments in Support of Adjustable Frequency Modulation

In references 13 and 14, DeLong and Hofstetter describe an adaptive transmission strategy that uses the above-mentioned scheme of burst waveforms. Their detailed theoretical study was in two parts: the first, published in 1967, investigated the use of adjustable pulse amplitudes, and the second, published in 1969, focused on the use of adjustable pulse phases with limited dynamic range. For both studies, they used the signal-to-interference ratio as the index of performance, with interference being composed of clutter and receiver noise. For performance optimization, they used a procedure based on the Karush–Kuhn–Tucker theorem [15]. The conclusions reported by DeLong and Hofstetter in those two early, but still very much valid, papers may be summarized as follows:

- Adjustment of phase is a more practical approach than amplitude for the design of adaptive radar transmitters.

5. The selection of waveforms to be used for adaptive radar transmission is application-dependent (as well as technology-dependent). In the waveform-selection procedure described in reference 11, the focus is on chirp-coded waveforms (stepped phase and frequency), the use of which is motivated by the relative ease of synthesizing constant-amplitude waveforms with fixed time–bandwidth product.

- The use of constant amplitude, quadratic-phase burst waveforms provides a significantly better ambiguity pattern than the corresponding constant-amplitude, zero-phase burst waveforms.

These findings have been confirmed in simulation experiments reported in reference 16, using the following system parameters:

- Burst waveform: 32 subpulses
- Single-pulse signal-to-interference ratio: 5 dB
- Clutter-to-target cross-section ratio: 100

In reference 16, the DeLong–Hofstetter algorithm using constant-amplitude, square-phase burst waveforms was found to reach a peak signal-to-interference ratio of 16.9 dB after 30 iterations of the algorithm; the performance of this system is almost equivalent to perfect clutter suppression in an environment highly dominated by clutter.

2.6 ECHO-LOCATION IN BATS

From the introductory section, we recall that a cognitive radar system embodies three fundamental ingredients:

- learning from the environment through experience,
- adjustment of the transmitted signal in an intelligent manner, and
- feedback from the receiver to the transmitter to make this adjustment possible.

All these three features are part and parcel of the echo-location system of a bat. Accordingly, there is much that we can learn from the echo-location system of a bat [17–20]. Most echo-locating bats are blind.[6] To see the world around it, the bat uses sonar, which is an active echo-location system. In addition to providing information about how far away a target (i.e. flying insect) is, the bat's sonar conveys information about the relative velocity of the target, the size of various features of the target, and the azimuth and elevation of the target [19–21]. The complex neural computations needed to extract all this information from the target echo occur within a brain the size of a plum. Indeed, an echo-locating bat can pursue and capture its target with a facility and success rate that would be the envy of a radar engineer. How then does the bat perform all these remarkable tasks? The answer to this fundamental question lies in the fact that soon after birth, the bat uses its innate hard-wired brain to build up rules of behavior through what we usually refer to as experience, hence the remarkable ability of the bat for echo-location.

6. We say "most," because not all species of bats are blind. Note also that not all bats use echo-location.

The bat uses its mouth (or nose) to broadcast echo-location sounds and uses its auditory system as the sonar receiver. The emitted sounds consist of burst waveforms whose characteristics are highly diverse, varying with both species and being situation-specific. The transmitted sound characteristics are summarized as follows:

- Duration: 0.3–300 ms
- Frequency: 12–200 kHz
- Structure: Frequency-modulated (FM) component, or constant-frequency (CF) component followed by FM component

The constant-frequency component can be single or multiple harmonic. The FM component can be of a downward or upward kind, the FM sweep varying linearly or nonlinearly with time. The use of FM is intended to improve the echo-location system's resolution capability for the bat. (It is noteworthy that an echo-location bat's emitted sounds consist of burst waveforms just as the adaptive transmission strategy used in the DeLong–Hofstetter algorithm also consists of burst waveforms.)

Broadly speaking, the adaptive behavior of bats may be categorized as follows [19]:

- *Velocity-dependent adaptation*, which involves adjustment of the transmitted sound frequency; this form of adaptation is most salient in species of constant frequency–frequency modulation (CF-FM) bats. These CF-FM bats also appear to make adjustments in temporal patterning as they close in on their targets.
- *Range-dependent adaptation*, which involves adjustment of the emitted-sound duration, bandwidth, and repetition rate; this second form of adaptation is most salient in bats using only frequency modulation (FM). These bats also appear to make adjustments in the transmitted sound bursts during target approach.

Echoes from targets (i.e. insects) are represented in the auditory system by *neuronal activities* that are sensitive to different combinations of acoustic inputs produced in response to the transmitted sound bursts. In particular, three principal dimensions of the bat's auditory representation have been identified [17]:

1. *Echo frequency*, which is initially encoded in the auditory periphery cochlea by place in the cochlear;
2. *Echo amplitude*, which is encoded by the neuronal responses under [1] and other neurons tuned to different dynamic ranges in the central nervous system;
3. *Echo delay*, which is encoded through neuronal computations that produce target-range tuning responses.

There are two principal (neuronal) computations that are performed by the bat's brain for image-forming purposes. One is the *spectrum* of the incoming echo, which is intended for the extraction of target shape, which is a particularly noteworthy point in light of the spectral processing performed by the radar scene analyzer in

Section 3. The other is *delay* in the received echo with respect to the transmitted sound bursts, which is intended for the extraction of target range. To carry out these computations, frequency-based information contained in the incoming echo spectrum is converted into estimates of the spatial (time) structure of the target.

In short, the echo-location system of a bat is very plastic, in that the parameters of the transmitted sound bursts can be changed considerably during the different phases of the target-pursuit sequence. We are therefore justified in viewing the echo-location system of a bat as physical proof (albeit in neurobiological terms) of cognitive radar.

2.7 DISCUSSION

Three important conclusions can be drawn from the presentations made in this chapter:

1. *Intelligence is a necessary requirement for the radar to be cognitive.* A striking difference is discernible between the presentations we have made on adaptive radar illumination and echo-location in bats. Simply put, in signal-processing terms, the echo-location systems of bats are far more plastic than the adaptive radar systems that are currently in use or being contemplated. This important point is best illustrated by the spectograms shown in Fig. 2.4, which were produced by four different bat species in their respective target (insect)-pursuit sequences. The significant characteristic that is immediately

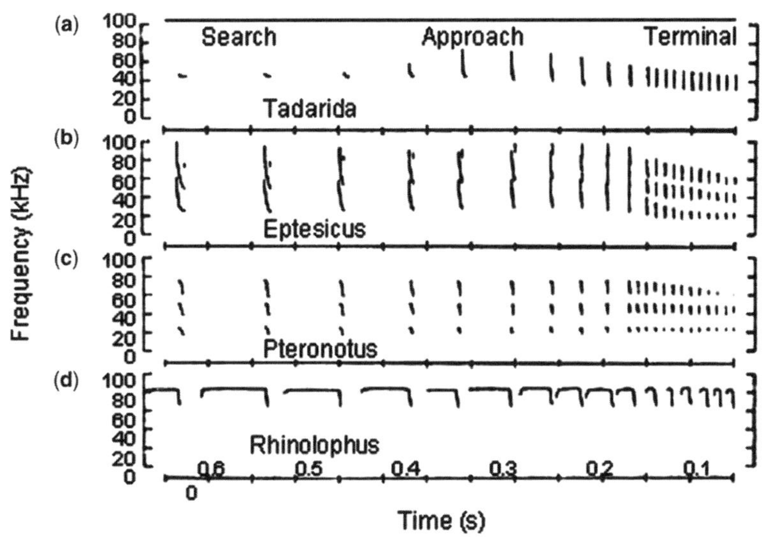

Figure 2.4 Spectrograms of sonar signals produced by four different species of bats as they advance from the search to approach and finally to the terminal phase of insect pursuit. (Reproduced from reference 20, with permission of the University of Chicago Press.)

apparent from this figure is that the transmitted signal duration decreases and the burst repetition rate increases as the bat gets closer to its target. In doing this, the bat is using acquired knowledge of the distance from its target to adjust the parameters of its transmitted sound bursts. For a radar system to be cognitive, therefore, it is a fundamental necessity for the radar transmitter to do the following:

> Learn from continuing interactions with the environment and intelligently use the information extracted by the receiver on targets under surveillance, all of this being done on the fly during the different phases of the target-track sequence.

2. *Feedback from the receiver to the transmitter is a facilitator of intelligent signal processing*. We say feedback is a facilitator of intelligence, because it is through feedback from the receiver to the transmitter that cognitive radar is enabled to learn from interactions with the environment. More is said on this issue in Section 2.7.1.
3. *The preservation of information in radar returns is of crucial importance to receiver performance, which is realized by the Bayesian approach to target detection through tracking*. The results presented in Section 2.4 on the Bayesian target-tracker emphasize the signal-processing power of the Bayesian approach. This approach is the only statistical approach in which a model of the received signal accounts for two factors contributing to the specification of information:

   - The statistical nature of interference (i.e. radar clutter and noise);
   - The explicit statistical structure of the radar environment (i.e. outside world), including targets.

   In the past, the Bayesian approach has been criticized for requiring a model that includes a statistical structure of the radar environment. In response to such criticism, we merely have to emphasize that if we are to account for the physical realities that are responsible for the generation of radar returns, then the inclusion of a statistical structure of the radar environment is a fundamental requirement for preserving the information content of the received signal.

   Most importantly, referring to the closed-loop feedback system of Fig. 2.1, encompassing the radar transmitter, the propagation medium, the radar receiver and the feedback channel back to the transmitter, there is much tat we can learn from the way in which this system operates. In this context, when we speak of the cognitive signal-processing cycle, the iterative processing of radar signals proceeds on a cycle-by-cycle basis inside the feedback loop, with each cycle corresponding to a frame of radar pulses produced by the transmitter. Moreover, within each cycle, there is processing being performed on a pulse-by-pulse basis. We thus have two forms of iterative processing that are performed side by side:

   - Processing on a pulse-by-pulse, which is performed in the receiver
   - Processing on a cycle-by-cycle basis, which is performed in the transmitter

2.7.1 Learning

Throughout this article, we have emphasized that learning is a basic ingredient of cognitive radar. In a generic sense, the learning process can take two different forms: off-line and on-line.

Through *off-line learning*, knowledge is acquired about the environment and then embedded in the receiver. In the radar context, an established way of accomplishing this acquisition is to collect real-life data by conducting ground-truthed experiments on the environment under varying conditions. Then, by performing statistical analysis on the radar data and formulating models on clutter and targets, the acquisition of knowledge of the environment is accomplished (see, for example, reference 21). In any event, the off-line learning takes place through the intervention of the experimenter.

Among the many different *on-line learning* procedures, *reinforcement learning* [22] stands out as a procedure well suited for cognitive radar. In the modern approach to reinforcement learning, also referred to as *approximate programming*, Bellman's dynamic programming (rooted in control theory) provides the theoretical foundation of the procedure. Bellman's dynamic programming suffers from the "curse of dimensionality," which limits its practical utility. The use of approximate dynamic programming (ADP) overcomes this limitation. In reference 23, different methods for implementing ADP are described.

2.7.2 Applications

A discussion of cognitive radar would be incomplete without some applications where it has the potential to make a difference. In what follows, we address two applications of cognitive radar, one dealing with multi-function radars that are expensive, and the other dealing with noncoherent radars that are inexpensive.

2.7.2.1 Multifunction Radars Thanks to continuing advances and improvements on two fronts, namely phased-array antennas and computers, multifunction radars are fast becoming, if not already, the norm in building sophisticated radar systems. For example, the radar may have to deal with a "fading target" due to the presence of multipath produced by close proximity of the target (e.g. a sea-skimming missile) to the sea surface in a hostile marine environment. One way of mitigating the fading problem is to increase the dwell time in order to track the target with adequate accuracy. In such an environment, we may identify two problems that require serious attention:

1. *Agility*, which mandates the use of phased-array antennas oriented to provide 360° coverage (e.g. four arrays at 90° with respect to each other).
2. *Fast response*, which is attained by using powerful computers that enable the radar to adapt its transmission waveforms so as to detect, track, and paint the target rapidly enough for the engagement to occupy a range of no more than 30 s to a couple of minutes.

Typically, while attending to the fading target, the radar is also required to handle other threatening targets. The radar is therefore faced with a new problem — *resource management*. Neurodynamic programming provides a "principled" approach for a solution (suboptimal, but perhaps adequate) to the resource management problem.[7]

2.7.2.2 Noncoherent Radar Network

For an entirely different application that could benefit from the use of cognitive radar, consider the international border security problem. To be more specific, consider the Great Lakes St. Lawrence Seaway. There are two challenging problems with this large open border between the United States of America and Canada [26]:

- the protection of assets and populations of people from terrorism and
- the prevention of illegal crossings across the border.

A cost-effective, all-weather, and all-day solution to both of these challenges is a *cognitive noncoherent radar network*. The network would be made up of inexpensive commercial off-the-shelf marine radars, which are distributed across the border. The only discriminant available for surveillance with such simple radars is amplitude, which severely limits the capability of the radar to detect noncooperative targets with small radar cross-section in the presence of lake clutter. To mitigate this serious problem, Weber et al. [26] depart from conventional radar signal processing by purposely setting low detection thresholds. Naturally, the false-alarm rates are raised to levels higher than a conventional processor, but, most importantly, the small noncooperative targets are now detectable. Then, through the use of a sophisticated tracking algorithm, the real targets are extracted and the false-alarm rates are reduced to an acceptable level.

Given a network of such noncoherent radars, which also incorporates a central base station, the real-target tracks computed by the component radars are transmitted by a communication channel (wireline or wireless) to that station. Consequently, we have yet another new problem, namely, multisensor fusion. Given the limited computing resources at the base station, the challenge here is how to design a cognitive radar network that produces a map in real-time for the entire Great Lakes St. Lawrence Seaway, which identifies the tracks of all noncooperative targets operating therein and does so in the most reliable manner possible.

In both of the applications addressed herein, another extremely challenging issue is that of knowing how to define a metric by means of which it can be said that the task in question has been accomplished? Stated in another way, what is the essence of

7. In reference 25, Krishnamurthy and Evans describe a beam-scheduling procedure for an agile radar. The procedure, rooted in reinforcement learning, enables the radar to adaptively decide which one of several targets to illuminate at each time.

The allocation of time resources in a multifunction radar is also addressed in reference 25. In this latter paper, Miranda et al. pursue an approach radically different from the Krishnamurthy–Evans procedure; their study embodies a comparative evaluation of two resource-scheduling procedures, both of which are a KB approach.

the description of the environmental scene that is under surveillance? The traditional radar specifications, based on the probability of detection and the problem of false alarm (which are never measured anyway in a real-time setting) are unsuitable. Rather, we need a new metric that addresses specifically what the end-user needs to see. The formulation of this metric is further exasperated when the application at hand involves several tasks and the tasks have to be prioritized. Here again, a cognitive metric that learns over time may well provide an answer, as is often the case with humans (private communication, Prof. C. J. Baker, 17 January 2004).

ACKNOWLEDGMENTS

Permission of the IEEE to reproduce an earlier version of this chapter that appeared in the *Signal Processing Magazine*, 2006, is gratefully acknowledged. Also, permission of the University of Chicago Press to reproduct Fig. 2.4 is deeply appreciated.

REFERENCES

[1] S. Haykin. Cognitive radar: A way of the future. *IEEE Signal Processing Magazine* 2006; 23:30–41.

[2] R. Bakker, T. Kirubarajan, B. Currie, S. Haykin. Adaptive radar detection: A Bayesian approach. In: EPSRC and IEE Workshop on Nonlinear and Non-Gaussian Signal Processing, Peebles, Scotland, 8–12 July 2002.

[3] R. Bakker, G. Lopez, S. Haykin. Bayesian approach to the direct filtering of radar targets in clutter. ASL Report 02-02. Adaptive Systems Laboratory, McMaster University, Canada, August 2002.

[4] M.B. Priestly. *Spectral Analysis and Time Series*. Academic Press; 1981.

[5] T.R. Field, R.J.A. Tough. Stochastic dynamics of the scattering amplitude generating *K*-distributed noise. *J. Math. Phys.* 2003;44:5212–5223.

[6] M.G.S. Bruno, J.M.F. Moura. Multiframe detector/tracker: Optimal performance. *IEEE Trans. Aerospace and Electronic Systems* 2001;37:925–944.

[7] A. Viterbi. Wireless digital communication: a view based on three lessons learned. *IEEE Communications Magazine* 1991;29(9):33–36.

[8] D.C. Knill, W. Richards, editors. *Perception as Bayesian Inference*. Cambridge University Press; 1996.

[9] B. Currie, S. Haykin. Bayesian detector evaluation and comparison. Final Report prepared for Defence Research Development-Ottawa (DRDC) under contract W7714-020683/001/SV, April 2004.

[10] D.T. Gjessing. *Target Adaptive Matched Illuminate Radar: Principles and Applications*. IEE Electromagnetic Waves Series 22. Peter Peregrinus Ltd: London; 1986.

[11] M.R. Bell, S. Monroq. Diversity waveform signal processing for delay-Doppler measurement and imaging. *Digital Signal Processing* (Special issue on defense applications) 2002; 12:329–346.

[12] P. Weber, S. Haykin. Non-linear FM pulse compression radar systems. CRL Report #120. Communications Research Laboratory, McMaster University, Canada, 1983.

[13] D.F. DeLong, E.M. Hofstetter. On the design of optimum radar waveforms for clutter rejection. *IEEE Trans. Information Theory* 1967;IT-13:454–463.

[14] D.F. DeLong, E.M. Hofstetter. The design of clutter-resistant radar waveforms with limited dynamic range. *IEEE Trans. Information Theory* 1969; IT-15:376–385.

[15] D.P. Bertsekas. *Nonlinear Programming*. Athenas Scientific, 1995.

[16] S. Haykin, T. Kirubarajan, B. Currie. Adaptive radar for improved small target detection in a maritime environment. ASL Report 03-01, 2003. Simulation setup for optimum energy-constrained and amplitude-constrained waveform design. ASL Report 03-02, 2003. Literature search on adaptive radar transmit waveforms. ASL Report 02-01, 2002. Adaptive Systems Laboratory, McMaster University, Canada. (An edited version of these reports is reproduced in Chapter 6 of reference 2.)

[17] J.A. Simons. The resolution of target range by echo-locating bats. *Journal of the Acoustical Society of America* 1973;54:157–173.

[18] N. Suga. Cortical computation maps for auditory imaging. *Neural Networks* 1990; 3:3–21.

[19] J.A. Thomas, C.F. Moss, M. Vater. *Echolocation in Bats and Dolphins*. Chicago: The University of Chicago Press; 2004.

[20] G. Schuller, C.F. Moss. Vocal control and acoustically guided behavior in bats. In: J.A. Thomas et al., editors. *Echolocation in Bats and Dolphins*. Chicago: University of Chicago Press; 2004. p. 3–16.

[21] S. Haykin, editor. *Adaptive Radar Signal Processing*. Wiley; 2007.

[22] S. Haykin. *Neural Networks: A Comprehensive Foundation*. 2nd ed. Prentice-Hall; 1999.

[23] J. Si, A.G. Barto, W.B. Powell, D. Wunsch II. *Handbook of Learning and Approximate Dynamic Programming*. Wiley; 2004.

[24] V. Krishnamurthy, R.J. Evans. Beam scheduling for electronically scanned array tracking systems using multi-arm bandits. *IEEE Trans. Signal Processing* 2001;49:2893–2908.

[25] S.L.C. Miranda, C.J. Baker, K.D. Woodbridge, H.D. Griffiths. Multifunction radar resource management. Chapter 10, this volume.

[26] P. Weber, A. Premji, T.J. Nohara, C. Krasnor. Low-cost radar surveillance of inland waterways for homeland security applications. *2004 IEEE International Radar Conference*, Philadelphia, PA, 26–29 April 2004.

3

KNOWLEDGE-BASED RADAR SIGNAL AND DATA PROCESSING: A TUTORIAL OVERVIEW[1]

Gerard T. Capraro, Alfonso Farina, Hugh D. Griffiths, and Michael C. Wicks

Radar systems are an important component in military operations. In response to increasingly severe threats from low flying aircraft hidden in foliage, and in environments with large numbers of targets, knowledge-based (KB) signal- and data-processing techniques offer the promise of significantly improved performance of all radar systems. Radars under KB control can be deployed to utilize valuable resources such as airspace or runways more effectively, and to aid human operators in carrying out their missions. As battlefield scenarios become more complex with increasing numbers of sensors and weapon systems, the challenge will be to use already available information effectively to enhance radar performance, including positioning, waveform selection, and modes of operation. Knowledge-based processing addresses this need and helps meet the challenge.

This chapter begins with a short introduction to radar, covering its evolution and taxonomy, followed by an overview of the signal- and data-processing chain of modern surveillance radar. Section 3.2 provides a brief overview of artificial intelligence (AI) and an explanation of why knowledge bases and robotics are two main areas of emphasis for bringing KB into fielded radar systems, why KB radar systems do not require thousands of rules, and how Semantic Web technologies have a role in KB radar systems. The third section describes an end-to-end radar signal and data processing architecture for an

Knowledge-Based Radar Detection, Tracking, and Classification. Edited by Fulvio Gini and Muralidhar Rangaswamy

airborne surveillance radar and describes its over-arching KB processing and control. This section also contains information concerning a variety of KB algorithms and their results, illustrating the benefits and demonstrated performance enhancements within the filtering, detection, and tracking stages of a radar's processing chain. The fourth section contains our view of the future of KB radar research, including waveform diversity and intelligent sensor systems. References are provided, allowing the reader to find the details of the work discussed and references to similar efforts that researchers will find informative.

3.1 RADAR EVOLUTION

Radar (RAdio Detection And Ranging) dates back to Hertz of Germany in 1885–1888. Hertz experimentally verified the prediction of Maxwell's theory of electromagnetic (EM) fields published in 1864 using an apparatus operating at around 455 MHz, similar to today's pulsed radar. Hertz demonstrated that radio waves are reflected from metal objects. In 1904 Hülsmeyer, in Germany, saw a practical application to Hertz's work and developed and patented a monostatic pulsed radar for the purpose of preventing collisions at sea. However, it was not until 1920 that Marconi of Italy observed the radio detection of targets in his experiments. Shortly thereafter (1922), Holt and colleagues from the United States (US) Naval Research Laboratory observed a fluctuating signal at a receiver when a ship passed between that receiver and a transmitter located on opposite sides of a river. This was probably the first occurrence of a bistatic continuous wave (CW) radar. A decisive step in the development of radar as we know it today was the 1940 British invention (University of Birmingham) of a high power microwave (μw) magnetron for use in higher frequency radars. The two British researchers were Harry Boot and John Randall. The development of this magnetron permitted the detecting of targets at great distances, with radars using small antennas.

The development of radar accelerated during World War II (WWII), offering long-range detection and performance that was not degraded by darkness, nor for the most part by weather conditions. In the United States, over 300 units were delivered to the Navy and Army in 1941 alone. These radars were used primarily for long-range air searching for aircraft. One of these radars detected Japanese bombers en route to Pearl Harbor. In Germany, by the end of 1940, three major operational radars had been developed: the 125 MHz *Freya* air search radar, the *Wurzburg* fire control radar at 565 MHz, and the *Seetack* ship-borne radar. In the Union of Soviet Socialist Republics (USSR), *RUS*-1, a bistatic CW radar at 75 MHz with 35 km separation between the transmitter and receiver, and the *RUS*-2, a monostatic pulse system at 75 MHz [1, 2] had been built. Elsewhere in Europe during this same time period, the first operating radar at 30 MHz for the Chain Home system had been developed in the United Kingdom (1938) and the Italian Professor U. Tiberio had made a substantial contribution, the development of the radar equation [3]. In addition, there were significant contributions during WWII by Japan, France, and the Netherlands. More details on the history of radar can be found in reference 1, from which most of the previous notes are taken.

Moving Target Indication (MTI) in pulsed radar is used to separate desired target echoes from undesired clutter returns. High power stable amplifiers (klystron, traveling wave tube, solid-state transistors) allow better exploitation of the Doppler effect. Sophisticated waveforms and high power transmitters have been developed. Highly accurate angle tracking is achieved with monopulse radar. Pulse compression has been developed to permit the use of long duration waveforms to obtain high energy and, at the same time, achieve the resolution of a short pulse by the internal modulation of the long pulse and suitable processing of received echoes. In addition to detecting targets, radars are also used to create images. Synthetic Aperture Radar (SAR) produces high resolution map-like images of the ground or ocean, while Inverse SAR (ISAR) has been adopted for imaging of targets and planets. For a comprehensive introduction to SAR, see reference 4. Interferometric SAR (InSAR) techniques and their application to remote sensing and ground moving target detection are also valuable. A good tutorial on InSAR may be found in reference 5. Airborne MTI and pulse Doppler radar were developed to detect a target in the midst of heavy ground clutter. Electronically steered phased-array antennas permit rapid beam steering without mechanical movement. High frequency (HF), over the horizon radar detects targets up to 2000 nautical miles away. Non Cooperative Target Recognition techniques based on high range resolution radar, ISAR, jet engine modulation, polarimetry, and a combination of the above techniques have been invented. Radar has been exploited by meteorologists for predicting weather, storms, and to enable safe and efficient air travel (i.e. air traffic control). More recently, space-based radar has been developed for surveillance and remote sensing of the globe and for planetary exploration.

Rapid advances in digital technology have made many theoretical capabilities practical with digital signal and data processing. Adaptive antennas for jammer cancellation were invented by Howells (1957) and Applebaum (1959). Practical Space Time Adaptive Processing (STAP) was demonstrated by F. Dickey, F. Staudaher, and M. Labitt in 1991 for enhancing clutter cancellation using measured radar data. Advances in digital and software technologies during the 1980s allowed the US Air Force Sensor's Directorate to investigate the embedding of KB technologies in radar systems, and time will tell the significance and value of this endeavor. A team headed by Carol Shilepsky from Digicomp and Vincent Vannicola from the Air Force investigated the use of rule-based systems in wide area surveillance radar for resource allocation. Their approach focused on situation assessment for the applications of counter measures and focused on clutter map based control for the engagement of moving target indication (MTI) filters. At that same time, Gerard Capraro and Michael Wicks assembled a team to investigate the use of expert systems in the control of parameters in adaptive signal processing algorithms. Initial demonstrations in the laboratory focused on expert system control of constant false alarm rate (CFAR) algorithms for radar detection processing in nonhomogeneous interference environments. Applications to adaptive filtering and training data selection were also investigated. This research resulted in numerous technical papers and patent applications (e.g. US patents 5,499,030 and 5,706,013).

3.2 TAXONOMY OF RADAR

A radar system can be characterized according to any of its many features:

- *Location of the radar*: ground-based, ship-borne, airborne, spaceborne.
- *Functions performed by the radar*: surveillance, tracking, reconnaissance, imaging, data link.
- *Application of the radar*: air traffic control (terminal area, en route, collision avoidance, apron), monitoring of surface traffic in airports (taxi radar), air defense, antitheatre ballistic missile defense, vessel traffic surveillance, remote sensing (application to crop evaluation, hydrology, geodesy, archaeology, astronomy, defense), meteorology (hydrology, rain/hail measurement), atmospheric science (detection of micro-burst and gust, wind profilers), spaceborne altimetry for measurement of sea surface height, acquisition and tracking of satellites in re-entry, monitoring of space debris, collision avoidance, ground penetrating radar (geology, gas pipe detection, archaeology, detection and location of mines, and so on).
- *Frequency band of operation.*
- *Radar main beam scanning*: fixed beam, mechanical scan (rotating, oscillating), electronic scan (phase control, frequency control, and mixed in azimuth/ elevation), mixed (electronic-mechanical) scan, multibeam configurations.
- *Types of data obtained from a radar*: range (delay time of echo), azimuth (antenna beam pointing), power (amplitude of echoes), elevation (only for 3D multifunctional and tracking radars), height (derived by range and elevation), intensity (echo power), radar cross-section (RCS; derived by echo intensity and range), radial speed (measurement of differential phase due to the Doppler effect requiring coherency), polarimetry (phase and amplitude of echo in the polarization channels: HH, horizontally transmitted, horizontally received; HV, horizontally transmitted, vertically received, VH, VV), RCS profiles along range and azimuth (high resolution along range, imaging radar).
- *Radar configuration*: monostatic (co-located transmit and receive — same antenna, mono-radar/multi-radar), bistatic (separated transmitter and receiver — two antennas), multistatic (one or more transmitters/receivers spatially dispersed).
- *Transmitted waveform*: continuous, pulsed, digitally synthesized.
- *Radar signal and data processing*: coherent (MTI/MTD/pulse-Doppler/super-resolution/SAR/ISAR, and so on), noncoherent (integration of envelope signals, moving window, adaptive threshold, e.g. constant false alarm rate (CFAR)) and mixed.
- *Major radar hardware technologies*: antenna (reflector plus feed, array (planar, conformal), corporate feed/air-coupled/lens), transmitter (magnetron, klystron, travelling wave tube (TWT), mini TWT, solid-state) and receiver (analog and digital technologies, base band, intermediate frequency sampling, and so on; the relevant parameters of the receiver are noise figure, bandwidth, and dynamic range).

3.3 SIGNAL PROCESSING

Radar signal processing enables the extraction of desired information while rejecting unwanted interference. In particular, a surveillance radar reports on the presence or absence of targets while cancelling radar echoes from clutter, radio frequency interference, and noise sources. An airborne radar accomplishes the same job in spite of Doppler spread clutter returns. A tracking radar, in addition to detection, is concerned with an accurate estimation of a target's kinematic parameters. The list could be extended to other radar systems such as low probability of intercept, synthetic aperture, space-based, and multistatic. Whatever the radar system, the basic operations performed by the signal and data processors are the detection of targets, if any, and the extraction of information from the received waveform to determine a wealth of relevant parameters about the targets (such as position, velocity, shape, and electromagnetic signature). The first step of the design process can be recognized in the formulation of mathematical models more adherent to the real environment in which the radar operates. Several major areas of research and development can be singled out in connection with radar detection: theory of optimum detection, adaptive detection theory, processing of non-Gaussian signals, multidimensional processing, and super-resolution. Some of these are extensively described in the literature [6]. Many of these techniques have been successfully demonstrated in fielded radar systems.

MTI, MTD, and pulse-Doppler radar concepts are successful processing schemes to reject clutter echoes and detect targets. Significant research has focused on accurate statistical modeling of clutter designed echoes. A remarkable result is the collection and processing of data from vegetated ground clutter organized by MIT–Lincoln Laboratory, probably the most extensive collection of ground clutter data. Some of these data have been processed with a variety of algorithms, the results being described in reference 7. Furthermore, under certain conditions (radar operating at low grazing angles and/or with a high resolution) the clutter echoes could be described as spherically invariant random processes that are clearly non-Gaussian in nature.

Clutter returns and/or jamming can be much larger than receiver noise. As a consequence, the detection threshold can be exceeded and false alarms can occur. This could overload an automatic detection and track system. CFAR automatically raises the threshold level thus avoiding overload of the automatic tracker with extraneous target reports [1]. CFAR is achieved at the expense of a lower probability of detection of desired targets. Cell averaging (CA) CFAR, due to Finn and Johnson, is optimum when the statistic of the envelope is Rayleigh distributed. It uses an adaptive threshold whose level is determined by the clutter and/or noise in the vicinity of the radar echo. For a review of other CFAR algorithms, see reference 8.

Adaptive signal processing applies to three different types of radars or application areas:

1. Ground-based or ship-borne radars for clutter cancellation;
2. Ground-based or ship-borne radars equipped with a multichannel phased-array antenna for jamming cancellation; and

3. Airborne early warning (AEW) radar equipped with a multichannel phased-array antenna for clutter and jammer cancellation.

A linear combination of (assumed independent and Gaussian) received signals is performed in all three radars. The filter output is envelope-detected and compared against a suitable detection threshold set so as to maximize the detection probability (P_d) and to obtain a prescribed probability of false alarm (P_{fa}). Let N be the number of spatial and/or temporal degrees of freedom used in adaptive processing, that is, the dimension of the snapshot of radar data. In application area [1], ground-based or ship-borne radar, N is the number of echoes (T seconds apart, where T is the radar pulse repetition time, PRT) captured by the radar receiver commensurate with a train of N coherent pulses transmitted. The clutter interference is cancelled by an adaptive filter, operating in the Doppler frequency domain, which rejects the Doppler frequency interval occupied by the clutter spectrum. The filter sets a peak at the Doppler frequency that is expected from the anticipated threat target. In application area [2], multichannel ground-based or ship-borne radar, N is the number of subarrays (in which the radar antenna is decomposed) and associated receiving channels. Jamming and radio frequency interferences are cancelled by adaptively shaping the received antenna pattern to place deep nulls in the directions of arrival of interference. In application area (3), AEW radar, N is the product of the number of received radar echoes, T seconds apart, and the number of subarrays into which the array aperture is decomposed. Both the clutter and jammer are cancelled by synthesizing a two-dimensional filter that operates in the domains of Doppler frequency and direction of arrival. This type of filtering is also referred to as STAP [9]. The common problem of all three applications is the calculation of the weights to be used in the linear combination for the derivation of the adapted filter output. While the desired signal can be assumed to be known a priori, the interference is not known and is changing with time and in space. This means that the interference characteristics (represented by a covariance matrix) have to be estimated in real time.

A direct implementation by means of Sample Matrix Inversion of the adaptive filter is not recommended. One reason is related to the poor numerical stability in the inversion of the interference covariance matrix, especially when a large dynamic range signal is expected during the operation. Another difficulty is the very high computational cost. Much attention today is placed on reduced-dimension STAP with the intent to limit the computational burden and the number of secondary data for adaptivity. Common transformations include beam forming and Doppler processing steps. The Factored Time–Space (FTS) algorithm is a post-Doppler method suitable for long coherent dwells and high radial velocity targets. The FTS method essentially involves spatial notching of the clutter in a given Doppler filter, but it does not exhibit joint domain adaptivity. To enhance performance with only a modest increase in required sample support and computational burden, Di Pietro introduced the Extended Factored Algorithm (EFA). The EFA method involves adaptively combining several adjacent Doppler filters (typically three or five) and all spatial channels. The EFA method often exhibits performance very close to the theoretical joint-domain space–time bound. To provide diversity in spatial and temporal degrees of freedom, Wang and Cai [10] developed the Joint Domain Localized (JDL) technique, which is a post-Doppler, beam space method. Basically, the processor forms multiple beams, then Doppler processes each beam and finally selects a collection of adjacent angle-Doppler bins over which to adapt the filter response. JDL provides good performance with very low training data requirements and very modest computational burden. Three adjacent beams by three adjacent Doppler bins form a typical localized JDL processing region.

Reduced-rank STAP methods also involve data-dependent transformation and selection. The principal component is one such method. A benefit of this approach is a reduction in training data support. However, computational burden remains high, as the processor must compute eigenvalues and eigenvectors of the interference covariance matrix.

From a practical point of view, the first (nonadaptive) Displaced Phase Center Antenna (DPCA) experiment involving an array antenna was carried out by Tsandoulas in the late 1950s. More recent experiments conducted in the United States adopt linear side-looking arrays. The Multi-Channel Airborne Radar Measurement (MCARM) and Mountaintop programs, started in 1990 to study advanced processing techniques and technologies to support the requirements of future airborne early warning radar platforms, studied the effect of geometry on performance.

3.4 DATA PROCESSING

Target tracking and identification are portions of the radar processing chain that are also referred to as data processing. In this chapter we only address target tracking. The tracking filter processes the target measurements (e.g. range, azimuth, elevation, and range rate) in order to reduce the measurement errors by means of a suitable time average, estimate the velocity and acceleration of targets, and predict future target position. The tracker can be considered as an application of stochastic filtering theory, which is an important branch of the modern theory of dynamic systems. The latter is characterized by the dynamic evolution of system state variables, optimal control under well-defined disturbances and inputs, use of stochastic processes to model noise-corrupted data, and uncertain parameters. The concept of a dynamic system is introduced to obtain a mathematical description of the input-output behavior of a physical object of interest, such as the time history of the position of an aircraft. Deterministic system theory is not sufficient for the practical design of operational systems. First, no mathematical model of a system is ever complete. Approximations, uncertainties, and neglected or misunderstood effects are inherent ingredients. Second, dynamic systems are driven not only by input commands but also by disturbances from the environment and imperfections in the actuator's ability to deliver commanded controls. These are uncontrollable effects for which there are usually no adequate deterministic models. Finally, sensors that provide data about the system may deliver only partial information about the system state. They introduce latency and new dynamics and are always noise-corrupted. These considerations justify the extension of dynamic system concepts to the stochastic case where the aforementioned uncertainties and approximations are modelled as random. A fundamental problem then arises in that it is necessary to estimate dynamic states on the basis of mathematical models and noisy data. The solution is given by optimal filtering theory. Important breakthroughs in this theory have been the Wiener filter for stationary processes and the Kalman–Bucy filter, which represent the optimal filter when both the dynamic state and the measurement equations are linear and the forcing and measurement noises are independent and Gaussian.

Textbooks and monographs, such as reference 11, are available that discuss tracking problems in detail. Recently, the need to perform more accurate estimates in a non-linear, non-Gaussian environment has prompted the conception of new approximations to optimum filtering that effectively go beyond the Kalman–Bucy theory [12].

3.5 INTRODUCTION TO ARTIFICIAL INTELLIGENCE

Current signal processing systems are built assuming Gaussian interference and are optimized for their processing requirements, whether the systems are mounted on aircraft, missiles, spacecraft, or at a ground-based site. The algorithms are "hardwired" into a computer architecture in order to meet real-time requirements demanded by the sensor's operating parameters, for example, update rate and number of sensor elements. This approach to building radar systems is being assessed today by the radar research and development community because of its rigidity and high costs, and changes will evolve slowly. This evolution will manifest itself such that different algorithms and/or their parameters will be modified by the radar's software as the environment changes. Some of the most progressive and earliest work in employing artificial intelligence (AI) techniques has been pursued by the US Air Force Research Laboratory's Sensors Directorate [13, 14]. Some of their original efforts have been in the CFAR portion of a radar's signal-processing chain.

This work was extended beyond the detection stage to the entire radar processing chain in the report in reference 14. This effort demonstrated the benefits of using outside data sources to affect the filtering, detection, and tracking stages of a surveillance radar sensor. Another program showed the benefits of using map data obtained from the US Geological Survey to improve the performance of STAP on an airborne radar by selecting range rings based on computed criteria rather than blindly choosing reference range rings surrounding the test range ring. This effort [15, 16] and others laid the groundwork for a new program called KASSPER (Knowledge Aided Sensor Signal Processing and Expert Reasoning). The objective of KASSPER was to investigate the use of outside data sources to dynamically change a radar's signal processing chain to enhance a radar's performance.

This section provides an overview of KB technologies that are needed for building radar systems that dynamically change their algorithms as required to enhance performance. Current KB algorithms only exercise simple rule-based logic, but are showing great promise. However, in the near future, we will be developing KB radar systems that will not only change algorithms dynamically, but will also be able to explain their actions and learn from their gathering of data, information, and by measuring performance.

3.5.1 Why Robotics and Knowledge-Based Systems?

Modern day AI has been around since the 1950s. Rich [17] defines AI as "the study of how to make computers do things at which, at the moment, people are better." Barr and Feigenbaum [18] define AI as "the part of computer science concerned with designing intelligent computer systems, that is, systems that exhibit the characteristics we associate with intelligence in human behavior." It is conjectured that there are varied definitions of AI because the field contains many subfields or areas of interest within its general domain. Some of these areas are planning, robotics, speech recognition, natural language processing, and expert systems.

In the areas of AI the most applicable to the radar domain are expert systems and robotics. Robotics is relevant because we want radar signal processing to change both on transmit and receive, depending upon its goals and position in the environment, and operate autonomously. Expert systems and knowledge bases are important because we want to change the transmit waveform and receive signal processing in response to a changing environment in an "expert" fashion. However, we do not have an expert; that is, there is no human that currently modifies the signal-processing software chain in real time based upon the changing environment. The radar community is just beginning to develop the rules or heuristics for determining how and when the processing chain should be changed. This is why the research community emphasizes the knowledge base and not the expert system or robotics areas of AI. As the community becomes more knowledgeable and proven techniques are obtained, more robotic-based solutions will be adapted. First, a KB approach will be built that operates with human intervention. Then we will capture the human intervention portion of the system and imbed this capability into an expert system. When this approach is proven for multiple sensors, even operating on airborne platforms, then these sensors may operate autonomously as robots.

3.5.2 Knowledge Base Systems (KBS)

A Knowledge Base Systems (KBS) has three main components: the user interface, the knowledge base, and the inference engine. This chapter will briefly discuss knowledge bases and inferencing.

The design and documentation of the knowledge base is the most important part of building a KB system, along with choosing how to describe it, that is, predicate logic, semantic nets, or frames. This choice is dependent upon the domain of interest, the purpose of the knowledge base, and the inference engine tool that is best for building the solution. Most KBS implementations are very large and contain numerous rules and facts. The embedded KB techniques we plan on implementing within a radar are not foreseen to require thousands of rules. Current research has shown significant results can be obtained by only implementing dozens of rules for each KB instantiation. This is because we are using KB or AI tools to choose one algorithm or parameter over another. We are not trying to supplant the current mathematical bases of radar systems, but to apply them appropriately. Our domain space is minimal and controlled. We do, however, expect to build systems in the future that will learn from its choices and evolve rules and an improved understanding as the technology and research progresses.

3.5.3 Semantic Web Technologies

In addition to leveraging KB and robotics, there are additional technologies being developed to enhance Web technologies that allow software to be written that understands the contents of Web pages. This technology can help multiple sensor systems share data and information and enhance their respective functions.

The World Wide Web Consortium (W3C) "develops interoperable technologies (specifications, guidelines, software, and tools) to lead the Web to its full potential. W3C is a forum for information, commerce, communication, and collective understanding." (See www.w3c.org.) They, along with the Defense Agency Research Project Agency (DARPA) Agent Markup Language (DAML) program, are building the next generation Internet or the Semantic Web. Leveraging these technologies (e.g. ontologies, XML, resource description framework) will allow for a platform of sensors to share and fuse data and information between sensors on a platform, multiple platforms located nearby, or miles away within a command center.

3.6 A GLOBAL VIEW AND KB ALGORITHMS

The performance of sensor systems can be enhanced by dynamically controlling algorithms based upon understanding the changing environment. A radar system can perform better if it knows where potential jammers are located and their characteristics. If an airborne radar system knows about certain features of the Earth (e.g. land–sea interfaces), terrain type and its surroundings, then it can use this information intelligently and improve performance. The sharing of information in real time with other sensor systems is very desirable, for example, sharing information between the surveillance radar system, navigation system, and the electronic warfare system on board an aircraft.

However, if an airborne radar is going to share and receive information from multiple sources, then it must be able to communicate and understand the information. A solution for the exchange of information between heterogeneous sensors is for each sensor to publish information based upon shared ontologies. In this manner, when a sensor publishes its track data, multiple sensors receiving this information will be able to interpret its contents without ambiguity. Each transmission of information between sensors must depict date/time, their unique identifier, their coordinates, pitch, yaw, and roll of the sensor's platform, and metadata describing the transmission.

In addition, if sharing track or target data, it must specify, for example their identifiers, RCS, velocity, and position. Unique identifiers will allow the receiving sensor to acquire, within its resident database, all of the sender's radar characteristics. The description of these data can be defined by ontologies such that all the sensor platforms will correctly understand the information provided, for example, sensor and platform characteristics. We require ontologies to define these data, and rules to explain the relationships among the data, so that the information published by any sensor can be understood correctly by the receiving sensor in order to perform functions such as fusion, track correlation, and target identification.

Sharing information between sensors on the same platform is also required, especially if one or more sensors are adaptively changing their waveform parameters to meet the demands of a changing environment. Figure 3.1 depicts a hypothetical intelligent sensor system. Each of the sensors has its own signal and data-processing functional capability. In addition to this capability we have added an intelligent processor to address fusion and communications between sensors and control of the sensors. Our goal is to be able to build this processor so that it can interface with any sensor and communicate with the other sensors using ontological descriptions via the intelligent platform network. This intelligent network will be able to coordinate the communications between the onboard and off-platform systems. There are also communications issues that need to be addressed for the sharing of information and for minimizing the potential of electromagnetic (EM) fratricide. The intelligent platform should determine if there is EM interference when a sensor varies its antenna's look direction or changes its waveform parameters, which may thereby

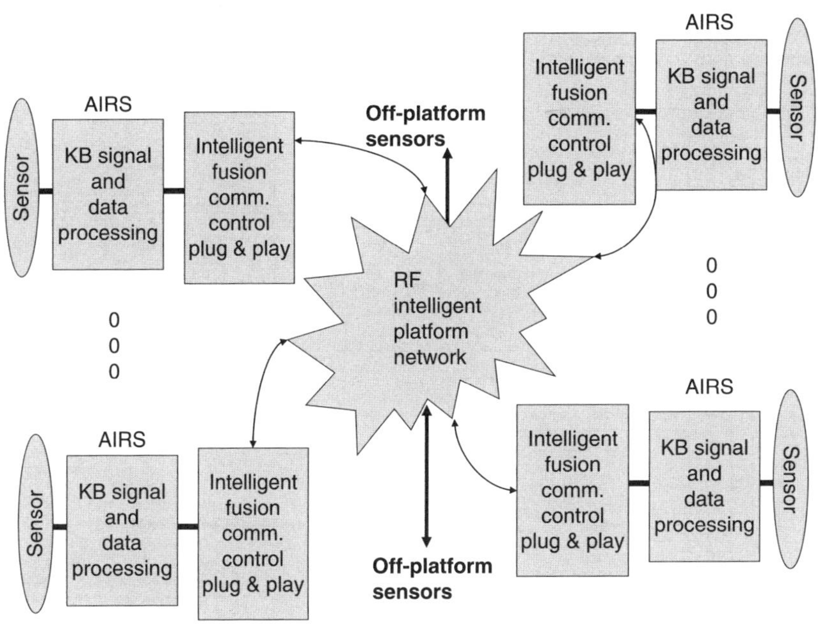

Figure 3.1 An intelligent sensor system.

cause additional interference. Rather than permitting each sensor on a platform to operate independently we designed a system of sensors with multiple goals controlled by an intelligent platform network that manages the dynamics of each sensor in order to meet the common goals of the platform.

3.6.1 An Airborne Autonomous Intelligent Radar System (AIRS)

The KB signal and data processing blocks shown in Fig. 3.1 represent one radar sensor. If this radar sensor is built using KB techniques, then processing is intelligently controlled. A modified design obtained from the KBSTAP effort [14] is shown in Fig. 3.2. The KB controller serves as the major integrator for communications and control of individual processors. These processors operate both independently and cooperatively. Each can be implemented on a separate processor, or on the same computer, and operate as separate software processes. It is also envisioned that each processor, using Semantic Web technologies, may some day be able to communicate with other processors on the same or different platforms. The knowledge base controller (KBC) receives information from many sources. Data about the radar, its frequency of operation, antenna configuration, where it is located on the aircraft, and so on, are provided by the block labeled "configuration information" (Fig. 3.2). Map data is preloaded prior to each mission to aid in the estimation of

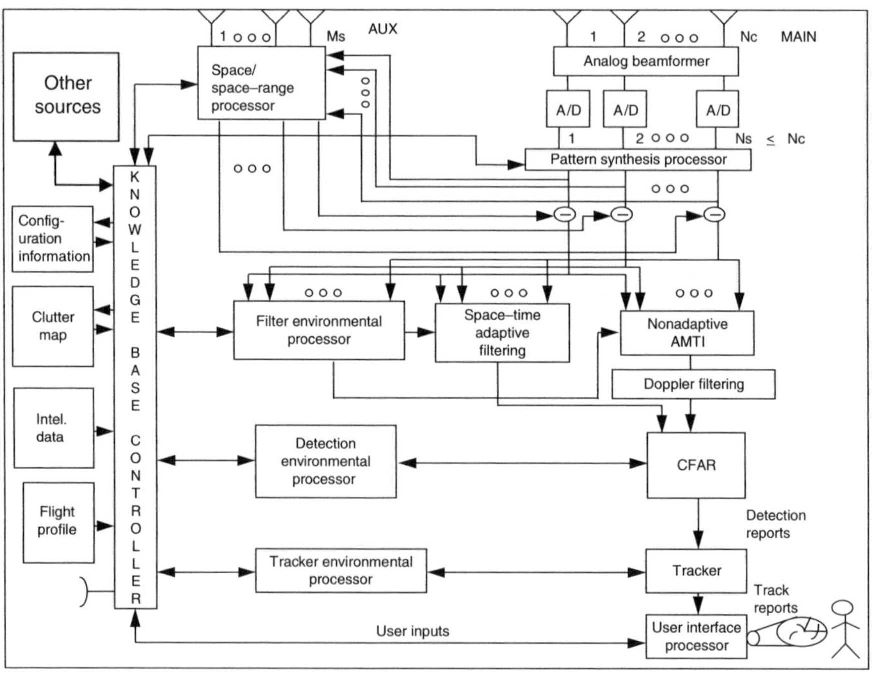

Figure 3.2 An AIRS architecture.

clutter statistics and to register platform location relative to the Earth and to other sensors. It is also preloaded with flight profile data and is updated continually by the platform's navigation system. It could receive intelligence information, both before and during a mission. In flight, the KBC will receive information about weather, jammers, requests for information, discretes, sensor fusion, and so on. The radar system is assumed to be aboard a surveillance aircraft flying a known and repeatable path over the same terrain. Therefore it can learn by monitoring the performance of different algorithms over repeatable passes of terrain. In the future, a truly intelligent sensor system may even optimize its flight path based upon analysis of real-time and archival data.

The KBC performs the overall control functions of the AIRS. It assigns tasks to all processors, communicates with external resources, and optimizes the system's global performance. Each individual processor optimizes its individual performance measures, for example, waveform parameters and probability of detection. The tracker in conjunction with the KBC, for example, optimizes the number of correct target tracks and minimizes the number of missed targets, incorrect tracks, and lost tracks. The KBC handles all interrupts from the User Interface Processor, assigns tasks to the individual processors based upon external requests, generates information gathered from sources to enhance the performance measures of the individual processors, works with other sensors and outside sources for target identification, and provides the User Interface Processor with reports for answering queries and requests from users. A more detailed description of each processor and how they change over time can be found in reference 19.

AIRS may operate within the Net-Centric data strategy and the global information grid (GIG) as shown in Fig. 3.3. One of the major driving forces of the Net-Centric approach is to provide information and instructions to as many users and resources as quickly as possible. Analyses of the data can be performed in parallel by numerous users and organizations rather than waiting for the developer of the information to

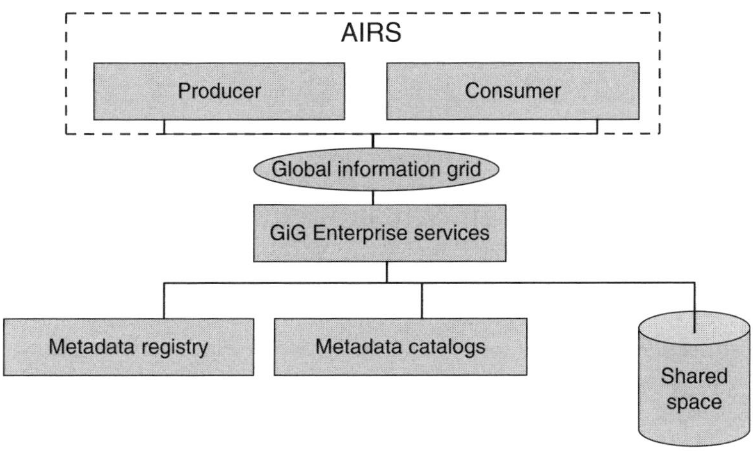

Figure 3.3 AIRS and the Net-Centric approach.

process data completely before reporting. There are three types of users in a Net-Centric strategy: the developer, the producer, and the consumer of data and information. The developers are those people that understand the GIG enterprise services and their capability along with knowledge of the domain of interest, in our case a radar system. The producers and consumers are users and/or processes that either produce or consume data and information that are published in a shared space, and produce or use the metadata in the catalog. Consumers can search the catalogs for data and information, or if they are aware of existing data as defined in the metadata catalog, they can periodically poll the shared data space, or they can subscribe to publications that are made available whenever published.

3.6.2 Filtering, Detection, and Tracking Algorithms and KB Processing

The AIRS architecture could be tailored for a ground, air, or space-based radar. In this section, algorithms are presented where the use of KB technologies, or information other than the radar data, have improved performance. The work presented is based on data obtained from the Multi-Channel Airborne Radar Measurement (MCARM) program except for the detection and track algorithms.

The pre-filtering and filtering stages of a radar can utilize outside information to enhance processing. If one knew where a jammer was located, then nulls could be placed in an antenna's beam pattern to limit the degradation of the radar. Having map data correlated with the ground may also be used to enhance a radar's performance. Antonik et al. [20] have shown that the use of map data can be used to place nulls in either the spatial or temporal domains of a radar's response. The authors showed marked improvement when placing nulls on road traffic, which affected the signal-to-interference-plus-noise ratio (SINR) of processed MCARM data. A test target was injected in the data near three intersecting roads. Each of the roads can be expected to provide traffic with the same Doppler of interest within the radar's sidelobes. Results of using classical STAP without pre-nulling yielded multiple false targets and an SINR of 3.7 dB. When spatial pre-nulling is implemented, the average SINR increases to 13.2 dB and the number of false targets are reduced. This work shows the benefits of using map data to help determine where and when to change the gain pattern of an airborne antenna.

KB techniques in the filtering process may involve eliminating heterogeneous test cells before filtering, because STAP assumes that the training cells are independent and identically distributed. The authors of reference 21 demonstrate the need to test the radar return data before applying STAP and thereby eliminating training cells based on their heterogeneity. An approach that they propose is presented in references 19 and 21 where, depending upon the clutter return data, different STAP algorithms will be applied. More information on this approach is presented in reference 22.

In a separate effort Bergin et al. [23] also looked at the effects of using map data to determine where roads are located and removing these cells from STAP training data.

Bergin also developed a simulation model of the MCARM test area, and obtained good correlation between measured data analysis and simulation results. Bergin demonstrated an improvement of the order of 15 dB. He used digital line graph (DLG) data to determine where roads are located and used this knowledge to improve STAP.

Researchers within the KASSPER program investigated the use of land use land cover (LULC) data and DLG to help in the selection of training data. Efforts in references 15 and 16 have shown improvements in post-Doppler STAP using LULC data only. Figures 3.4 and 3.5 present MCARM data analysis with returns from a moving target simulator located at range bin 450. Figure 3.4 presents analysis using training rings symmetrically around the test cell. Figure 3.5 shows results using STAP training data containing similar LULC characterizations and Doppler as the cell intersecting the test ring. The preferred performance measure (PPM) shown in both figures represents the ratio of the target's modified sample matrix inversion (MSMI) statistic to the average MSMI for all ranges at the Doppler of interest. There is a 3 dB improvement in target detection using this approach and, depending upon where the target threshold is set, there is a marked improvement in the false alarm control.

The detection stage of signal processing was introduced earlier and results are presented below. A common algorithm used in most radar systems is CA-CFAR. This algorithm is optimum if the clutter is Rayleigh distributed. However, a ground-looking radar mounted on an aircraft will not always encounter this ideal

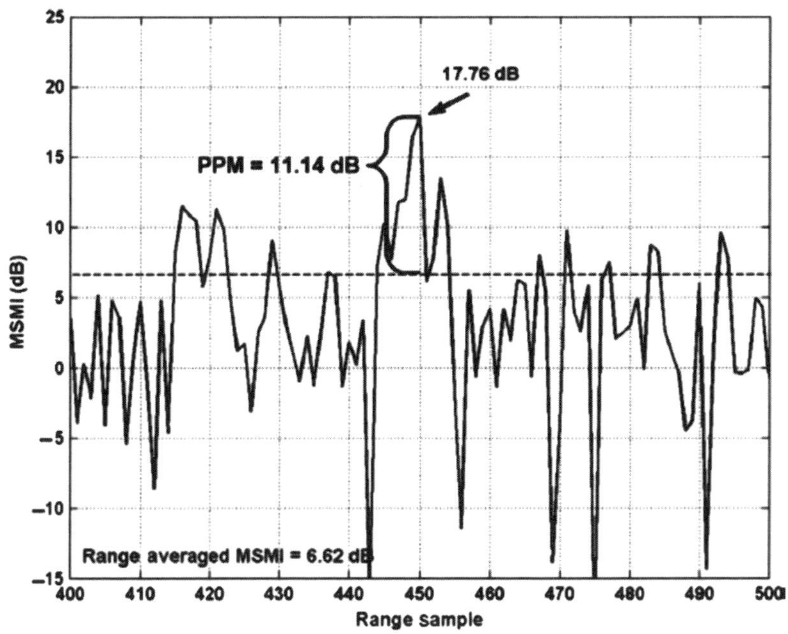

Figure 3.4 Sliding window STAP results for half array.

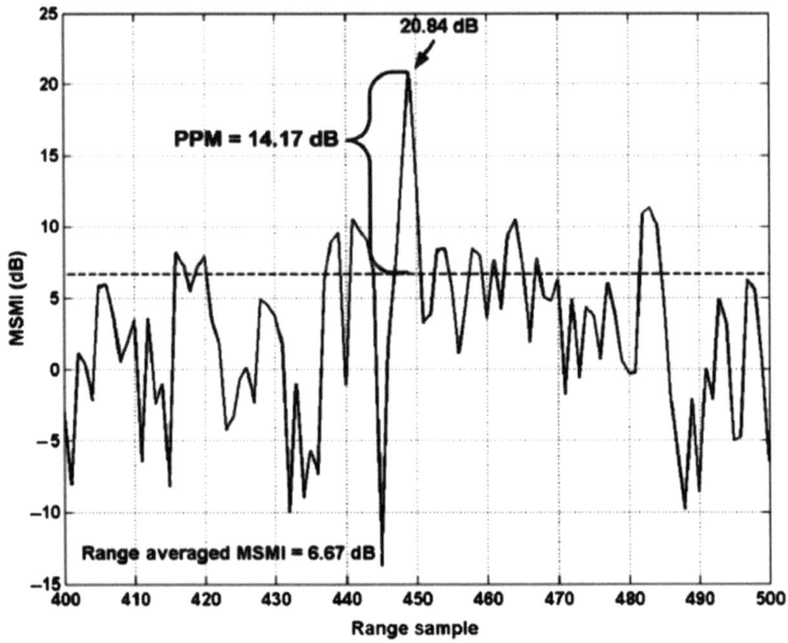

Figure 3.5 STAP results with similar LULC data for half array.

situation. Where the distribution is not Rayleigh and when some of the surrounding cells are of a different clutter type, such as at a land–sea interface, researchers have developed other algorithms such as trimmed mean CFAR and ordered statistic CFAR. The US Air Force began a research program to determine which of these algorithms would be best given the type of terrain an airborne radar was illuminating. A set of rules were developed along with a voting scheme based upon five CFAR algorithms: CA, TM, GO, SO, and OS. An expert system (ES) tool was developed and simulations were performed that showed that the ES approach outperformed CA-CFAR in heterogeneous terrain [24]. The authors took this research one step further by comparing it with actual data from an E-3A radar. The results of that experiment are provided in Fig. 3.6, which compares ES-CFAR to CA-CFAR and OS-CFAR.

The tracking stage of the signal processing chain also uses terrain/mapping data to obtain information for improving performance. A simulation was developed [25] that evaluated different modeling approaches for tracking aircraft traveling on the ground, from the runway, taxiways, and to the terminal. The modeling approaches used were as follows:

- The first is an Extended Kalman Filtering (EKF) without constraints.
- The second is an Interactive Multiple Model (IMM) with four models — three constrained and matched respectively to a runway, two taxiway and an unconstrained model for maneuvers.

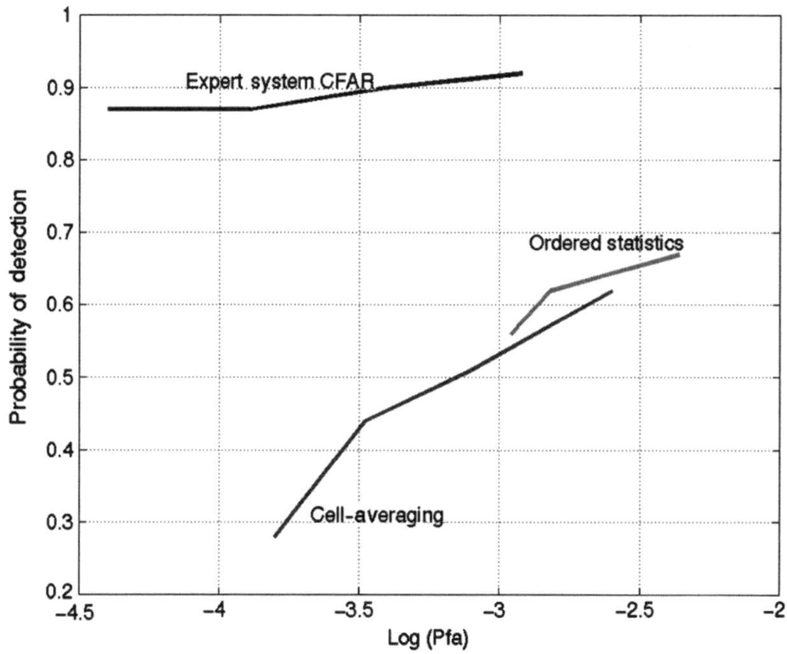

Figure 3.6 ES-CFAR measured E-3A results.

- The third is an IMM with seven models — three constrained to a runway, two taxiway; two constrained and matched to the trajectory of an aircraft during maneuvers; a constrained model matched to a centripetal maneuver that the airplane will not perform ("not performed maneuver"); and, finally, an unconstrained model.
- Finally, a Variable Structure–IMM (VS-IMM) with the same seven models is considered. The set of models and the transition probabilities of the VS-IMM are updated taking into account a map of the airport and the track estimation in the previous step.

The results of simulations plotting the standard deviation versus frame (or time) of the target position are shown in Fig. 3.7. It can be noted that the IMM with four models and the VS-IMM estimation are more accurate than the EKF due to the use of the constraints and to the correct selection of the models provided via knowledge of the runways.

This VS-IMM approach can also be used for GMTI for tracking targets on and off roads from an airborne sensor and where the topography of the terrain is considered. A topography-based variable structure mechanism is developed that eliminates the need for carrying all the possible models throughout the entire tracking period as in the standard IMM estimator. The specification of a road map with visibility conditions can be tabulated including road segment, the waypoints, the visibility condition, and indication whether it is possible to enter or exit a road segment.

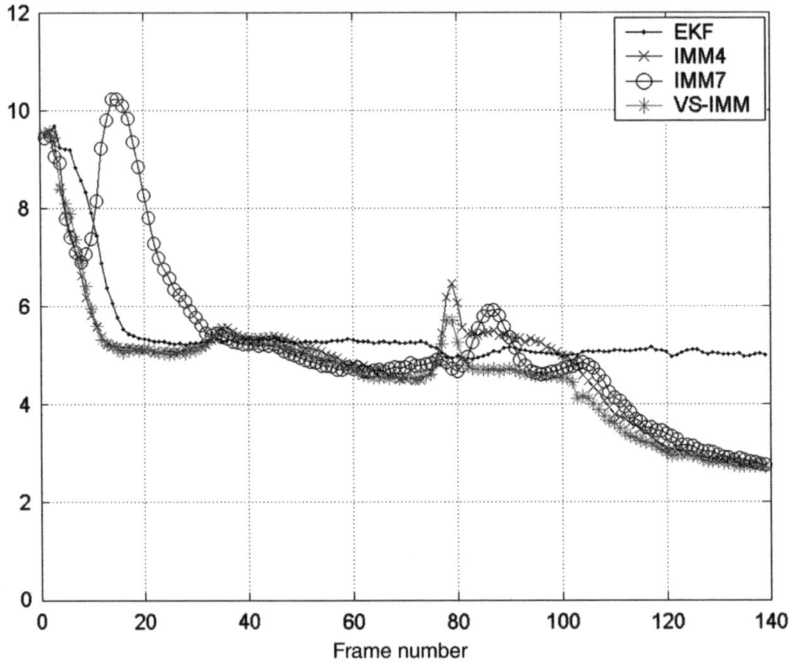

Figure 3.7 Standard deviation (*m*) of the target position error along the *X*-axis. Results are averaged over 10,000 Monte Carlo trials. (*Source*: reference 25.)

Unlike an off-road-capable target, which is free to move in any direction, the motion of an on-road-only target is highly constrained. To handle motion along the road the concept of directionally dependent noise is introduced [26]. The standard motion model assumes that the target can move in any direction and, therefore, uses equal process noise variances in both the x and y directions. This means that for off-road targets the motion uncertainties in both directions are equal. For on-road targets, the constraint means more uncertainty along the road than orthogonal to it. Thus, the IMM module representing on-road motion consists of process noise components along and orthogonal to the road, rather than along x and y directions as in the standard model. Extensive simulations have been performed [26] that analyzed four algorithms: IMM with two and three dimension assignments and VS-IMM with two and three dimension assignments. Analysis shows that the best performance is obtained using VS-IMM with three dimension assignment followed by the VS-IMM with two dimension assignment, and was able to handle the on-road/off-road transitions and the change from one road to another more smoothly than the fixed IMM by anticipating target dynamics. Also, once the target begins to move along a particular road, the VS-IMM, which uses a model matched to the road, yields a better course estimate than the fixed IMM, which uses an open field model.

The VS-IMM concept is combined with a particle filter (PF) technique giving rise to an algorithm called the variable structure multiple model particle filter

(VS-MMPF) in reference 27. The key features of the VS-MMPF are the number of models active at any particular time, and the state transitions vary depending on the current state and topography. Simulation results compare the VS-IMM and the VS-MMPF for a single target moving on a network of roads. The VS-MMPF provides a 65% improvement in tracking accuracy with respect to the VS-IMM for road targets; with knowledge of speed constraints the improvement in mean square error is ~77% [27]. These models and simulations realize performance increases through the use of map data and the ability to look ahead and realize which models are best.

A unique use of knowledge in tracking was presented in reference 28 where the authors showed that by knowing the possible kinematics of an aircraft they could maintain track while the maneuvering aircraft tried to force a drop in track by flying in and out of the radar's Doppler blind zone (DBZ). Problems with tracking targets occasionally hidden in the blind Doppler of a radar are investigated in reference 28. The authors demonstrate that by using prior knowledge to the limits of the DBZ, one can design a tracker that will perform better, in terms of track continuity, than one that ignores this prior information. The concept has been proven using particle filter theory, mainly because nonstandard information (such as blind Doppler) can be easily incorporated into the framework of a sequential Monte Carlo estimator.

3.7 FUTURE WORK

As the techniques described in this chapter mature, and as future radar systems become more and more complex, we can expect greater use of KB techniques. Some particular techniques and potential applications include the following.

3.7.1 Target Matched Illumination

The original concept of the matched filter illumination by Gjessing [29] and Bell [30] considered the optimum waveform for the detection of a target of a given range profile against noise. The target is characterized in terms of its impulse response as a function of delay time (i.e. range), which will also be a function of aspect angle (and in practice would require a library of target impulse responses versus aspect angle). The concept has been extended by Guerci and Pillai [31] to include detection and discrimination of multiple targets in colored noise.

3.7.2 Spectral Interpolation

This technique allows interpolation between two (or more) spectral bands to give the effect of a signal of very high bandwidth. The technique was demonstrated by Cuomo et al. at MIT Lincoln Labs [32]. The scheme may be regarded as a variant of super-resolution processing, and essentially involves modeling signals in each sub-band with an all-pole filter, adjusting the poles until they cohere, and then interpolating the missing spectrum.

3.7.3 Bistatic Radar and Passive Coherent Location

KB techniques may have several applications in geometrically diverse radar systems. In passive coherent location, broadcast or communications signals are used as the illuminating sources rather than dedicated radar transmissions. However, it is found that the ambiguity performance of the signals depends significantly on bistatic geometry and (for analog modulation, at least) on the instantaneous modulation of the signal [33, 34]. Both of these dependencies are deterministic, suggesting that the optimum sources to be used to track a particular target can be dynamically selected in real time.

3.7.4 Synthetic Aperture Radar

High resolution synthetic aperture radar is now widely used for military surveillance as well as in geophysical remote sensing applications. Although huge volumes of image data are readily generated, one of the major challenges is to extract information from those images in a reliable and automated manner. The use of prior information in a KB approach is clearly attractive. Blacknell [35] has considered the use of 'context' in the detection of vehicles in SAR images, making use of the fact that military vehicles are more likely to be found in groups close to hedges and the edges of woods rather than individually in open areas. His results show a consistent improvement of the order of 13% in the number of targets detected.

3.7.5 Resource Allocation in a Multifunction Phased Array Radar

A multifunction phased array radar is able to perform several different functions such as tracking, surveillance, missile guidance, kill assessment, all sharing the same antenna system. The greatest benefit of such a radar is its flexibility to undertake all these functions, changing the radar parameters such as waveform, frequency, pulse compression, dwell time and beam shape, in order to cope well with all the different environments and operational scenarios. The functions consist of a set of individual tasks that are competing for radar resources at any given instant. Because of this the allocation requirement radar resources must be executed efficiently by a resource manager to optimize the performance of the overall system. Two important processes must be undertaken to allow optimum resource management. The first is the prioritization of the tasks that must be performed. The second is the scheduling that consists of forming a queue of tasks in a time line to be executed by the radar.

Knowledge-based approaches have been used in recent years to design efficient scheduling algorithms. Researchers have proposed numerous approaches:

1. Scheduling processes to interleave tasks using neural networks;
2. Scheduling tracking and surveillance tasks based on operations research theory and on temporal logic using artificial intelligence;

3. Algorithms using the concept of on-line coupled-task scheduling where the time between the pulse transmission and reception are used for interleaving new tasks; and
4. The use of fuzzy logic to introduce concepts such as dangerous and friendly in the scheduling process in order to resolve conflicts between tasks when the radar system is operating in an environment that leads to an overload situation (these techniques are described in greater detail in reference 36).

3.7.6 Waveform Diversity and Sensor Geometry

This paper has covered the results of using KB techniques on the receiver portion of a radar. There is clearly scope for exploiting prior knowledge and using adaptivity in the transmitter portions of the radar, too. This constitutes the subject of waveform diversity, in which the sensor may transmit different waveforms via different transmit radiation patterns, and receive via different receive patterns, all adapted dynamically on the basis of the information obtained from the target scene and the set of tasks to be accomplished [37]. This, combined with the ability to dynamically reconfigure flight paths, will offer significant improvements in performance achievable only through the application of KBS to radar.

ACKNOWLEDGMENTS

The authors would like to acknowledge the support and encouragement of their employers. The authors would like to specifically thank Mr Gerard Genello of the US Air Force Research Laboratory (SNRT) and Dr Joseph Guerci, formerly of DARPA, for their financial support and technical assistance. This work was inspired by NATO RTO Lecture Series 233 held across Europe in 2003, 2004 and 2006.

REFERENCES

[1] M.I. Skolnik. *Introduction to Radar Systems*. 3rd ed. McGraw Hill; 2001. New York City, NY, p. 14–19.

[2] V.S. Chernyak. *Fundamentals of Multisite Radar Systems*. Gordon and Breach Science Publishers; New York City, NY, 1998.

[3] U. Tiberio. *L'insegnamento e l'opera di* (in Italian), 24 October 1998. (Fifty years after publication of his paper in *Proc. IRE.*)

[4] *IEEE Aerospace and Electronic Systems Magazine* 2000;10:119.

[5] F. Gini, F. Lombardini. Multibaseline cross-track SAR interferometry: A signal processing perspective. *IEEE Trans. Aerospace and Electronic Systems*, Tutorials II. 2005;20(8):71–93.

[6] A. Farina. *Antenna Based Signal Processing Techniques for Radar Systems*. Norwood, MA, Artech House; 1992.

[7] P. Lombardo, M.V. Greco, F. Gini, A. Farina, J.B. Billingsley. Impact of clutter spectra on radar performance prediction. *IEEE Trans. Aerospace and Electronic Systems*, 2001; AES-37[3]:1022–1038.

[8] A. Farina, F.A. Studer. A review of CFAR detection techniques in radar systems. *Microwave Journal* 1986;29(9):115–128.

[9] R. Klemm. Space time adaptive processing: principles and applications. In: *IEE Radar, Sonar, Navigation and Avionics* 9. IEE Press; London, England, 1998.

[10] H. Wang, L. Cai. On adaptive spatial-temporal processing for airborne surveillance radar systems. *IEEE Transactions on AES* 1994;30(3):660–669.

[11] A. Farina F.A. Studer. Radar data processing. Introduction and Tracking (vol. I). Research Studies Press (England), John Wiley and Sons (USA), 1985.

[12] B. Ristic, S. Arulampalam, N. Gordon. Beyond the Kalman filter: Tracking applications of particle filters. Norwood, MA: Artech House; 2004.

[13] W. Baldygo, M. Wicks, R. Brown, P. Antonik, G. Capraro, L. Hennington. Artificial intelligence applications to constant false alarm rate (CFAR) processing. *Proc. IEEE 1993 National Radar Conference*, Boston, MA, April 1993.

[14] Knowledge base applications to adaptive space–time processing. AFRL-SN-TR-2001 Final Technical Report, Volumes I–VI, July 2001.

[15] C. Capraro, G. Capraro, D. Weiner, M. Wicks, W. Baldygo. Improved STAP performance using knowledge-aided secondary data selection. *Proc. 2004 IEEE Radar Conference*, Philadelphia, PA, April 2004.

[16] C.T. Capraro, G.T. Capraro, D.D. Weiner, M. Wicks. Knowledge based map space time adaptive processing (KBMapSTAP). *Proc. 2001 International Conference on Imaging Science, Systems, and Technology*, Las Vegas, Nevada, June 2001.

[17] E. Rich. *Artificial Intelligence*. New York, NY: McGraw-Hill; 1983.

[18] A. Barr, E.A. Feigenbaum. *The Handbook of Artificial Intelligence*, 3 vols. Los Altos, CA: William Kaufman; 1981–1982.

[19] A. Farina, H. Griffiths, G. Capraro, M. Wicks. Knowledge-based radar signal & data processing. NATO RTO Lecture Series 233, November 2003.

[20] P. Antonik, H. Schuman, P. Li, K.W. Melvin, M. Wicks. Knowledge-based space–time adaptive processing. *Proc. 1997 IEEE National Radar Conference.* Syracuse, NY, May 1997.

[21] R.S. Adve, M.C. Wicks, T.B. Hale, P. Antonik. Ground moving target indication using knowledge-based space–time adaptive processing. *Proc. 2000 IEEE National Radar Conference.* Alexandria VA, May 2000.

[22] M.C. Wicks, M. Rangaswamy, R. Adve, T.B. Hale. Space–time adaptive processing. *IEEE Signal Processing Magazine* 2006;23(1):51–65.

[23] J.S. Bergin, P.M. Techau, W.L. Melvin, J.R. Guerci. GMTI STAP in target-rich environments: site-specific analysis. *Proc. 2000 IEEE National Radar Conference*, 2002.

[24] M.C. Wicks, W.J. Baldygo, R.D. Brown. Expert system constant false alarm rate (CFAR) processor. U States patent 5,499,030. 12 March 1996.

[25] A. Farina, L. Ferranti, G. Golino. Constrained tracking filters for A-SMGCS. *Proc. Fusion 2003*, 8–11 July 2003, Cairns, Queensland, Australia.

[26] T. Kirubarajan, Y. Bar-Shalom, K.R. Pattipati, I. Kadar. Ground target tracking with variable structure IMM estimator. *IEEE Trans. Aerospace and Electronic Systems* 2000; 36(1):26–46.

[27] M.S. Arulampalam, N. Gordon, M. Orton, B. Ristic. A variable structure multiple model particle filter for GMTI tracking. *Proc. Fifth International Conference on Information*

Fusion. FUSION 2002, Sunnyvale, CA, (IEEE cat. no. 02EX5997). *Int. Soc. Inf. Fusion* 2002;2:927–934.

[28] N.J. Gordon, B. Ristic. Tracking airborne targets occasionally hidden in the blind Doppler. *Digital Signal Processing* 2002;12(2/3):383–393.

[29] D.T. Gjessing. *Target Adaptive Matched Illumination Radar: Principles and Applications*. Peter Peregrinus; London, England, 1986.

[30] M. Bell. Information theory and radar waveform design. *IEEE Trans. Information Theory* 1993;39(5):1578–1597.

[31] S.U. Pillai, H.S. Oh, D.C. Youla, J.R. Guerci. Optimum transmit-receiver design in the presence of signal-dependent interference and channel noise. *IEEE Trans. Information Theory* 2000;46(2):577–584.

[32] K.M. Cuomo, J.E. Piou, J.T. Mayhan. Ultra-wideband coherent processing. Special issue of The Lincoln Laboratory. *Journal on Superresolution* 1997;10(2):203–221.

[33] H.D. Griffiths, C.J. Baker, H. Ghaleb, R. Ramakrishnan, E. Willman. Measurement and analysis of ambiguity functions of off-air signals for passive coherent location. *Electronics Letters* 2003;39(13):1005–1007.

[34] D.D. Weiner, M.C. Wicks, G.T. Capraro. Waveform diversity and sensors as robots in advanced military systems. First International Conference on Waveform Diversity and Design, Edinburgh, November 2004.

[35] D. Blacknell. Contextual information in SAR target detection. *IEE Proc. Radar, Sonar and Navigation* 2001;148(1):41–47.

[36] S. Miranda, C.J. Baker, K. Woodbridge, H.D. Griffiths. Knowledge-based resource management for multi function radar. Special Issue of *IEEE Signal Processing Magazine on* Knowledge Based Systems for Adaptive Radar Detection, Tracking and Classification.

[37] *Proc. First International Conference on Waveform Diversity and Design*. Edinburgh, Scotland, November 2004.

4

AN OVERVIEW OF KNOWLEDGE-AIDED ADAPTIVE RADAR AT DARPA AND BEYOND[1]

Joseph R. Guerci[2] and Edward J. Baranoski

For the past several years, the Defense Advanced Research Projects Agency (DARPA) has been pioneering the development of the first ever real-time knowledge-aided adaptive radar architecture. The impetus for this program arises from the increasingly complex missions and operational environments encountered by modern radars and the inability of traditional adaptation methods to address rapidly varying interference environments. The DARPA Knowledge-Aided Sensor Signal Processing and Expert Reasoning (KASSPER) program has as its goal the demonstration of a high performance embedded computing (HPEC) architecture capable of integrating high- fidelity environmental "knowledge" (i.e. "priors") into the most computationally demanding subsystem of a modern radar: the adaptive space–time beamformer. This is no mean feat, as environmental knowledge is a "memory" quantity that is inherently difficult (if not impossible) to access at the rates required to meet radar front-end throughput requirements. In this chapter we will provide an overview of the KASSPER program highlighting both the benefits of knowledge-aided (KA) adaptive radar, key algorithmic concepts, and the breakthrough "look-ahead" radar scheduling approach that is the keystone to the KASSPER HPEC architecture. In addition, we will introduce the notion of extending KA processing to the adaptive Multi-Input Multi-Output (MIMO) radar problem.

1. Reprinted, in part, with permission, from "Knowledge-aided adaptive radar at DARPA: an overview," *IEEE Signal Processing Magazine* 2006;23(1):41–50.
2. J.R. Guerci is an independent consultant.

Knowledge-Based Radar Detection, Tracking, and Classification. Edited by Fulvio Gini and Muralidhar Rangaswamy

4.1 INTRODUCTION

4.1.1 Background on STAP

Even for a nonspecialist it is not hard to qualitatively imagine the enormous challenges presented to modern airborne radar systems when attempting to separate ephemeral ground target echoes (associated with moving vehicles for example) from overwhelming land clutter returns, which can be eight or more orders of magnitude stronger than the desired target, especially in highly complex and nonstationary clutter environments (e.g. urban clutter) [1–4]. The lion's share of this signal separation/detection task falls to the real-time space–time adaptive processor (STAP) [5–7].

STAP attempts to filter out the clutter and noise returns by an adaptive multidimensional finite impulse response (FIR) filter structure as depicted in Fig. 4.1, consisting of M time taps to support Doppler filtering and N spatial taps to support angle processing. By judicious setting of the complex weights $\{w_{i,j}\}$, a two-dimensional angle–Doppler adaptive antenna pattern can be obtained that maximizes the return from a desired target resolution cell, while simultaneously minimizing the returns from clutter and possibly jamming. A notional example of this process is depicted in Fig. 4.2a and b, where the clutter plus noise power spectrum of Fig. 4.2a is filtered by the pattern in Fig. 4.2b. Note that an airborne radar is assumed as evidenced by the angle–Doppler coupled clutter ridge, which is the result of aircraft motion [7].

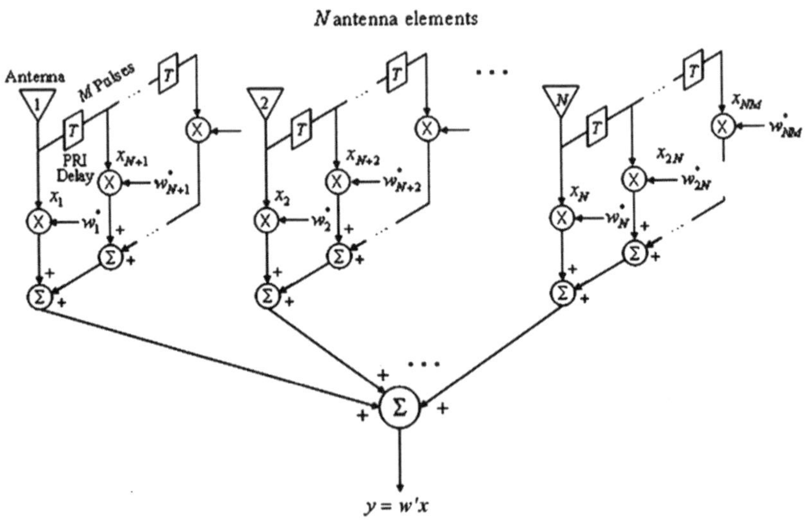

Figure 4.1 Space–time (angle–Doppler) beamformer consisting of N independent antenna channels and M time taps (pulse returns) comprising the coherent processing interval (CPI). A specific angle–Doppler pattern is obtained by judicious selection of the complex linear combiner weights. For other space–time variations of the above architecture (e.g. post-Doppler beamspace) see references 6 and 7.

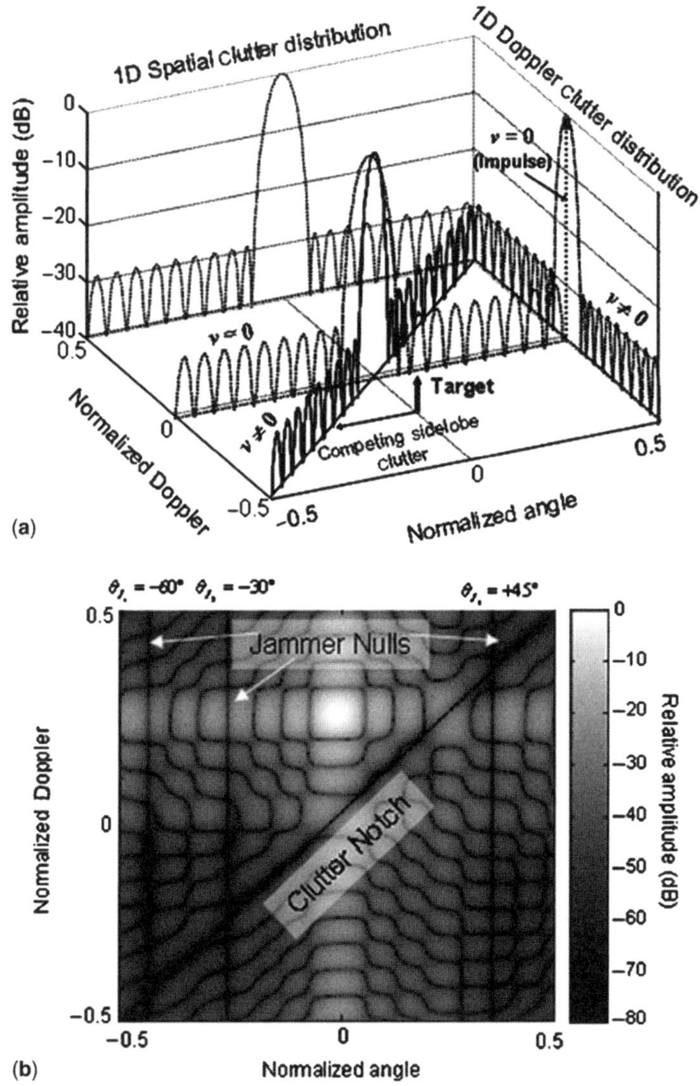

Figure 4.2 (**a**) Notional clutter spectrum for an airborne MTI radar. (**b**) The optimal 2D space–time filter response for a boresight aligned target with a normalized Doppler of 0.25. (*Source*: reference 7.)

The multidimensional adaptive FIR filter thus consists of a total of NM taps (or spatio-temporal degrees of freedom, DoFs). The spatial DoFs could be the outputs of individual antenna array elements (as part of an electronically scanned array (ESA) for example), subarrays, or outputs of a beamformer [7]; and the temporal DoFs could be the sampled pulse returns or Doppler bins [7]. The optimum STAP weights vector, $w \in C^{NM}$, which maximizes the signal-to-interference-plus-noise ratio (SINR), satisfies the matrix Weiner–Hopf equation

$$w = \mathbf{R}^{-1}s, \tag{4.1}$$

where $s \in C^{NM}$ is the desired target signal of interest (i.e. "steering" vector [7]), and $\mathbf{R} \in C^{NM \times NM}$ is the clutter-plus-jamming-plus-noise covariance matrix [7], which is guaranteed to be positive definite in practice due to the omnipresent receiver noise [7].

A consideration of Equation 4.1 in the context of practical real-world considerations leads one to immediately recognize a set of serious fundamental issues [7]:

1. $\mathbf{R}$ is potentially very large (due to the product NM).
2. $\mathbf{R}$ cannot be known a priori, as it depends on the clutter and jamming environment. It must therefore be estimated, at least in part, on the fly.
3. Owing to the unknown statistics of the radar interference, $\mathbf{R}$ must be determined adaptively from the data.
4. The need for immediate results will require Equation 4.1 to be implemented on the aircraft in real time.

A first-generation solution to these issues essentially consists of the following three elements [5–7]:

1. Selection of the STAP domain (e.g. pre-Doppler element space, post-Doppler beamspace [7]).
2. Selection of a rank-reduction method, as a full-dimensional implementation of Equation 4.1 is not only very difficult/costly, but often not a good idea because adding adaptive DoFs (ADoFs) can actually hurt performance in nonstationary environments [1–3, 7].
3. Selection of a covariance estimation scheme (implicit or explicit) [5–7].
4. A real-time computing architecture [8].

Although many ingenious solutions to the first, second and fourth elements have been devised, they have been predicated on a single basic approach to element 3: sample covariance estimation derived from local radar measurements [5–7]. In practice, this estimation is generally implicit, because the solution to the fourth element generally consists of a data domain implementation (e.g. a QR-factorization implemented via a parallel pipelined computing architecture [8, 9]) in which a covariance matrix is not explicitly formed. It is, nonetheless, mathematically equivalent to replacing $\mathbf{R}$ in Equation 4.1 with a sample estimate $\hat{\mathbf{R}}$ generally of the form

$$\hat{\mathbf{R}} = \frac{1}{L} \sum_{i \in \{L \text{ datasamples}\}} x_i x_i', \tag{4.2}$$

where $x_i \in C^{NM}$ is a space–time snapshot [7] data vector for the ith range bin measured by the radar in the domain selected in element 1, and L data samples are used in the formation of Equation 4.2. Ideally, if the snapshot data samples were drawn from a Gaussian independent and identically distributed (i.i.d.) stochastic process, then Equation 4.2 corresponds to the maximum likelihood covariance estimate, that is,

$$\hat{\mathbf{R}} = \underset{\{\mathbf{R}\}}{\arg\max} f(x_i : i = 1, \ldots, L|\mathbf{R}), \tag{4.3}$$

where $f(x_i : i = 1, \ldots, L|\mathbf{R})$ denotes the probability density function (pdf) of L snapshots conditioned on $\mathbf{R}$. Reed, Mallet and Brennan (RMB) have shown that in this case, the expected SINR loss ρ, due to finite sample estimation, is [10]

$$\rho = \frac{L - NM + 2}{L + 1}, \qquad L \geq NM, \tag{4.4}$$

where it is assumed that there are at least NM samples. A useful rule of thumb derived from Equation 4.4 is that to achieve a loss no greater than 3 dB, at least on order of $2NM$ i.i.d. samples are required. For an $N = 16$, $M = 16$, case, this corresponds to an assumption that at least 512 i.i.d. samples are available that share identical statistics with the cell under test.

In practice, the L training samples are generally derived from a subset of the total space–time data cube measured during a coherent processing interval (CPI) [5]. One such commonly discussed scheme is depicted in Fig. 4.3. Unfortunately, as will be made readily apparent in the next section, real-world clutter often violates any such stationarity assumptions.

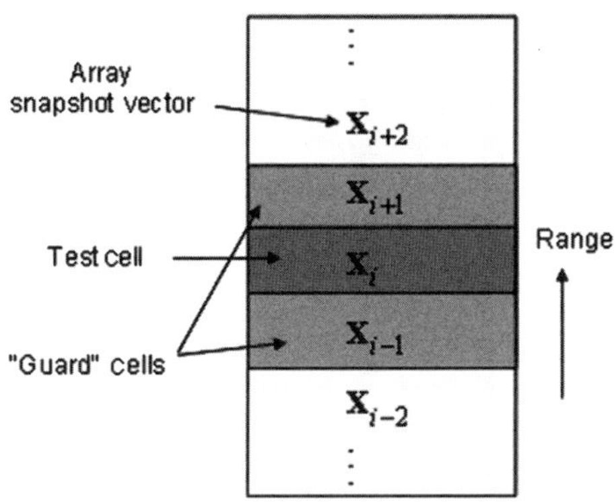

Figure 4.3 An example data selection strategy for estimating a sample space–time covariance matrix. L data samples are selected from range bins in proximity to a range cell under test (ith range bin in this example). The actual test cell, and cells adjacent to it (guard cells), may be excluded to avoid contamination from the actual target signal (5).

4.1.2 Examples of Real-World Clutter

It is obvious that real-world ground clutter is not well modeled by a homogenous stationary stochastic process [1–3]. Variations in underlying terrain, foliage, land–sea interfaces, urban/manmade structures, as well as nonlinear array responses (e.g. circular and/or tilted arrays [25, 26]), and so on, all contribute to stationarity violations. Consequently, significant deviations from predicted ideal performance are to be expected — their exact nature of course depends intimately on the environment [3].

Figure 4.4 shows a comparison between actual measured radar returns for the DARPA Mountain Top experimental UHF radar (11) located at the White Sands Missile Range in New Mexico (Fig. 4.4a), and what would have been measured if the terrain were homogeneous (Fig. 4.4b). This form of inhomogeneity leads to either over- or undernulling of the clutter [7], with resulting poor detection or false-alarm rate performance, respectively.

Figure 4.5 shows data from an experimental X-band airborne radar [12]. Clearly evident are distinct bright clutter discretes, which will not be adequately nulled if an averaging process like Equation 4.2 is simply applied.

For details on the many deleterious effects of nonstationary, non-Gaussian real-world clutter, the reader is referred to the proceedings of the KASSPER workshops [13] available on the Internet at www.darpa.mil/sto/programs/kassper.htm.

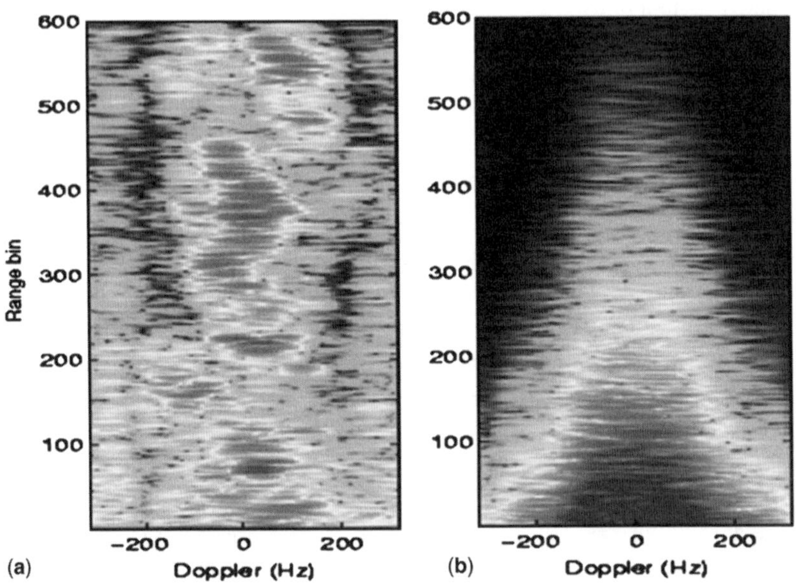

Figure 4.4 Comparison between real-world clutter (**a**) from the DARPA Mountain Top radar (7), and returns assuming homogeneous clutter (7) (**b**).

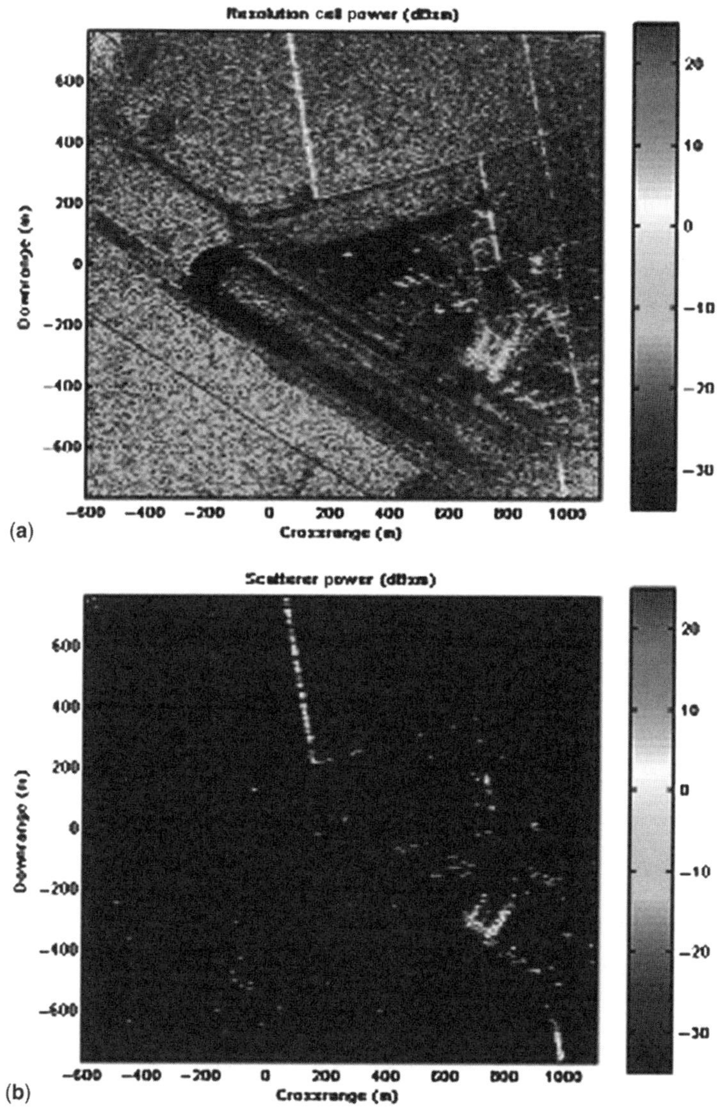

Figure 4.5 Example real-world X-band radar measurements corresponding to a geographical location with discrete clutter (see reference 14 for details). (**a**) High-resolution geo-registered reflectivity image. (**b**) Corresponding discrete map.

4.2 KNOWLEDGE-AIDED STAP (KA-STAP)

4.2.1 Knowledge-Aided STAP: Back to "Bayes-ics"

In this section we describe in some detail basic ways in which prior knowledge can be incorporated into the most demanding component of a modern MTI radar: the space–time adaptive beamformer. Although radar centric, the methods discussed below were specifically chosen because of their general applicability in many other adaptive sensor signal processing systems such as sonar, lidar, and other multidimensional sensor arrays where environmental clutter (as opposed to say random thermal receiver noise) is a dominant source of interference. We will describe two broad categories of KA processing: Case I, Intelligent Training and Filter Selection, and Case II, Bayesian Filtering and Data Pre-Whitening. The former consists primarily of *indirect* exploitation of prior knowledge sources (such as training data selection) and the latter entails *direct* filtering of the incoming multidimensional data stream based on prior information.

4.2.1.1 Case I: Intelligent Training and Filter Selection (ITFS) In the Intelligent Training and Filter Selection (ITFS) approach, prior knowledge of the interference environment is used to optimize two adaptive filtering processes: [1] the filter selection and [2] the filter training strategy. In the case of radar clutter, this is accomplished by first conducting an environmental segmentation analysis based on whatever prior terrain/clutter database is available. Everything from Digital Terrain and Elevation Data (DTED) to Land Cover Land Use (LCLU), to synthetic aperture radar (SAR) imagery, or even hyperspectral imagery can be used [1]. Land clutter tends to be “clumpy”; that is, it tends to be locally similar, but with distinct and often abrupt boundaries (see Fig. 4.6, for example). Clearly, from physical principles, an adaptive filter should not attempt to lump all these regions together and

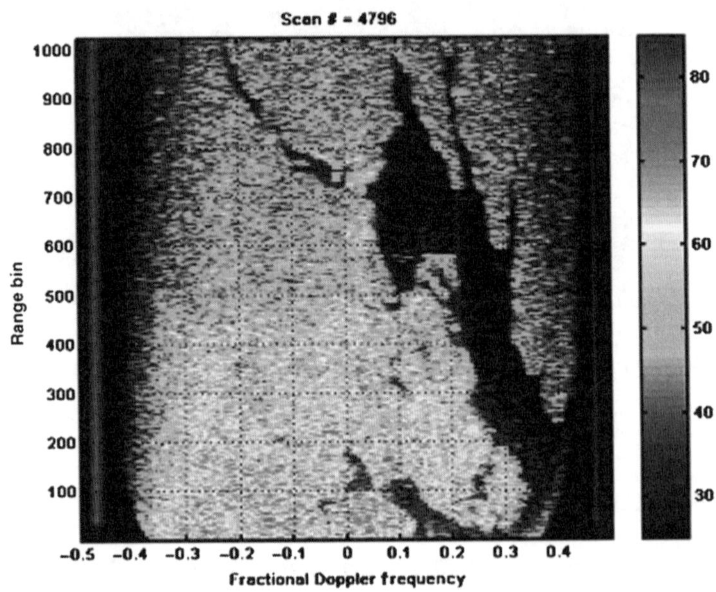

Figure 4.6 Measured range–Doppler returns for the MCARM data set (12) showing the highly segmented nature of radar clutter returns.

apply a single filtering strategy. Instead, a segmentation analysis should be performed and an adaptive filter tailored to that region should be applied.

Generally speaking, the filter selection stage determines what type of adaptive filter is best suited to a given segmented region. In the case of STAP filtering for clutter suppression in radar, a pivotal step is the domain in which the actual filtering is performed (e.g. pre- or post-Doppler, element or beamspace [5–7]) and the number of adaptive degrees of freedom (ADoFs), which manifests itself ultimately in the "size" of the adaptive filter. For example, in the case of the Principal Components (PC) method, the number of adaptive DoFs refers to the number of significant eigenvectors to be included in the adaptive weight calculation. Similarly, for the Multi-Stage Weiner Filter (MWF), the number of "stages" is the metric for ADoFs [15]. What is critical is that the number of ADoFs be matched to the available training data (and of course the real-time computing architecture). A general rule of thumb, which has its origins in the RMB result of Equation 4.4, is that there be of order $2k$ i.i.d. samples available for training the adaptive weights (e.g. sample covariance estimation), where k is the number of effective ADoFs.

Once a basic filtering structure has been selected, a training strategy can be selected and optimized for that choice. In essence, all or a subset of the samples from the locally stationary region are utilized in the weight training stage. In the case of PC, all of the range bins — including the cell under test — might be included, as it has been shown that this approach is robust to target cancellation. In contrast, a multi-bin post-Doppler approach might need extra care and the introduction of exclusion and guard cells to prevent target signal cancellation [5].

Figure 4.7 illustrates the impact that ITFS can have when applied to real-world data. As described in reference 16, the MCARM data set included a number of significant highways, that is, "moving clutter"! If one simply applied traditional sample averaging techniques such as those previously described, one could suffer significant detection losses at "roadway" speeds [16]. Using an intelligent training and adaptation scheme that essentially took account of the road networks, a significant improvement in detection was achieved.

4.2.1.2 Case II: Bayesian Filtering and Data Pre-Whitening

In the Bayesian approach to radar STAP, prior knowledge is used directly by the filter to aid in adapting to nonstationary clutter. A convenient pedagogical framework for this approach is the Bayesian covariance estimation approach of T. W. Anderson [17].

Wishart [18] established that the elements of a sample covariance matrix $[L\hat{\mathbf{R}}]_{i,j}$ formed from an outer product sum of L Gaussian i.i.d. samples, that is,

$$\hat{\mathbf{R}} = \frac{1}{L}\sum_{i=1}^{L} x_i x_i' \tag{4.5}$$

obey a Wishart distribution (actually complex Wishart [19]) of degree L, that is, $\hat{\mathbf{R}} \approx W(L\hat{\mathbf{R}}, L)$.

If a prior estimate of the covariance matrix exists, $\hat{\mathbf{R}}_0$, it is not unreasonable to assume it too is Wishart distributed [17]. The rationale for this is simple. If it is

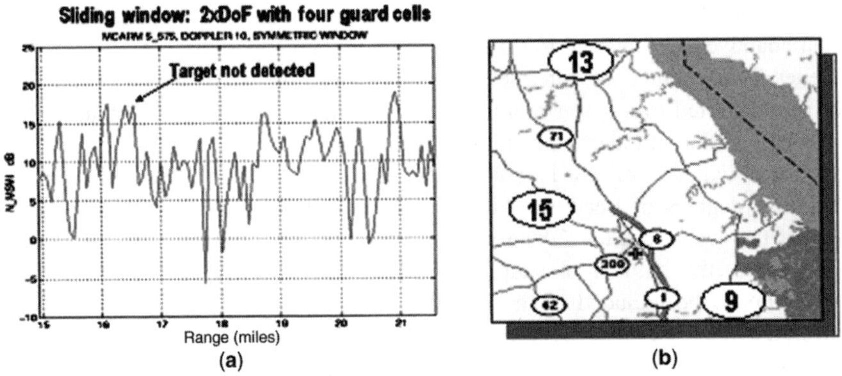

Cells excluded	Rationale
13.5 – 14 miles	Rt. 15
16 – 17.5 miles	Rt. 13
19 – 20 miles	Rt. 9 in sidelobe region, Radial alignment
21 – 22 miles	Rt. 9 in mainlobe

(c)

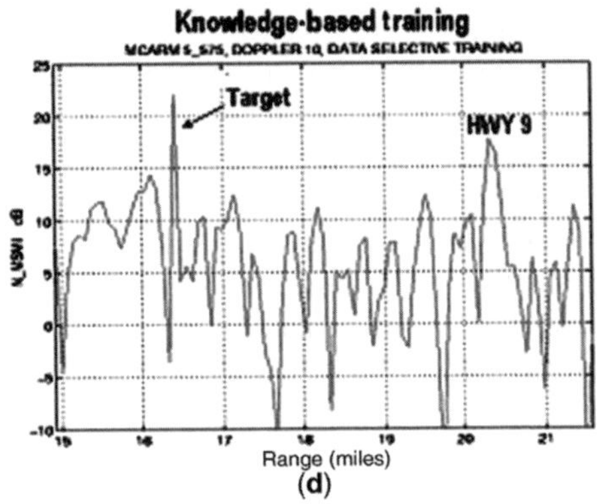

Figure 4.7 Illustration of the impact that prior knowledge (in this case prior road network data) can have on improving detection performance for the MCARM data set (16). (**a**) STAP filter residue without knowledge-aided processing. (**b**) Local map of region indicating locations of road networks. (**c**) A training cell exclusion rule based on the map data. (**d**) The target, which was previously undetected, is clearly visible after intelligent training. (*Source*: reference 16.)

based on prior radar observations, then it is also of the form of Equation 4.5. The corresponding Bayesian (maximum a posteriori) estimate, which combines $\hat{\mathbf{R}}$ and $\hat{\mathbf{R}}_0$, is easily derived. Let L_1 and L_1 denote the degrees of $\hat{\mathbf{R}}$ and $\hat{\mathbf{R}}_0$, respectively, which are further assumed to be i.i.d. [17]. Then $\hat{\mathbf{R}}$ and $\hat{\mathbf{R}}_0$ are collectively sufficient statistics for $L_0 + L_1$ i.i.d. samples $\{x_i : i = 1, \ldots, L_0 + L_1\}$. Thus, the maximum a posteriori solution of $\hat{\mathbf{R}}$ given prior $\hat{\mathbf{R}}_0$ is equivalent to the maximum likelihood solution based on $\{x_i : i = 1, \ldots, L_0 + L_1\}$; that is,

$$\begin{aligned}\hat{\mathbf{R}} &= \max_{\mathbf{R}} f(x_i : i = 1, \ldots, L_1 | \mathbf{R}) f_0(\mathbf{R}) \\ &= \max_{\mathbf{R}} f(x_i : i = 1, \ldots, L_0 + L_1 | \mathbf{R}) \\ &= \frac{1}{L_0 + L_1} (L_0 \hat{\mathbf{R}}_0 + L_1 \hat{\mathbf{R}}_1),\end{aligned} \tag{4.6}$$

where $f_0(\mathbf{R})$ denotes the prior pdf associated with the prior covariance estimate $\hat{\mathbf{R}}_0$ based on L_0 samples, and thus is $W(L_0\hat{\mathbf{R}}_0, L_0)$, and $\hat{\mathbf{R}}_1$ denotes the maximum likelihood (ML) estimate based on L_1 samples. Equation 4.6 has an obvious intuitive appeal: the a posteriori covariance estimate is formed as a weighted sum of the prior and current estimates with weighting factors proportional to the amount of data used in the formation of the respective sample covariances.

An obvious yet useful generalization of Equation 4.6 is given by

$$\hat{\mathbf{R}} = \alpha\hat{\mathbf{R}}_0 + \beta\hat{\mathbf{R}}_1, \qquad \alpha + \beta = 1, \tag{4.7}$$

which is the familiar "colored loading" or "blending" approach of references 20 and 21. The practical advantages of Equation 4.7 relative to Equation 4.6 are many. For example, the data used to form the prior covariance might lose its relevance with time — the so-called stale weights problem [22]. In that case, even though $\hat{\mathbf{R}}_0$ might have been formed from L_0 samples, it effectively has less information and should be commensurately "de-weighted." A common method for accomplishing this, borrowed from Kalman filtering, is the fading memory approach in which case α, in Equation 4.7, is given by

$$\alpha = e^{-\gamma t}L_0, \tag{4.8}$$

where t is the time elapsed since the prior covariance estimate was formed, and the positive scalar γ is the fading memory constant [23].

In a more general setting, the blending parameters (α, β) could be chosen based on the relative confidence in the estimates. For example, $\hat{\mathbf{R}}_0$ could be derived from a physical scattering model of the terrain. In which case it is also typically of the form of Equation 4.5, with the distinction that the outer products represent clutter patch steering vectors weighted by the estimated reflectivity [7, 24]; that is,

$$\hat{\mathbf{R}}_0 = \frac{1}{N_\mathrm{c}}\sum_{i=1}^{N_\mathrm{c}} G_i v_i v_i'. \tag{4.9}$$

N_c clutter "patches" have been utilized in the formation of $\hat{\mathbf{R}}_0$ (typically corresponding to a particular iso-range ring [7]) where $v_i \in C^{NM}$ is the space–time (angle–Doppler) steering vector corresponding to the ith clutter patch and G_i is its corresponding power [7]. Such information could be available a priori from SAR imagery [25] (essentially a high-resolution clutter reflectivity map) or physics-based models [24].

Although the confidence metric to apply, in the form of the weighting parameter α, is difficult to ascribe in practice because the quality of the a priori estimate is vulnerable to a number of error sources, a straightforward remedy is to choose α adaptively so as to maximally "whiten" the observed interference data. For example;

$$\min_{\{\alpha\}} Z_L(\alpha), \tag{4.10}$$

where

$$Z_L(\alpha) = \left\| \sum_i y_i y_i' - I \right\| \tag{4.11}$$

and where

$$y_i = (\alpha \hat{\mathbf{R}}_0 + \beta \hat{\mathbf{R}}_1)^{-\frac{1}{2}} x_i. \tag{4.12}$$

In Equations 4.10–4.12, x_i is the space–time snapshot vector for the ith range bin, $(\alpha\hat{\mathbf{R}}_0 + \beta\hat{\mathbf{R}}_1)^{-1/2}$ is the whitening matrix corresponding to a particular α, y_i is the vector residue with $\dim(y_i) = \dim(x_i)$, and the summation in Equation 4.11 is performed over a suitable subset of the radar observations for which $\hat{\mathbf{R}}_0$ is believed valid. If an a priori covariance estimate is available for each range bin, then the vector residue in Equation 4.12 can be replaced with

$$y_i = (\alpha \hat{\mathbf{R}}_0(i) + \beta \hat{\mathbf{R}}_1)^{-\frac{1}{2}} x_i, \tag{4.13}$$

where $\hat{\mathbf{R}}_0(i)$ is the a priori estimate for the ith range bin.

The above adaptive α approach is but a special case of an entire class of direct filtering methods incorporating prior information, that is, *data pre-whitening* (or simply data de-trending). In a more general setting, the space–time vector residues $\{y_i\}$ can be viewed as a "de-trended" vector time series using prior knowledge in the form of a covariance-based whitening filter. The major potential advantage of this is to remove (or attenuate) the major quasi-deterministic trends in the data (e.g. clutter discretes,

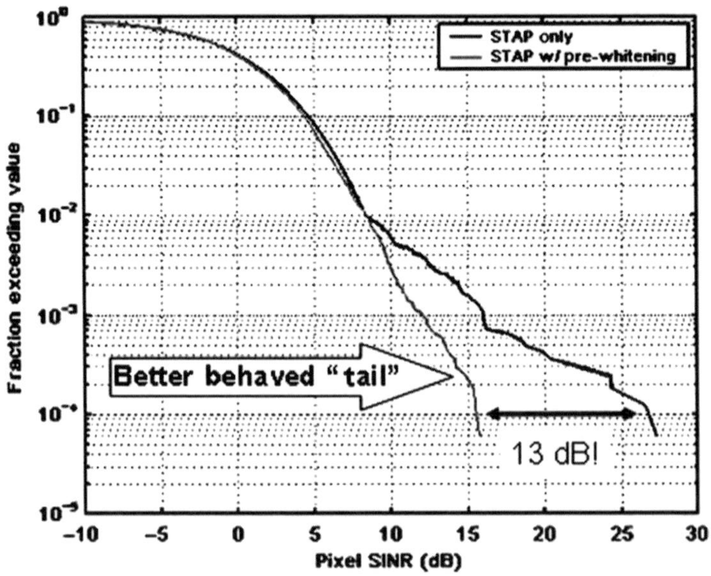

Figure 4.8 Illustration of the effectiveness of the pre-whitening approach on real-world data (25). Pre-whitening the data corresponding to Fig. 4.5 resulted in a significant reduction in the tails of the clutter residue.

mountains) so that the resulting residue vector time series is less nonstationary or inhomogeneous.

An interesting example of this can be found in Equation 4.14. In this pre-whitening example, a "CLEAN" algorithm was applied to the APTI data set of Fig. 4.5a, resulting in the discrete map of Fig. 4.5b. A deterministic covariance [24] was then formed as in Equation 4.9, from which a square-root whitening filter matrix could be derived. Figure 4.8 shows a log exceedance plot of the difference between the unwhitened data and the pre-whitened data. Note the presence of spiky clutter as evidence by the so-called "fat-tails" in the unwhitened data.

In the next section, we address the seemingly daunting challenge of incorporating prior knowledge — an inherently memory-intensive process — into a high performance embedded computer.

4.3 REAL-TIME KA-STAP: THE DARPA KASSPER PROGRAM

4.3.1 Obstacles to Real-Time KA-STAP

As mentioned previously, ingenious real-time computing architecture solutions have been devised to implement the sample matrix-based (maximum likelihood) solutions to STAP. In particular, to achieve the enormous throughput burden of a modern multichannel STAP radar, highly parallel HPEC systems based on so-called data domain reformulations of the Weiner–Hopf equation (Equation 4.1) have been devised [8, 9]. Figure 4.9 shows one such architecture based on a QR-factorization solution to Equations 4.1 and 4.5.

With such architectures, tens or hundreds of GFLOPs (Giga floating point operations) of real-time compute power can be achieved in hardware that can fit onto an airborne radar aircraft. Although marvels of modern technology, these machines are cyber savants — they can solve Equations 4.1 and 4.5 at blinding speeds in a strict pipelined fashion, but grind to a snail's pace if the data flow is disrupted for nonpipelined operations. This is a major fundamental obstacle to implementing knowledge-aided or general Bayesian approaches, which are inherently memory-intensive (prior information needs to be stored). Figure 4.10 shows the order-of-magnitude time-scales for accessing different memory storage devices. Thus, to create a real-time knowledge-aided HPEC (KA-HPEC, pronounced "K-PEC") architecture, a major breakthrough in memory management must be achieved as much of the prior information (e.g. terrain maps, road networks, discrete maps) will reside on mass storage (and thus slow) media.

4.3.2 Solution: Look-Ahead Scheduling

The key KA-HPEC breakthrough in the DARPA KASSPER project is based on a basic fundamental insight:

> There is a significant degree of determinism — and thus predictability — to radar clutter returns, particularly if the prediction horizon is only on the order of seconds.

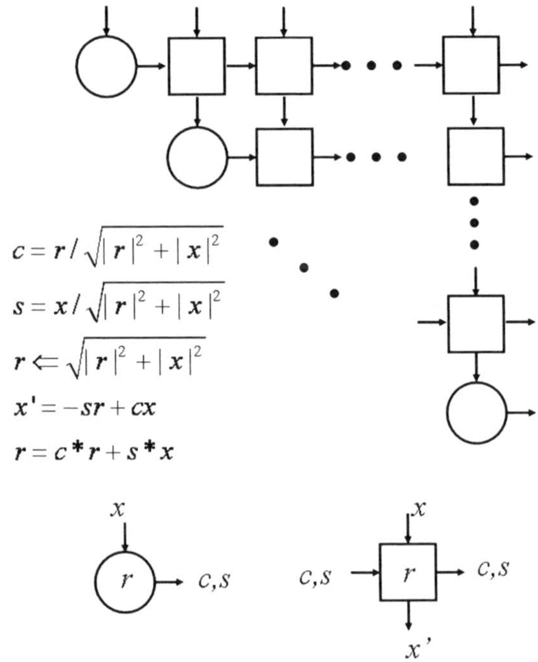

Figure 4.9 An example of a computer array implementing a data domain reformulation of the sample matrix-based Weiner–Hopf equation. (*Source*: reference 8.)

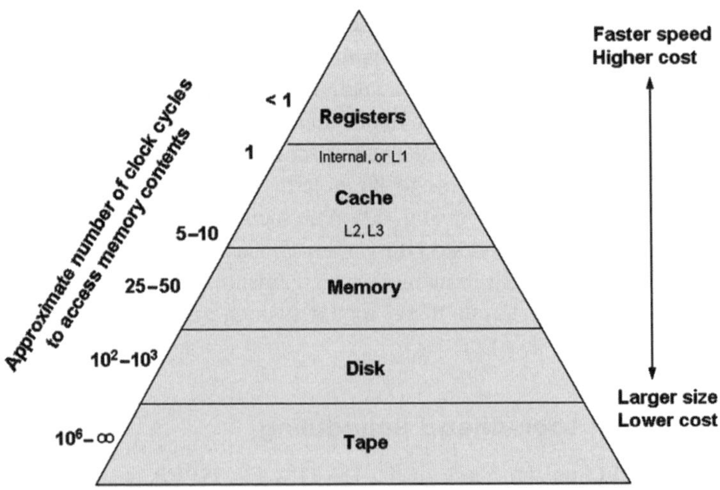

Figure 4.10 Illustration of the time-scales involved in accessing different memory storage media. (*Source*: Dr D. Patterson, Graduate Computer Architecture Course, University of California, Berkeley, Spring 2001.)

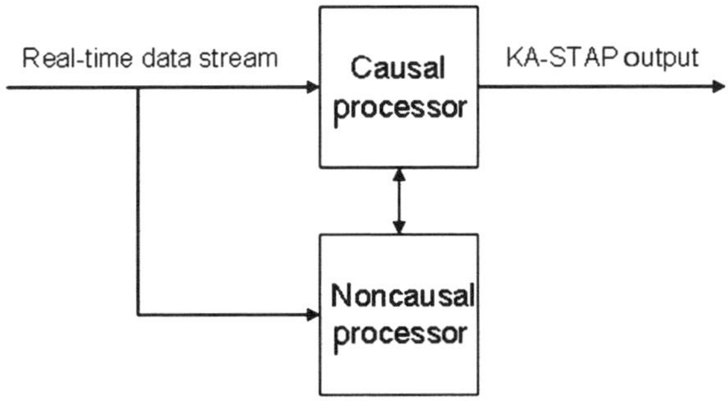

Figure 4.11 Example of a KA-HPEC architecture exploiting the high degree of radar determinism with look-ahead time-scales on the order of seconds. The noncausal processor, running in parallel with a more conventional HPEC STAP processor, is used to look ahead and detect regions of the radar field of regard requiring knowledge-aided processing, and thus modifications to the normal adaptive weights calculations.

For example, let t_0 denote the present time of the airborne radar depicted in Fig. 4.11. Let $t_0 + \delta t$ denote a time slightly in the future, say $\delta t = 1$ s. Then, in practice, the following are true:

1. The location of the aircraft at $t + \delta t$ can be predicted to a very high degree of accuracy assuming that no radical maneuvering is occurring.
2. The future state of the radar (look-direction, frequency, PRF, and so on) at $t_0 + \delta t$ is also known to a very high degree of accuracy.

The justification for the first assertion is simply that given the full kinematic state vector of the aircraft (position, speed, heading, and so on), Newtonian mechanics ensures fairly deterministic behavior — particularly for just a few seconds into the future. Justification for the second assertion arises from the simple fact that modern airborne radar systems typically utilize a radar scheduler. As the radar is computer controlled, it must have a tasking schedule. The scheduler is highly deterministic when considering a future time horizon on the order of seconds.

Why are the above assertions so critical to solving the memory access problem described in Section 4.3.1? Simple: they allow for "look-ahead scheduling." More specifically, they allow for noncausal processing, with a prediction horizon commensurate with the memory access delays! To see how this can be exploited by a KA-HPEC architecture, consider Fig. 4.11. In this instantiation, a noncausal look-ahead computer is running in parallel with a more conventional causal STAP HPEC processor. The noncausal processor is used to spot trouble before it occurs, and perform the necessary memory retrieval and pre-computations to ensure that the right weight modification scheme is ready to go when the data appear.

Figure 4.12 shows the MIT Lincoln Laboratory KASSPER HPEC system, a real-time 96-node parallel processing architecture implementing the noncausal look-ahead scheduling scheme of Fig. 4.11 [27]. The system has the capability of receiving

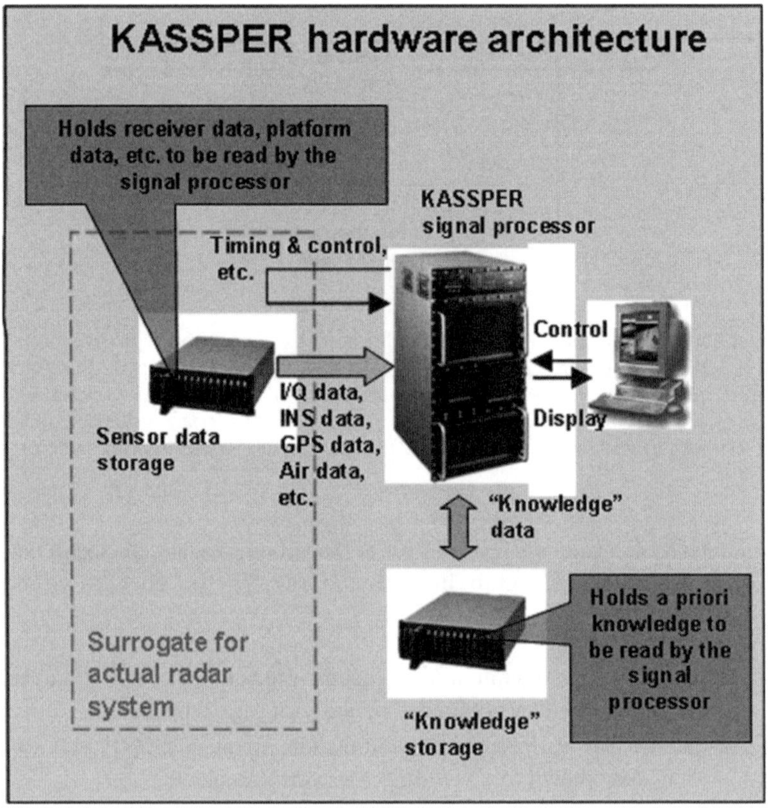

Figure 4.12 The MIT Lincoln Laboratory 96-node real-time KASSPER HPEC system.

real-time I&Q (In-phase and Quadrature) digitized samples from multiple receive channels over the full range extent of a radar and implementing a variety of knowledge-aided algorithms throughout the entire radar signal processing chain from STAP to CFAR (Constant False Alarm Rate).

Although when it comes to real-time HPEC the "devil" is most certainly in the details, Fig. 4.13 gives the basic idea of how the look-ahead scheduling is implemented. As the aircraft moves, a "sliding window" of data is migrated from a mass storage medium (e.g. disk drives) to a more readily accessible location (e.g. RAM). Depending on the particulars of the radar tasking, a "first-pass" decision is made as to what regions require knowledge-aided processing. For example, if the radar is scheduled to point in a direction where a major road network is known to exist, essential details regarding this road network (orientation, range extent, and so on) are extracted and exploited in potentially several stages of the radar signal processing chain. Given the look-ahead time buffer, this is all accomplished prior to the actual radar event. The exact extent of the sliding window depicted in Fig. 4.13 depends (of course) on the particular radar parameters (min/max range, altitude, and so on).

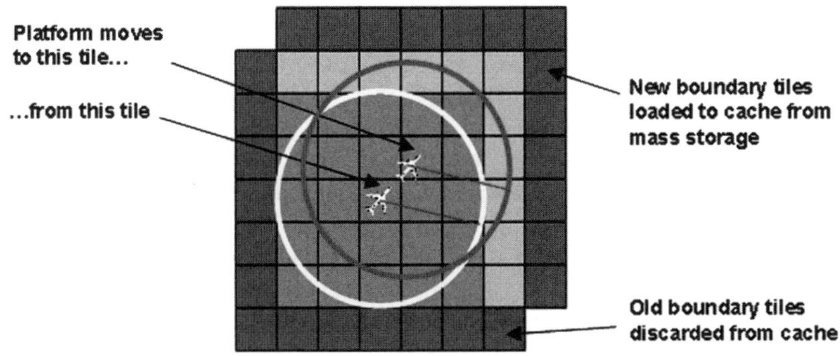

Figure 4.13 Illustration of the environmental database manipulation illustrating the "sliding window" approach to migrating data from mass storage to RAM and ultimately to cache.

4.4 APPLYING KA PROCESSING TO THE ADAPTIVE MIMO RADAR PROBLEM

To date, KA processing has been applied to the receive processing chain — that is, the receive "channel." However, if the transmit function (waveform selection for example) is allowed to adapt to the environment, then KA processing could be employed to aid in the estimation of the "transmit channel" [28].

Figure 4.14 shows the basic signal slow diagram for a MIMO radar. It has previously been shown [28–30] that the optimum MIMO transmit waveform $s_{\text{opt}} \in C^q$ corresponds to the eigenfunction with maximum eigenvalue of the following

$$(\mathbf{H}'\mathbf{H})s = \lambda s, \tag{4.14}$$

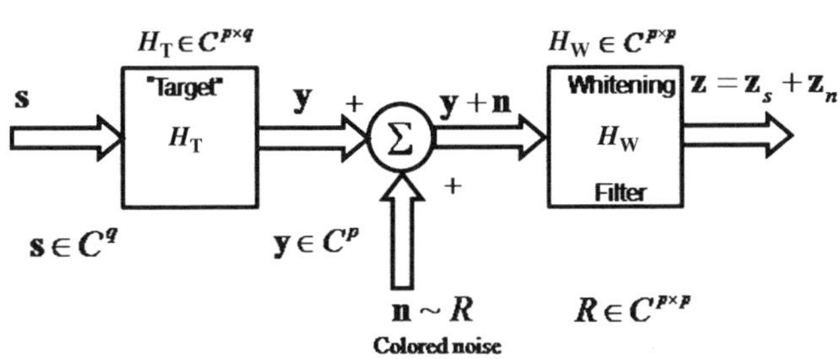

Figure 4.14 The basic multi-input, multi-output (MIMO) radar signal flow block diagram used to analyze the optimum transmit MIMO input (e.g. waveforms). Note that the dimensions of the transmit input are completely general and could include fast time, slow time, polarization, and even location (such as for the sparse transmit array case) (29).

where $\mathbf{H} \in C^{p \times q}$ denotes the total composite channel transfer function (target and receive whitening filter [28]), that is,

$$\mathbf{H} = \mathbf{H}_W \mathbf{H}_T. \tag{4.15}$$

Note that the dimensions of the transmit and receive channels are completely arbitrary and could include fast time, slow time, polarization, and even location for sparse arrays for example.

The KA processing described in this chapter (and elsewhere in this book) was developed to estimate $\mathbf{H}_W \in C^{p \times q}$ (i.e. the total interference covariance matrix and its corresponding whitening filter). It can thus immediately be applied to adapting the multi-input transmit function to $\mathbf{H}_W$. Of course, KA processing can likewise play a role in estimating the so-called "target" transfer function as well [28]. For further details regarding optimum MIMO radar, the reader is referred to references 28–30.

4.5 THE FUTURE: NEXT-GENERATION INTELLIGENT ADAPTIVE SENSORS

The heart of the DARPA KASSPER program is an architecture for performing knowledge-aided/Bayesian adaptive sensor signal processing. It is *not* a specific set of algorithms. Indeed, this chapter and the entire monograph are merely scratching the surface of potential instantiations not only for radar, but for any sensor interacting with the environment. For example, multichannel sonar systems face an analogous set of problems when attempting to detect small target echoes or emanations in a background of largely environmentally induced noise. If environmental databases were available (bathymetry, sea state, and so on), KA processing could be employed. The same is true of automated target detection sensors working in the EO/IR (electro-optical/infrared) regime (e.g. hyperspectral, lidar). Although radar centric, it is hoped that this paper will spark interest in others outside of radar, and usher in, or should we say, "re-usher" in real-time Bayesian adaptive sensor signal processing.

REFERENCES

[1] J.R. Guerci. DARPA KASSPER overview. *Proc. 2004 DARPA Workshop on Knowledge-Aided Sensor Signal Processing and Expert Reasoning (KASSPER)*, Clearwater, FL, 5–7 April 2004.

[2] A.O. Steinhardt, J.R. Guerci. STAP for RADAR: what works, what doesn't, and what's in store. *Proc. IEEE Radar Conference*, 2004, 26–29 April 2004.

[3] W.L. Melvin. Space–time adaptive radar performance in heterogeneous clutter. *IEEE Trans. Aerospace and Electronic Systems* 2000;36(2):621–633.

[4] W.L. Melvin. Introduction to the special section on STAP. *IEEE Trans. Aerospace and Electronic Systems* 2000;36(2):508–509.

[5] J. Ward. *Space–Time Adaptive Processing for Airborne Radar*. MIT Technical Report 1015, MIT Lincoln Laboratory, December 1994.

[6] R. Klemm. *Principles of Space–Time Adaptive Processing*. London: IEE Press; 2002.

[7] J.R. Guerci. *Space–Time Adaptive Processing for Radar*. Norwood, MA: Artech House; 2003.

[8] A. Farina. *Antenna-Based Signal Processing for Radar Systems*. Norwood, MA: Artech House; 1992.

[9] A.O. Steinhardt, C. Rader. Householder transforms in signal processing. *IEEE Signal Processing Magazine* 1988;5(3):4–12.

[10] I.S. Reed, J.D. Mallet, L.E. Brennan. Rapid convergence rate in adaptive arrays. *IEEE Trans. Aerospace and Electronic Systems* 1974;AES 10.6:853–863.

[11] G.W. Titi. An overview of the ARPA/Navy Mountain Top program. *Proc. IEEE Adaptive Antenna Systems Symposium*, November 1994.

[12] D.J. Zywicki, W.L. Melvin, G.A. Showman, J.R. Guerci. STAP performance in site-specific clutter environments. *2003 IEEE Aerospace Conference*, Big Sky, MT, USA, 8–15 March 2003.

[13] Proc. KASSPER Workshops, 2002–2005. Updated URL last accessed Feb 2, 2008. http://www.darpa.mil/STO/space/kassper_workshop2002.html.

[14] D.J. Zywicki, W.L. Melvin, G.A. Showman, J.R. Guerci. STAP performance in site-specific clutter environments. *Proc. 2003 IEEE Aerospace Conference*, 8–15 March 2003. Last accessed Feb 2, 2008.

[15] J.R. Guerci, J.S. Goldstein, I.S. Reed. Optimal and adaptive reduced-rank STAP. Special Section on Space–Time Adaptive Processing. *IEEE Trans. Aerospace and Electronic Systems* 2000;36(2):647–663.

[16] W. Melvin, M. Wicks, P. Antonik, Y. Salama, L. Ping, H. Schuman. Knowledge-based space–time adaptive processing for airborne early warning radar. *IEEE Aerospace and Electronic Systems Magazine* 1998;13(4). April 1998:37–42.

[17] T.W. Anderson. *An Introduction to Multivariate Statistical Analysis* 2nd ed. John Wiley & Sons; New York, 1984.

[18] J. Wishart. The generalized product moment distribution in samples from a normal multivariate population. *Biometrika* 1928;20A:32–52.

[19] S.U. Pillai. *Array Signal Processing*. New York: Springer-Verlag; 1989.

[20] J.S. Bergin, C.M. Teixeira, P.M. Techau, J.R. Guerci. STAP with knowledge-aided data pre-whitening. *Proc. 2004 IEEE Radar Conference*, April 2004, p. 289–294.

[21] J.D. Hiemstra. Colored diagonal loading. *Proc. 2002 IEEE Radar Conference*, April 2002, p. 386–390.

[22] J.R. Guerci. Theory and application of covariance matrix tapering for robust adaptive beamforming. *IEEE Trans. Signal Processing* 1999;47(4):977–985.

[23] A. Gelb. *Applied Optimal Estimation*. Cambridge MA: MIT Press; 1984.

[24] P.M. Techau, J.R. Guerci, T.H. Slocumb, L.J. Griffiths. Performance bounds for hot and cold clutter mitigation. *IEEE Trans. Aerospace and Electronic Systems* 1999; 35(4):1253–1265.

[25] M. Zatman. Circular array STAP. *Proc. 1999 IEEE Radar Conference*, 20–22 April 1999, p. 108–113.

[26] G.K. Borsari. Mitigating effects on STAP processing caused by an inclined array. *Proc. 1998 IEEE Radar Conference*, 11–14 May 1998, p. 135–140.

[27] *Proc. 23rd Systems and Technology Symposium (DARPA Tech)*, Anaheim, CA, March 2004. Available at www.darpa.mil.

[28] J.R. Guerci. Knowledge-aided adaptive waveform design. Presented at the *2004 IEE/IEEE Workshop on Waveform Diversity*, Edinburgh, UK, November 2004.

[29] J.R. Guerci. Next-generation intelligent radar. *Proc. 2007 IEEE Radar Conference*, Waltham, MA, USA, 17–20 April 2007.

[30] J.R. Guerci, P. Grieve. Optimum matched illumination waveform design process. US patent 5,121,125, 1992.

5

SPACE–TIME ADAPTIVE PROCESSING FOR AIRBORNE RADAR: A KNOWLEDGE-BASED PERSPECTIVE

Michael C. Wicks, Muralidhar Rangaswamy,
Raviraj S. Adve, and Todd B. Hale

This chapter provides an overview of radar space–time adaptive processing (STAP) from its inception to state-of-the-art developments. The topic is treated from both intuitive and theoretical aspects. A key requirement of STAP is knowledge of the spectral characteristics underlying the interference scenario of interest. However, the spectral characteristics are seldom known in practice and must be estimated using training data. The collection of training data for a given scenario is limited by the scale of change of the interference phenomenon with respect to space and time as well as by system considerations such as bandwidth. Increasingly complex interference scenarios give rise to stressful conditions of training data support. Consequently, the choice of training data becomes a crucial component of the adaptive process. Additional issues of importance in STAP include the computational cost of the adaptive algorithm as well as the ability to maintain a constant false alarm rate (CFAR) over widely and often rapidly varying interference statistics. This chapter treats these topics, from a knowledge-based (KB) perspective.

Knowledge-Based Radar Detection, Tracking, and Classification. Edited by Fulvio Gini and Muralidhar Rangaswamy

5.1 INTRODUCTION

Signal detection using adaptive processing in spatial and temporal domains offers significant benefits in a variety of applications including radar, sonar, satellite communications, and seismic systems [1]. The focus of this chapter is signal processing for radar systems using multiple antenna elements that coherently process multiple pulses. An adaptive array of spatially distributed sensors, which processes multiple temporal snapshots, overcomes the directivity and resolution limitations of a single sensor. Specifically, using STAP, that is, joint adaptive processing in the spatial and temporal domains, creates an ability to suppress interfering signals while simultaneously preserving gain on the desired signal. Using *training* to estimate interference statistics, this suppression is possible despite lack of knowledge of the interference spectral properties. Training, therefore, plays a pivotal role in adaptive systems. This chapter focuses on several aspects of this crucial phase from a *knowledge-based perspective*.

Consider the operation of an airborne phased array radar with J elements. The radar transmits a pulse in a chosen direction. The goal is to look for a target in this direction (the look angle). This transmitted pulse reflects off (possibly) a target (the desired signal) and the ground (or other clutter interfering sources). On receive, the radar samples this reflected wave at a high rate, with each of the R samples corresponding directly to reflections from a specific range. The sampled signal may also include other interfering effects of electronic counter-measures (ECM), such as jamming. This process is repeated for N pulses transmitted at a rate of the pulse repetition frequency (PRF). The entire received data can therefore be organized in a $J \times N \times R$ datacube [2, 3].

The problem at hand is to detect and locate targets, if they exist, within this datacube. This location is in terms of range (at a *primary range cell*) and Doppler (velocity) with the angle set to the look angle. In practice, the interference statistics and the target complex amplitude are unknown; thus the detection problem is equivalent to the problem of statistical hypothesis testing in the presence of unknown nuisance parameters. From another point of view, the Doppler–wavenumber or angle–Doppler spectrum per range cell provides a unique representation of a signal in a three-dimensional (3D) plane. Hence, the STAP problem may also be viewed as spectrum estimation where the two-dimensional (2D) adaptive spectral transform of spatio-temporal data affords separation of the desired target from interference. Indeed, in spatially and temporally white noise, the 2D Fourier transform is optimal.

In the context of STAP, for each range cell, the interference spectral characteristics correspond to the spatio-temporal covariance matrix of the $JN \times 1$ complex data vector under the target-free condition. The presence of these unknown parameters precludes use of a uniformly most powerful test for hypothesis testing [4]. This is because joint maximization of a likelihood ratio over the domain of unknown parameters becomes mathematically intractable and computationally expensive. Hence, ad hoc approaches have been proposed to overcome this problem. Nevertheless, present-day computing power permits the use of well-known tools from statistical detection and estimation theory in the radar problem.

The optimal STAP algorithm assigns a complex weight vector to each spatio-temporal degree of freedom (DOF) one range cell at a time. The complex weight vector corresponds to the transfer function of a linear finite impulse response (FIR) filter, which maximizes the signal-to-interference-plus-noise ratio at its output. For Gaussian interference statistics, the filter output has the interpretation of a pre-whitening operation (related to inversion of the interference covariance matrix) followed by matching (calculation of an inner product), and constitutes the optimal processor. In practice, the $JN \times JN$ covariance matrix is unknown and is therefore estimated from training data. Clearly, the statistics of this data must match that of the interference; that is, the training data must be target-free and homogeneous. Unfortunately, obtaining an accurate covariance matrix estimate requires a large number of homogeneous training samples, which are, generally, not available in practice. This is mainly because the training uses data from the *secondary range cells*, that is, range cells other than the primary range cell. Furthermore, even if they were available, the associated computation load makes the optimal approach impractical. This problem is exacerbated by the fact that the STAP process must be repeated for each Doppler and range bin of interest.

There are, therefore, two fundamental issues that limit the application of STAP algorithms in practice: the need for adequate homogeneous training data and the computation load of the algorithm. This chapter addresses these issues in some detail, drawing from the authors' extensive research in these areas. In the area of algorithms the exposition is devoted to salient developments in the field (including the authors' own contributions), as opposed to providing a comprehensive overview. We also discuss the important role of nonhomogeneity detection, covering the basics of ranking and selection theory, the theory of SIRPs, and the use of a nonhomogeneity detector tied to the STAP algorithm used for target detection. There is, unfortunately, no one "best" algorithm or approach. The chapter attempts to analyze by placing these algorithms using a KB perspective. We conclude with a preliminary algorithm wherein these issues are tied together in a combined approach that addresses all the critical issues mentioned above.

Section 5.2 discusses the STAP problem in some detail, covering early work on radar adaptive signal processing and developing a data model for the algorithms that follow. Section 5.3 discusses the issue of computation load while Section 5.4 discusses the issue of secondary data support. Section 5.5 then places the algorithms presented from a KB perspective. Section 5.6 concludes the paper.

5.2 PROBLEM STATEMENT

A radar is a sensor, in our case an antenna array on an airborne platform, which transmits and receives electromagnetic radiation. The transmitted electromagnetic signal impinges on various objects such as buildings, land, water, vegetation, and one or more targets of interest. The illuminated objects reflect the incident wave, which is received and processed by the radar receiver. The reflected signal includes desired signals (targets) but also undesired returns from extraneous objects, designated as clutter. Additionally, there could be one or more jammers, high powered noise-like signals transmitted as ECM, masking the desired target signals. Finally, the received returns include the ubiquitous background white noise caused by the radar receiver circuitry as well as by man-made sources and machinery. Typically, if it exists, the power of the desired signal return is a very small fraction of the overall interference power (due to clutter, jamming, and noise). The problem at hand is to detect the target, if it exists, within the background of clutter and jammer returns. The key to solving this problem is the availability of suitable models for targets, clutter, and jammers [2, 3, 5]. These models account for the angular position of the target in relation to the receiving array. If moving, the target signature includes the effect of the resulting Doppler frequency.

More precisely, the radar receiver front end consists of an array of J antenna elements, which receives signals from targets, clutter, and jammers. These radiations induce a voltage at each element of the antenna array, which constitutes the measured array data at a given time instant. Snapshots of the measured data collected at N successive time epochs give rise to the spatio-temporal nature of the received radar data. The spatio-temporal product $JN = M$ is defined to be the system dimensionality. Figure 5.1 uses the angle–Doppler space to illustrate the need for space–time (joint domain) processing. A target at a specific angle and traveling at a specific velocity (corresponding to a Doppler frequency) occupies a single point in this space. A jammer originates from a particular angle, but is temporally white (noise-like). The clutter, due to the motion of the platform, occupies a *ridge* in this 2D space [5]. A clutter patch in front of the moving aircraft has the highest Doppler frequency while one at broadside has zero Doppler (no relative velocity). The clutter spectrum reflects the two-way beam pattern of the transmitted signal.

The figure also illustrates the effect of strictly temporal (Doppler) or spatial processing (in angle). The former is equivalent to a projection of the two-dimensional target plus interference spectrum onto the Doppler plane; however, the target signal is masked by the temporally white jamming. The latter is equivalent to a projection of the said spectrum onto an angular plane, but as the clutter power is strongest at

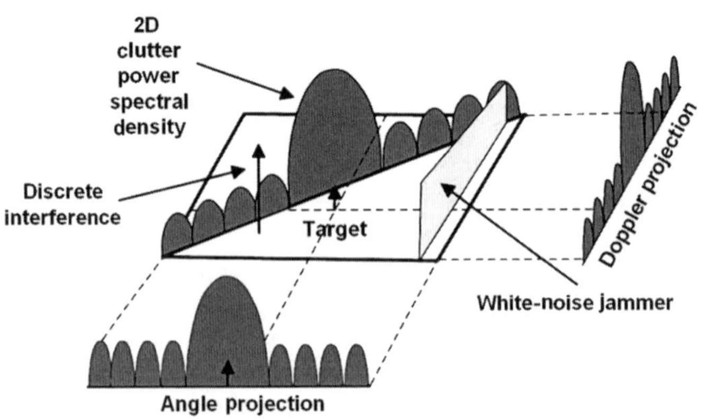

Figure 5.1 The target and interference scenario in an airborne radar.

the look angle, the target cannot be distinguished from clutter. However, joint domain processing identifies clear regions in the 2D plane, which affords recovery of the target from the interference background.

The detection problem can be formally cast in the framework of a statistical hypothesis test of the form

$$H_0: \quad \mathbf{x} = \mathbf{d} = \mathbf{c} + \mathbf{j} + \mathbf{n} \tag{5.1}$$

$$H_1: \quad \mathbf{x} = \alpha\mathbf{e}(\theta_t, f_t) + \mathbf{d} = \alpha\mathbf{e}(\theta_t, f_t) + \mathbf{c} + \mathbf{j} + \mathbf{n}, \tag{5.2}$$

where $\mathbf{x} \in C^{JN\times 1}$ denotes the received data under either hypothesis, $\mathbf{d}$ represents the overall interference being the sum of $\mathbf{c}$, the clutter vector, $\mathbf{j}$, the jammer vector, and $\mathbf{n}$, the background white noise. Finally, $\mathbf{e}$ is a known spatio-temporal steering vector that *models* the target return for a specific angle–Doppler and α is the unknown target complex amplitude. For the popular case of a linear array of equally spaced elements,

$$\mathbf{e} = \mathbf{e}_t \otimes \mathbf{e}_s \tag{5.3}$$

$$\mathbf{e}_t = \begin{bmatrix} 1 & z_t & z_t^2 \cdots z_t^{(N-1)} \end{bmatrix}^T, \tag{5.4}$$

$$\mathbf{e}_s = \begin{bmatrix} 1 & z_s & z_s^2 \cdots z_s^{(J-1)} \end{bmatrix}^T, \tag{5.5}$$

$$\mathbf{z}_s = e^{j2\pi f_s} = e^{(j2\pi \frac{d}{\lambda} \sin\phi_t)}, \qquad z_t = e^{j2\pi f_t/f_R}, \tag{5.6}$$

where ϕ_t and f_t represent the look angle, measured from broadside, and Doppler frequency respectively, $\otimes$ represents the Kronecker product of two vectors, f_R the pulse repetition frequency (PRF), and λ the wavelength of operation. The vectors $\mathbf{e}_t$ and $\mathbf{e}_s$ represent the temporal and spatial steering vectors, respectively. Note that from one pulse to the next and from one element to the next the steering vectors represent a *constant phase shift*.

Adaptive algorithms generally determine a weight vector $\mathbf{w}$ to obtain a test statistic Λ; that is,

$$\Lambda = |\mathbf{w}^H\mathbf{x}|^2 \underset{H_0}{\overset{H_1}{\gtrless}} \lambda, \tag{5.7}$$

where the superscript H represents the Hermitian transpose of a vector/matrix and λ represents a *threshold* above which a target is declared present. This threshold determines the probability of false alarm, the rate at which a target is *detected* "by mistake." For Gaussian interference statistics, the optimum processing method, corresponding to the case of a known interference covariance matrix $\mathbf{R}_d$, is the whiten-and-match filter (MF) for detecting a rank-1 signal given by [6]

$$\mathbf{w} = \frac{\mathbf{R}_d^{-1}\mathbf{e}}{\sqrt{\mathbf{e}^H\mathbf{R}_d^{-1}\mathbf{e}}} \Longrightarrow \Lambda_{MF} = \frac{|\mathbf{e}^H\mathbf{R}_d^{-1}\mathbf{x}|^2}{\mathbf{e}^H\mathbf{R}_d^{-1}\mathbf{e}} \underset{H_0}{\overset{H_3}{\gtrless}} \lambda_{MF}, \tag{5.8}$$

which represents the matched filtering of the whitened data $\breve{\mathbf{x}} = \mathbf{R}_d^{-1/2}\mathbf{x}$ and whitened steering vector $\breve{\mathbf{e}} = \mathbf{R}_d^{-1/2}\mathbf{e}$. It can be readily shown that Λ_{MF} is simply the output signal-to-interference-plus-noise ratio (SINR) of the minimum variance distortionless response (MVDR) beamformer, which calculates the maximum likelihood *estimate* of the target complex amplitude. The relationship between Λ_{MF} and the MVDR beamformer output SINR thus provides a unified perspective of detection and estimation in the context of STAP.

In practice, the covariance matrix $\mathbf{R}_d$ is unknown and must be estimated. Early work on antenna arrays by Widrow (least-squares method) [7] and Applebaum (maximum signal-to-noise-ratio criterion) [8] suggest use of feedback loops to ensure convergence of iterative methods for calculating the weight vector. However, these methods were slow to converge to the steady-state solution. Fundamental work by Reed, Mallet and Brennan (RMB beamformer) [9] showed that the sample matrix inverse (SMI) method offered considerably better convergence. In the SMI approach, the basis for most modern STAP algorithms, the interference covariance matrix is *estimated* using K data ranges for training,

$$\hat{\mathbf{R}}_d = \frac{1}{K}\sum_{k=1}^{K}\mathbf{x}_k\mathbf{x}_k^H = \frac{1}{K}\mathbf{X}\mathbf{X}^H, \tag{5.9}$$

where $\mathbf{X} = \{\mathbf{x}_1, \mathbf{x}_2, \ldots, \mathbf{x}_K\}$ and the adaptive weights are obtained as $\mathbf{w} = \hat{\mathbf{R}}_d^{-1}\mathbf{e}$. A drawback of the RMB approach is the lack of a constant false-alarm rate (CFAR); that is, the false-alarm rate varies with the interference level, an important consideration in practical systems. Variants of the RMB beamformer to obtain CFAR, such as the Kelly GLRT [10], the adaptive matched filter (AMF) [6], and the adaptive coherence estimator (ACE) [11], were the focus of a number of efforts in the 1980s and early 1990s. The AMF test is also referred to as the modified sample matrix inverter (MSMI) in the literature. Interestingly, the whiten-and-match filter of Equation 5.8, with the true covariance matrix $\mathbf{R}_d$ replaced with the estimated covariance matrix $\hat{\mathbf{R}}_d$ has CFAR. There are, however, three fundamental problems with this approach when applied in the real world: the associated computation load, the need for an adequate number of training samples, and finally, most importantly, the heterogeneity of the available data.

The SMI algorithm and its variants mentioned above require the solution to a system of linear equations involving a $JN \times JN$ matrix in real time, an $\mathcal{O}(J^3N^3)$ operation. The fact that the algorithm must be executed for each range and Doppler bin of interest exacerbates the problem. Furthermore, to obtain performance within 3 dB of optimum, one requires approximately $K \geq 2JN$ training samples to estimate the $JN \times JN$ matrix $\mathbf{R}_d$. Such a large number of samples are generally not always available.

Finally and most importantly, the training data must be *homogeneous*, that is, statistically representative of the interference within the range cell of interest. This is generally impossible to obtain in practice due to limitations imposed by the spatio-temporal nonstationarity of the interference as well as by system considerations such as bandwidth and fast scanning arrays. For example with $J = 11$ and $N = 32$, the parameters for the KASSPER data set [12], the training data support for 3 dB performance is 704. Assuming an instantaneous RF bandwidth of 500 kHz, this calls for the wide-sense stationarity (homogeneity) over a 400 km range! The scarcity of training data is exacerbated by system errors such as aircraft crabbing and internal clutter motion [5] and environmental considerations such as strong clutter discretes [13], range-varying interference spectra and power levels [14], and outlier contamination of training data by target-like signals [15] occurring in dense target scenarios caused by flight formations.

These three issues are interlinked — the computation load is a function of the DOF in the adaptive process and the number of training samples are approximately twice the DOF. In other words, reducing the computation load also reduces the required training. Clearly, reducing the required training also addresses the heterogeneity problem, making it easier to acquire an adequate number of training samples.

As is clear from the above discussion, adequately and effectively training the adaptive filter is essential. The central theme of the following discussion is the use of pre-existing and the development of real-time knowledge bases to help in the training process. This knowledge base comprises many aspects — using a priori knowledge in choosing the secondary data, using real-time processing to identify homogeneous data samples, and choosing the most effective algorithm based on the available information. The use of KB processing has resulted in the development of the Knowledge Aided Sensor Signal Processing Expert Reasoning (KASSPER) program [12]. Using simulated and measured data the preliminary results, now appearing in the literature, show both the importance of and improvements from using knowledge-aided processing [15–19].

5.3 LOW COMPUTATION LOAD ALGORITHMS

Succinctly stated, the fully optimal STAP algorithm consists of the following steps:

1. Starting with a datacube, identify the cell under test (corresponding to the length-JN data vector $\mathbf{x}$) and form the target steering vector $\mathbf{e}$ for every Doppler bin of interest.
2. Select K representative training data from both sides of the cell under test, avoiding guard cells to account for target leakage and competing targets.
3. Form $\hat{\mathbf{R}}_{\mathrm{d}}$, the estimated interference covariance matrix, using the training data.
4. Calculate a weight vector $\mathbf{w} \propto \hat{\mathbf{R}}_{\mathrm{d}}^{-1}\,\mathbf{e}$ and apply the weight vector to test cell data to obtain a test statistic, $\Lambda \propto \mathbf{w}^{H}\mathbf{x}$.
5. Compare the test statistic to a threshold (corresponding to a specified false-alarm rate) and declare target presence when the test statistic exceeds the threshold.

To overcome the problems associated with this fully adaptive algorithm, researchers have developed alternative, partially adaptive, approaches that reduce the DOF with attendant reductions in required sample support and computation cost. Important works in this area include the Joint Domain Localized (JDL) processing algorithm [20], the Parametric Adaptive Matched Filter (PAMF) [21] (and references therein), the Multi-Stage Wiener Filter (MSWF) [22] and factored STAP methods [5]. Another important approach, not dependent on any statistical training, is the Direct Data Domain (D^3) approach [23]. This algorithm was then extended to include statistical processing in [24]. Several studies show that there is no "best" algorithm, but that an effective implementation would require use of the most effective from a library of algorithms. Other than the PAMF, D^3, and MWF (depending on the type of implementation) algorithms, all STAP methods require explicit formation and inversion of the interference covariance matrix; that is, the issue of homogeneous training data remains.

This section develops in some detail the most popular low computation load algorithms. These algorithms are the most popular for a specific reason — they all address the issue of computation load in innovative, though completely different, ways. A common framework for the reduced DOF process is that they all rely on a transformation of the steering vector and received data into a subspace of dimension $r < JN$. In the following, $\mathbf{T}$ denotes a general transformation matrix, and $\tilde{\mathbf{e}} = \mathbf{Te}$, $\tilde{\mathbf{x}} = \mathbf{Tx}$, and $\tilde{\mathbf{X}} = \mathbf{TX}$ denote the transformed steering vector, test cell data, and training data, respectively. The transformation matrix $\mathbf{T}$ can either be data dependent or data independent. The JDL algorithm is an example of a data-independent transformation, while the PAMF, MWF, and LRNAMF [15] are instances of data-dependent transformations.

5.3.1 Joint Domain Localized Processing

The JDL algorithm as developed by Wang and Cai [20] maps the received data to the angle–Doppler domain. The transformation to angle–Doppler localizes the target and interference to a few angle and Doppler bins, significantly reducing the required DOF, with corresponding reductions in required sample support and computation load. The authors assume the receiving antenna to be an equally spaced linear array of ideal, isotropic, point sensors. Based on this assumption, space–time data is transformed to the angle–Doppler domain using a 2D discrete Fourier transform (DFT). This approach is only valid in the ideal case under certain restrictions. The presentation here is for the generalized JDL algorithm valid for real-world antenna arrays as well [25].

The JDL algorithm begins with a transformation to the angle–Doppler space; that is, the angle–Doppler response of the data is obtained at the few angle and Doppler bins within the LPR. Mathematically, the angle–Doppler response of the data vector $\mathbf{x}$ at angle ϕ and Doppler f_d is given by

$$\tilde{x}(\phi, f_d) = \mathbf{e}^H(\phi, f_d)\mathbf{x}, \tag{5.10}$$

where the tilde above the scalar x denotes the transform domain. Repeating this process for η_a angles and η_d Doppler bins (corresponding to $\eta_a\eta_d$ space–time steering vectors) generates a length$-\eta_a\eta_d$ vector $\tilde{\mathbf{x}}$ of angle–Doppler domain data. These η_a angles and η_d Doppler frequencies are said to comprise the Localized Processing Region (LPR). Note that this scheme may be used in conjunction with real-world arrays where the space–time steering vector would include a measured spatial steering vector. The scheme reverts to the 2D DFT in the case of an idealized linear array of isotropic point sensors. The transformation matrix T is given by

$$\mathbf{T} = \left[\mathbf{e}_1, \mathbf{e}_2, \ldots, \mathbf{e}_{\eta_a\eta_d}\right], \qquad \tilde{\mathbf{x}} = \mathbf{T}^H\mathbf{x}, \tag{5.11}$$

where $\mathbf{e}_i$, $i = 1, \ldots, \eta_a\eta_d$ are the steering vectors corresponding to the angles and Dopplers in the LPR. In practice, the angle and Doppler points are chosen to be close to and symmetric around the look angle and Doppler. Note that the transformation matrix is independent of the data. As is usual with a Fourier transform, one could also use a taper, such as a Hamming or Kaiser window, to lower the transformation sidelobes. In the angle–Doppler domain the adaptive weights are given by

$$\tilde{\mathbf{w}} = \tilde{\mathbf{R}}_d^{-1}\tilde{\mathbf{e}}, \tag{5.12}$$

and

$$\tilde{\mathbf{R}}_d = \frac{1}{K}\sum_{k=1}^{K} \tilde{\mathbf{x}}_k\tilde{\mathbf{x}}_k^H. \tag{5.13}$$

That is, *the JDL algorithm is basically the original SMI algorithm, but using data in the angle–Doppler space.*

The steps in implementing the JDL adaptive processor are as follows:

1. Choose the size of the LPR, that is, η_a and η_d and the number of training data vectors that will be used to estimate the covariance matrix.
2. Set the angle bin to be the look direction. Choose a set of η_a angles centered around (and including) the look angle.
3. For each Doppler bin of interest, choose a set of η_d Doppler bins centered around (and including) the look Doppler. Use the set of angles and Doppler bins to form the transformation matrix **T** using Equation 5.11.
4. Transform the entire datacube to angle–Doppler space and find the transformed steering vector $\tilde{\mathbf{e}}$.
5. For each range of interest, estimate an angle–Doppler covariance matrix using Equation 5.13 and obtain the angle–Doppler weights using Equation 5.12 to obtain a decision statistic.

Comparing the decision statistic to the chosen threshold, as in Equation 5.8, completes the detection process. The key here is that K, the number of required homogeneous samples, is reduced to about $2\eta_a\eta_d$–$4\eta_a\eta_d$ (as opposed to $2JN$) and the computation load to $\mathcal{O}(\eta_a^3\eta_d^3)$. Common values of η_a and η_d, usually odd, are on the order of 3, 5, 7, resulting in enormous savings in computation load and required sample support.

5.3.2 Parametric Adaptive Matched Filter

The PAMF method is a case of reduced-dimension processing that relies on a decomposition of the $JN \times JN$ interference covariance matrix $\mathbf{R}_d$ of the form $\mathbf{R}_d = \mathbf{LDL}^H$, where $\mathbf{L}$ is a lower block-triangular matrix with $J \times J$ identity matrices along the main block diagonal, and $\mathbf{D}$ is a block diagonal matrix with Hermitian matrices $\mathbf{D}_i \in C^{J\times J}$, $i = 1, 2, \ldots, N$ [21]. Consequently, $\mathbf{L}^{-1}$ admits a representation of the form

$$L^{-1} = \begin{bmatrix} \mathbf{I}_J & \mathbf{O} & \mathbf{O} & \cdots & \mathbf{O} \\ \mathbf{A}_1^H(1) & \mathbf{I}_J & \mathbf{O} & \cdots & \mathbf{O} \\ \vdots & \vdots & \vdots & \vdots & \vdots \\ \mathbf{A}_{N-1}^H(N-1) & \mathbf{A}_{N-1}^H(N-2) & \cdots & \mathbf{A}_{N-1}^H(1) & \mathbf{I}_J \end{bmatrix} \tag{5.14}$$

where $\mathbf{A}_l(k)$, $k = 1, 2, \ldots, N$ denote the coefficients of the lth order multichannel forward linear predictor or multichannel (matrix) autoregressive (AR) linear predictor and $\mathbf{D}_i$ is the covariance matrix of the residual from the ith order multichannel linear predictor. Thus the transformation matrix $\mathbf{T}$ takes on the form $\mathbf{T}_{\text{PAMF}} = \mathbf{D}^{-\frac{1}{2}}\mathbf{L}^{H^{-1}}$ for the multichannel parametric method. The block form of the transformation is computationally expensive to implement due to the fact that it requires the calculation of all the matrix prediction coefficients of orders 1 to N. Consequently, a sequential method for implementing $\mathbf{T}_{\text{PAMF}}$ is developed in reference [21]. Furthermore, it has been found for a variety of simulated and measured radar data sets that a low order $p = 3$ or 4 multichannel linear predictor provides a good approximation to $\mathbf{R}_d$ [21]. When $\mathbf{R}_d$ is unknown, the adaptive algorithm directly estimates the coefficients of a pth order multichannel linear predictor and the associated block diagonal covariance matrix using training data from the data matrix $\mathbf{X}$. The PAMF test is given by

$$\Lambda_{\text{PAMF}} = \frac{|\tilde{\mathbf{e}}^H\tilde{\mathbf{x}}|^2}{\tilde{\mathbf{e}}^H\tilde{\mathbf{e}}} \underset{H_0}{\overset{H_1}{\gtrless}} \lambda_{\text{PAMF}}, \tag{5.15}$$

where $\tilde{\mathbf{e}}$ and $\tilde{\mathbf{x}}$ are the steering vector and observed data vector transformed by $\mathbf{T}_{\text{PAMF}}$. The low model order $p \ll JN$ enables significant reduction in the training data support requirements. Thus, explicit formation and inversion of the interference covariance matrix is avoided. Instead the coefficients of multichannel linear prediction and the associated residual error covariance matrices, succinctly embed the information contained in $\mathbf{R}_d$. For example, with both simulated and measured data sets, it was shown in reference [21] that the sample support for $JN = 128$; $p = 3$, and $K = 8$, the PAMF exhibited performance to within 0.5 dB of the optimal matched filter. The key to obtaining enhanced PAMF performance is the use of efficient parameter estimation algorithms for calculating the multichannel coefficients of linear prediction and the block diagonal error covariance matrix. A number of approaches for this purpose have been discussed in reference [26]. However, the method best suited for the STAP problem is the covariance method of linear prediction also known as the method of least squares. The computational cost underlying the algorithm is $\mathcal{O}(JN)2p)$, which provides an order of magnitude reduction in the computational cost for $p \ll JN$. However, when the training data is subject to outlier contamination, the PAMF performance is severely degraded. Mitigating strategies for this problem have been discussed in reference [27], with other PAMF extensions and variants presented in the references therein. Unfortunately, the detection and false-alarm probability for the PAMF test and its variants cannot be calculated using closed-form analytical expression. Consequently, these issues are studied using Monte Carlo simulations in references [21] and [27] and references therein.

5.3.3 Multistage Wiener Filter

The MWF is another reduced rank processing method, which relies upon a serial decomposition of the MVDR beamformer weight vector in the form of a generalized sidelobe canceller (GSC) [22, 28]. The GSC processor relies upon a projection of $\mathbf{x}$ onto the signal subspace, $d = \mathbf{e}^H\mathbf{x}$, and a projection onto an orthogonal complement subspace, $\mathbf{b} = \mathbf{B}\mathbf{x}$, where $\mathbf{B}$ is a blocking matrix with orthonormal columns that are orthogonal to $\mathbf{e}$. The GSC weight vector is given by $\mathbf{w}_{\mathrm{GSC}} = [1 \ \ \mathbf{v}]^T$, where $\mathbf{v} = -\mathbf{R}_{\mathrm{b}}^{-1}\mathbf{r}_{\mathbf{bd}}$ is a $1 \times M - 1$ row vector, with $\mathbf{R}_{\mathrm{b}} = \mathbf{B}\mathbf{R}_{\mathbf{d}}\mathbf{B}^H$ and $\mathbf{r}_{\mathbf{bd}} = \mathbf{B}\mathbf{R}_{\mathbf{d}}\mathbf{e}$. The error variance at the MVDR beamformer output can then be expressed as $\frac{1}{\mathbf{e}^H\mathbf{R}_{\mathbf{d}}^{-1}\mathbf{e}} = \mathbf{e}^H\mathbf{R}_{\mathbf{d}}\mathbf{e} - \mathbf{r}_{\mathbf{bd}}^H\mathbf{R}_{\mathrm{b}}^{-1}\mathbf{r}_{\mathbf{bd}}$. This form of the error admits a sequential representation in terms of a Rayleigh quotient and an inverse Rayleigh quotient. More precisely, let $\delta_i = \frac{\mathbf{v}_i^H\mathbf{R}_i\mathbf{v}_i}{\mathbf{v}_i^H\mathbf{v}_i}$ and $\xi_i = \frac{\mathbf{v}_i^H\mathbf{v}_i}{\mathbf{v}_i^H\mathbf{R}_i^{-1}\mathbf{v}_i}$, where $\mathbf{v}_0 = \mathbf{e}$, $\mathbf{R}_0 = \mathbf{R}_{\mathbf{d}}$, $\mathbf{v}_1 = \mathbf{R}_{\mathrm{b}}^{-1}\mathbf{r}_{\mathbf{bd}}$. Then, we have a recurrence relationship between the error residuals at the output of successive stages of the MWF. Specifically, for $i = 0, 1, \ldots, M - 1$,

$$\xi_i = ||\mathbf{v}_i||^2\delta_i - \frac{||\mathbf{v}_{i+1}||^2}{\xi_{i+1}}, \tag{5.16}$$

where $||\cdot||$ denotes the norm of a vector. This form of implementation provides a continued fraction expansion of the MVDR beamformer output variance, which results in a tridiagonal covariance structure for the transformed data [22, 28]. Such a form lends itself to an iterative calculation of the MVDR beamformer weight vector via the conjugate gradient method [29]. In sample support deficient scenarios, this method has been found to converge to the principal components inverse method [30]. Key features of this method are the absence of the formation and inversion of the full dimension covariance matrix. Additionally, the MWF implementation in references [22] and [28] is computationally expensive due to the need to calculate a sequence of matrix products to recombine the error residuals from the transformed data. However, this is greatly alleviated by the conjugate gradient method, which only requires a one-way computation [29]. Performance comparisons of the MWF with competing techniques can be found in references [22, 28], and [29] and references therein.

5.4 ISSUES OF DATA SUPPORT

As discussed in this chapter, an extremely important issue in space–time adaptive processing (STAP) is the formation and inversion of the covariance matrix underlying the disturbance. In practice, the unknown interference covariance matrix is estimated from a set of independent identically distributed (i.i.d.) target-free training data, which is assumed to be representative of the interference statistics in a cell under test. Frequently, the training data are subject to contamination by discrete scatterers or interfering targets [31]. In either event, the training data become nonhomogeneous. As a result, they are not representative of the interference in the test cell. Hence, standard estimates of the covariance matrix from nonhomogeneous training data result in severely undernulled clutter. Consequently, CFAR and detection performance suffer. Significant performance improvement can be achieved by employing pre-processing to select representative training data.

Figure 5.2, borrowed from reference [13], illustrates the importance of nonhomogeneous data. The figure plots the probability of detection P_d versus SINR for a false-alarm rate of $P_{fa} = 0.01$ and a clutter-to-noise ratio of 40 dB. The system uses $M = JN = 64$ and $K = 128$ range cells to estimate the interference covariance matrix. The matched filter (MF) curve is theoretical, corresponding to optimal performance in Gaussian clutter. The curve for the AMF, operating with homogeneous interference, has performance within 3 dB of optimal. The curve corresponding to nonhomogeneous data is obtained using Monte Carlo simulations. The training data are corrupted using 30 high amplitude discrete targets. As is clear, the nonhomogeneity

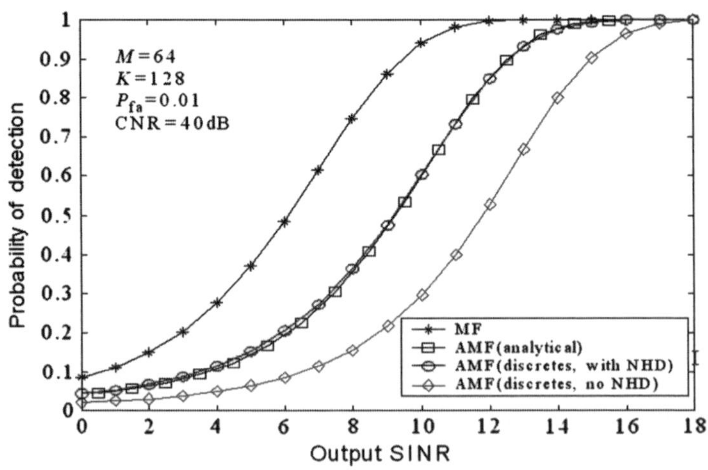

Figure 5.2 Impact of nonhomogeneous data on detection performance.

of the data significantly worsens detection performance by as much as 3–5 dB. Using a nonhomogeneity detector (NHD) [32], the performance of the AMF algorithm is restored.

In general, nonhomogeneity of training data is caused by environmental factors such as the presence of strong discrete scatterers, dense target environments, nonstationary reflectivity properties of the scanned area, and radar system configurations such as conformal arrays and bistatic geometries. A variety of robust adaptive signal-processing methods to combat specific types of nonhomogeneities have been developed in references [15, 31], and [33–35]. In this effort, we confine ourselves to the problem of selecting representative training data, when the training data are contaminated by outliers resembling a target of interest specifically, outliers sharing the same steering vector as a target of interest.

5.4.1 Nonhomogeneity Detection

The problem of outlier contamination of STAP training data assumes increased significance in dense target scenarios, where outliers resembling a target of interest contaminate the training data. This results in an incorrect threshold setting due to an erroneous estimate of the interference covariance matrix. Furthermore, the presence of outliers in the training data causes target cancelation, resulting in degraded output signal-to-interference ratio and perforce degraded detection performance. A common signal-processing method in this context is to excise outliers from the training data and use the resulting outlier-free training data for covariance matrix estimation. Several algorithms for outlier removal have been proposed in recent times [13–15, 17, 27, 32] in a variety of dense target environments. For the purpose of this section, and in practice, the columns of the data matrix $\mathbf{X}$ are no longer independent, identically distributed, and free from outlier contamination. The problem therefore is to classify the columns of $\mathbf{X}$ into groups sharing the same covariance matrix and thereby detect the presence of outliers, which have a deleterious impact on STAP performance when used in covariance estimation. When outliers are in the form of strong clutter discretes, the generalized inner product (GIP) method [13, 32] and references therein gives a method for outlier removal as summarized in the following:

1. First an initial estimate of the covariance matrix using an extended training data set is formed as $\tilde{\mathbf{R}} = \tilde{\mathbf{X}}\tilde{\mathbf{X}}^H/L$ where $\tilde{\mathbf{X}}$ is a data matrix with L columns, where $L \gg 2JN$. For example, L may be equal to all available ranges R.
2. Each column of $\tilde{\mathbf{X}}$ is used in a sliding window process to form a statistic $p_i = \mathbf{x}_i^H\tilde{\mathbf{R}}^{-1}\mathbf{x}_i$ for the available range of i. Note that $\tilde{\mathbf{R}}$ for each $\mathbf{x}_i$ is formed by excluding that column and a one column on either side of $\mathbf{x}_i$ (to allow for guard cells).
3. If the columns $\tilde{\mathbf{X}}$ shared the same covariance matrix, empirical realizations of p_i will conform to an F-distribution [13], whose theoretical mean μ is readily calculated.

4. The absolute value $|p_i - \mu|$ are calculated and sorted in increasing order and $K \simeq 2(\text{DOF})$ columns of $\tilde{\mathbf{X}}$ corresponding to p_i showing the least deviation from μ are retained for covariance matrix estimation. The remaining columns are discarded.

Approximately 3–5 dB of performance improvement in the AMF performance in heterogeneous clutter scenarios was demonstrated in reference [13] using simulated and measured data. However, such an approach relies on full-dimension STAP processing and therefore is not suited for conditions of limited sample support. Hence, an alternative reduced-dimension extension of this procedure known as the innovations power sort was developed in reference [27], wherein a multichannel linear predictor approximation to $\tilde{\mathbf{R}}$ is employed along the lines of the multichannel AR model described in the PAMF. This form of the estimator has been found to be extremely valuable in conditions of small sample support. The procedure for outlier removal therein is very similar to the GIP approach described above. Significant performance improvement over competing methods was demonstrated using measured radar data in reference [27]. When outliers resembling a target of interest contaminate the training data, it becomes imperative to use the steering vector in calculating the test statistic for use in outlier identification and removal. Motivated by this and the need to operate in conditions of limited sample support, the authors in reference [15] developed an eigen-based method, which relies upon the simple principle that the output of a matched filter peaks when data containing a desired target is passed through the filter. This fundamental idea is used in an iterative manner in reference [15] to identify the outliers in training data. An extension of this method is pursued in reference [17] from a KB perspective to significantly reduce the sample support for covariance estimation, while obtaining near clairvoyant STAP detection performance. Other approaches include the use of the adaptive process as a NHD [36].

Theoretical approaches to the problem of nonhomogeneity include use of spherically invariant random processes (SIRP). In other instances, there could be range-varying clutter power properties due to environmental and system considerations. In this instance, the clutter statistics depart from the Gaussian behavior, which leads to unacceptably large false-alarm rates. This calls into question a suitable model for these impulsive (heavy-tailed) clutter scenarios. There is no unique model for representing the joint probability density function (PDF) of a set of M correlated non-Gaussian random variables. However, a popular model for non-Gaussian radar clutter is the SIRP [14]. Every SIRP is equivalent to the product of a complex Gaussian process and a non-negative random variable, whose PDF is defined to be the first-order characteristic PDF of the SIRP. Consequently, every SIRP is uniquely determined by the specification of a mean vector, a covariance matrix, and a characteristic first-order PDF. As a result, the sample covariance matrix is no longer the maximum likelihood estimate for the SIRP covariance matrix. Furthermore, the covariance matrix estimate cannot be calculated in closed form [14, 37]. Instead the ML estimate is a weighted sample covariance matrix, which could be calculated iteratively using the expectation maximization (EM) algorithm. Key issues in this context include the convergence properties of the algorithm and the associated numerics. Having obtained the ML estimate of the covariance matrix (which is usually within a multiplicative constant of the covariance matrix of the Gaussian component of the (SIRP) a scale-invariant test statistic, such as the ACE, is called for. Using the statistics of the ACE test, a formal goodness-of-fit test is developed in reference [14] to detect and remove outliers. Performance of the approach is presented in reference [14] using simulated and measured data. The method outperforms all competing candidate algorithms. The extension of this method for sample support-starved scenarios is the focus of ongoing research.

5.4.2 Direct Data Domain Methods

Purely statistical algorithms, such as JDL and MWF, cannot suppress a discrete interference source within the primary range cell. For example, a large target within the test range cell but at a different angle and/or Doppler appears as a false alarm, through the sidelobes of the adapted beam pattern, at the look angle–Doppler domain. The secondary data cells do not carry information about the discrete nonhomogeneity and hence a statistical algorithm cannot suppress discrete (uncorrelated) interference within the range cell under test. This issue of adaptive processing within nonhomogeneous cells has led to the investigation of a new class of algorithms — nonstatistical, or direct data domain (D^3), algorithms [23, 24]. D^3 algorithms use data from the primary range cell only, and so bypass the problem of the required homogeneous secondary data support.

The basis of D^3 processing is that, as shown in Equation 5.6, given the look angle and Doppler, the steering vector determines the phase shift of the target signal from one antenna element/transmitted pulse to the next. The look angle and Doppler determine z_s, the phase shift of the target signal from one antenna element to the next and z_t, the phase shift from one pulse to the next. If $x_j(n)$ represents the total signal at the jth element and nth pulse, terms such as $x_j(n) - z_s^{-1}x_{(j+1)}(n)$ and $x_j(n) - z_t^{-1}x_j(n+1)$ should therefore contain only interference and noise terms. The D^3 approach minimizes the power in these terms while maximizing processing gain in the look direction constant. For example, to determine a set of spatial weights, define the $N \times (J-1)$ interference and noise matrix $\mathbf{A}$,

$$\mathbf{A} = \begin{bmatrix} x_0(0) - z_s^{-1}x_1(0) & \cdots & x_{(J-2)}(0) - z_s^{-1}x_{(J-1)}(0) \\ x_0(1) - z_s^{-1}x_1(1) & \cdots & x_{(J-2)}(1) - z_s^{-1}x_{(J-1)}(1) \\ \vdots & \vdots & \vdots \\ x_0(N-1) - z_s^{-1}x_1(N-1) & \cdots & x_{(J-2)}(N-1) - z_s^{-1}x_{(J-1)}(N-1) \end{bmatrix} \tag{5.17}$$

and the optimal weights $\mathbf{w}_s$ are the solution to the following optimization problem,

$$\mathbf{w}_s^{\text{opt}} = \arg \max_{\mathbf{w}_s, \mathbf{w}_s^H \mathbf{w}_s = 1} \left[|\mathbf{w}_s^H \mathbf{e}_{s,0:J-2}(\theta)|^2 - \kappa_s \mathbf{w}_s^H \mathbf{A}^T \mathbf{A}^* \mathbf{w}_s \right], \tag{5.18}$$

where the superscripts ^{T} and * represent the transpose and conjugation operators, respectively. This formulation is chosen to remain consistent with the notion that the conjugates of the weights multiply the data. The vector $\mathbf{e}_{s0:J-2}(\theta)$ represents the first $J-1$ entries of the length-J *spatial* steering vector. The use of only $J-1$ weights represents the DOF lost due to the subtraction operation in $x_j(n) - z_s^{-1}x_{(j+1)}(n)$.

The first term in Equation 5.18 represents the gain of the weight vector in the direction of the look angle while the second term represents the residual interference power after the data is filtered by the same weights. Hence, the optimal D^3 weights maximize the difference between the gain of the antenna at the look Doppler and the residual interference power. The term κ_s is chosen as a tradeoff between gain and interference cancellation. Using the method of Lagrange multipliers, it can be shown that the desired weight vector is the eigenvector corresponding to the maximum eigenvalue of the $(J-1) \times (J-1)$ matrix $\mathbf{a}_{0:J-2}\mathbf{a}_{0:J-2}^H - \mathbf{A}^T\mathbf{A}^*$. A temporal weight vector $\mathbf{w}_t$ can be found analogously and the overall weight vector is

$$\mathbf{w} = \begin{bmatrix} \mathbf{w}_t \\ 0 \end{bmatrix} \otimes \begin{bmatrix} \mathbf{w}_s \\ 0 \end{bmatrix}, \tag{5.19}$$

where $\otimes$ represents the Kronecker product and the zeros appended represent the loss of one DOF in space and time.

The steps in implementing the D^3 processor are as follows:

1. Choose the emphasis parameter κ and form matrix $\mathbf{A}$ using Equation 5.17 and data from within the range cell of interest *only*.
2. Find the eigenvector corresponding to its largest eigenvalue of $\mathbf{a}_{0:J-2}\mathbf{a}_{0:J-2}^H - \mathbf{A}^T\mathbf{A}^*$. This is $\mathbf{w}_s$.
3. Repeat Steps 1 and 2 to obtain a temporal weight vector and then the overall weights $\mathbf{w}$ using Equation 5.19.

Note that the adaptive weight vector in Equation 5.18 is obtained using data from the primary range cell *only*, without estimation of a covariance matrix. This property gives direct data domain processing its greatest advantage and its greatest disadvantage. The lack of an estimation of correlation allows use of D^3 processing in severely nonhomogeneous situations. In theory, it could be used by itself; however, the nonhomogeneous range cells have two components of interference — the discrete and the homogeneous components. By their very nature, D^3 algorithms are effective against discrete interference, but they are not as effective against the homogeneous component of the interference. This is because they ignore all statistical information.

5.4.2.1 *Hybrid Approach* We present here a hybrid technique, a two-stage process based on the D^3 and JDL algorithms, that combines the benefits of D^3 and statistical processing [24]. Consider the framework of any STAP algorithm. The algorithm processes receive data to obtain a complex weight vector for each range bin and each look angle/Doppler. The weight vector multiplies the primary data vector to yield a complex number. The process of obtaining a real scalar from this number for threshold comparison is part of the post-processing and not inherent to the algorithm itself. The adaptive process effectively estimates the signal component in the look angle/Doppler; that is, it is a 2D adaptive spectral estimate. The adaptive weights can therefore be viewed in a role similar to that of the nonadaptive steering vectors in JDL processing, used to transform the space–time data to the angle–Doppler domain.

The JDL processing algorithm begins with a transformation of the data from the space–time domain to the angle–Doppler domain. Statistical adaptive processing within a LPR in the angle–Doppler domain follows. The hybrid approach uses the D^3 weights, replacing the nonadaptive steering vectors used earlier. By choosing the set of look angles and Dopplers to form the LPR, the D^3 weights perform a function analogous to the nonadaptive transform. The D^3 algorithm is used repeatedly with the η_a look angles and the η_d look Doppler frequencies to form the LPR using the *same primary data*. This implies that there is a main look direction for the overall hybrid STAP process, but a set of auxiliary look directions for use with the D^3 algorithm.

The steps in implementing the hybrid adaptive processor are as follows:

1. Choose the size of the LPR (η_a and η_d), the number of secondary data vectors that will be used to estimate the covariance matrix (usually of the order of $2\eta_a\eta_d$–$4\eta_a\eta_d$) and the number of guard cells (usually 2–4).
2. Choose a set of η_a angles centered around (and including) the look angle.
3. For each range bin and Doppler bin of interest, choose a set of η_d Doppler bins centered around (and including) the look Doppler.
4. Using only the primary data, use the D^3 algorithm repeatedly ($\eta_a\eta_d$ times) with each combination of the chosen angles and Dopplers as the look direction.[1] These $\eta_a\eta_d$ weight vectors form the transformation matrix **T** as in Equation 5.11.
5. JDL processing continues as in Equations 5.12 and 5.13.

5.5 KNOWLEDGE-AIDED APPROACHES

The previous sections have addressed the three fundamental issues associated with practical adaptive processing for airborne radar: computation load, required sample support, and nonhomogeneity detection (including adaptive processing within heterogeneous ranges). Clearly, for each issue there exists an embarrassment of riches — this chapter has detailed only a few key schemes addressing each issue. An equally important issue that arises is therefore a scheme to pick within all these

1. Note, this implies that there is a main look angle/Doppler for the overall STAP process, but a set of auxiliary look directions for use with the D^3 algorithm.

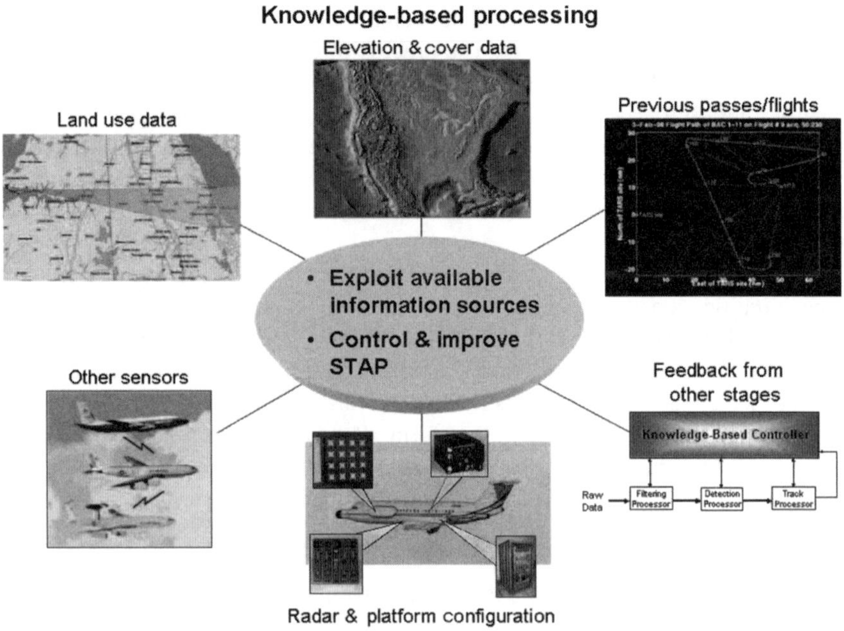

Figure 5.3 Sources informing a knowledge-based processor.

potential approaches. One should start with the fundamental notion that there is no "one-best" approach — different algorithms have their own advantages and disadvantages. This introduces the need for a *knowledge-aided approach* wherein a database informs the choice of algorithm, sample support both in terms of quantity and choice of range bins, the threshold level that sets the probability of false alarm, potentially even radar parameters such as frequency of transmission, PRF, and transmitted waveform.

Figure 5.3 illustrates the potential knowledge sources that could be exploited; it includes land-use and coverage data, information from earlier passes over the same terrain, radar parameters, and feedback from other stages in the detection and tracking process. However, clearly this requires a massively complex series of decisions to be made in real time. The figure therefore serves more to illustrate the long-term goal of knowledge-aided processing.

5.5.1 A Preliminary Knowledge-Based Processor

This section implements a very preliminary KB processor [38]. Knowledge-based processing best matches the adaptive processing algorithm to the interference scenario. The STAP technique is chosen using knowledge gained by processing the received data. In the KB processor of Fig. 5.4, each range cell is classified into one of only two types — homogeneous or nonhomogeneous — with different algorithms used for each type of cell. This classification is made using the NHD of Section

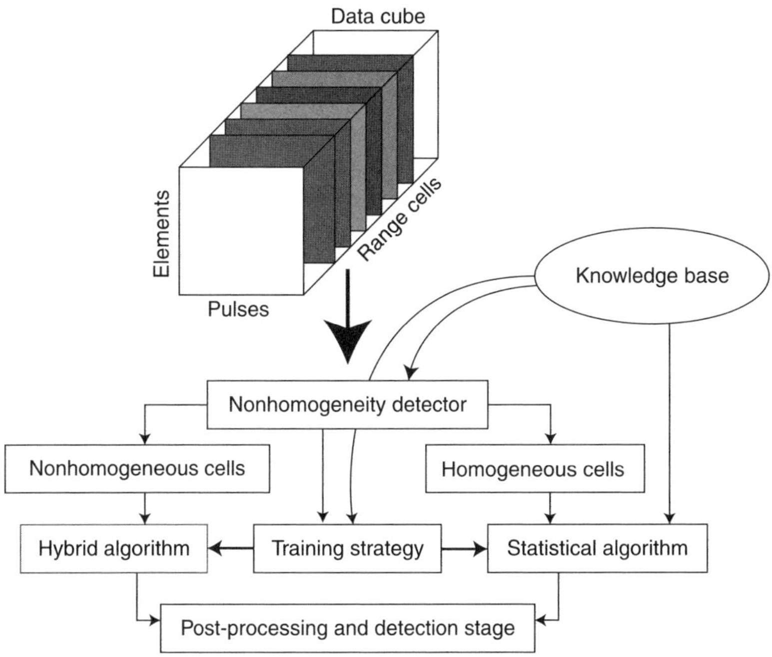

Figure 5.4 A preliminary KB process.

5.4.1 based on whether the JDL detection statistic crosses a chosen threshold. Within the range cells deemed nonhomogeneous, the interference is assumed to have discrete and homogeneous components and the hybrid algorithm is used for target detection. We use the JDL processor of Section 5.3.1. This choice of statistical processing allows for the use of the JDL algorithm in all three components of the KB processor. The only difference between processing in the homogeneous cells and in the nonhomogeneous cells is the choice of transformation matrix. Within the homogeneous cells, the transformation matrix is the nonadaptive transform of Equation 5.11. Within the nonhomogeneous range cells, the transformation matrix is given by the D^3 weights. In both cases, the secondary data used to estimate the angle–Doppler covariance matrix are chosen from range cells deemed homogeneous.

The steps in implementing the simple KB processor are as follows:

1. For each Doppler bin of interest, repeat the following steps.
2. For all range bins, identify homogeneous and nonhomogeneous cells using the JDL-NHD.
3. For each range cell of interest, if it is homogeneous, apply the JDL algorithm, but now using other homogeneous cells as sample support.
4. If it is nonhomogeneous, apply the hybrid algorithm, using other homogeneous cells as sample support.

Another KB processor is the fast maximum likelihood reiterative self-censoring adaptive power residue concurrent block processing two weight vector adaptive cosine estimator (FRACTA) [17, 39] which employs a priori information pertaining to the clutter covariance matrix. The FRACTA method demonstrates near-clairvoyant detection performance while employing 30% of the sample support needed in reduced rank STAP (for reduced rank STAP, the RMB rule requires $K = 2r$ (where r is the clutter rank; typically $r \ll M$) training data snapshots to obtain performance within 3 dB of the optimum). Performance analysis of the FRACTA algorithm is carried out using data from the KASSPER program. Due to constraints of space, the interested reader is referred to reference [17] for further details. Finally, the LRNAMF developed in reference [15] is another example of knowledge-aided adaptive processing, where a priori information about the clutter rank gained from system parameters such as platform speed, pulse repetition interval, array element spacing, number of antenna array elements, and number of pulses processed in a coherent processing interval is used to significantly reduce the training data support for covariance matrix estimation. Performance of the LRNAMF is benchmarked using data from the KASSPER program.

5.5.2 Numerical Example

The motivation for the KB processor is practical implementation of STAP in airborne radars for GMTI. With this in mind, we present here a result of using the KB formulation of Fig. 5.4 using measured data from the multichannel airborne radar measurements (MCARM) program [40]. The example chosen here uses the data from acquisition 575 on flight 5. Included with the data is information regarding the position, aspect, and velocity of the airborne platform and the mainbeam transmit direction. This information is used to correlate target detections with ground features.

While recording this acquisition, the radar platform was at latitude–longitude coordinates of (39.379°, −75.972°), placing the aircraft close to Chesapeake Haven, Maryland, USA. The plane was flying mainly south with velocity 223.78 mph and east with velocity 26.48 mph. The aircraft location and the transmit mainbeam are shown in Fig. 5.5. The mainbeam is close to broadside. Note that the mainbeam illuminates terrain of various types, including several major highways. Each data cube comprises 22 elements ($J = 22$), 128 pulses ($N = 128$) at a PRF of 1984 Hz and 630 range bins sampled at 0.8 μs (corresponding to 0.075 miles). The array is a 2×11 rectangular array. The array operates at a center frequency of 1.24 GHz.

To illustrate the effects of nonhomogeneities in secondary training data we inject two targets at closely spaced range bins. These artificial targets are in addition to the ground targets of opportunity on the roadways illuminated by the array. The artificial targets are injected in range bins 290 and 295. In this acquisition, the zero range is referenced to range bin 74 and so these injected targets are at ranges of 16.2 miles and 16.575 miles, respectively. The parameters of the injected targets are given in Table 5.1. These values are chosen to ensure that the targets cannot be distinguished

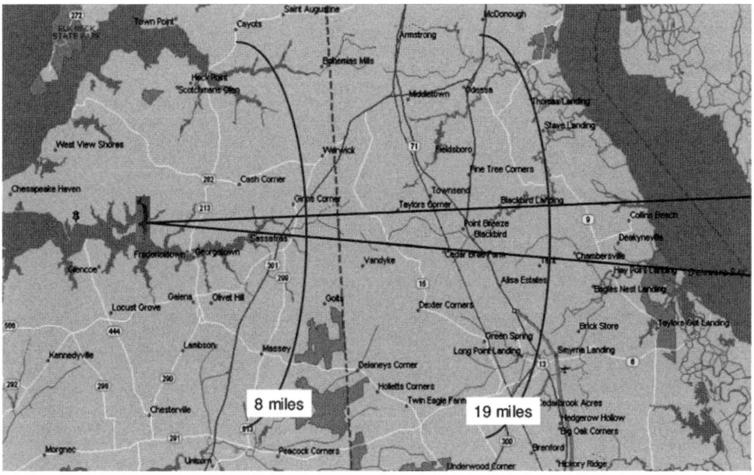

Figure 5.5 Location and transmit direction of the MCARM airplane during acquisition 575.

using nonadaptive, matched filter processing. Note that the two targets are at the same look angle and Doppler frequency and the second target is 20 dB stronger than the first.

This example is based on the JDL algorithm in all stages. The NHD uses the JDL-NHD discussed earlier and the statistical algorithm is the JDL algorithm using homogeneous range cells for sample support. The hybrid algorithm, as discussed earlier, is the JDL algorithm with an adaptive D^3 transform to the angle–Doppler domain. All stages use three angle bins and three Doppler bins (a 3×3 LPR). Thirty-six secondary data vectors are used to estimate the 9×9 angle–Doppler LPR covariance matrix. Two guard cells are used on either side of the primary data vector. Based on these numbers, without an NHD stage, range bin 295 would be used as a secondary data vector for detection within range bin 290, violating the homogeneity assumption of statistical STAP algorithms. The example compares the original JDL algorithm of reference [20] and the KB-STAP algorithm of Fig. 5.4.

Figure 5.6 plots the results of the original JDL algorithm without attempting to compensate for array effects or nonhomogeneities. The plot is of the MSMI statistic as a function of range and Doppler. The red spots correspond to higher statistics; that is, the red tends to correspond to target detections. The figure shows that targets are

TABLE 5.1 Parameters defining the injected targets.

	Target 1	Target 2
Amplitude	1×10^{-4}	1×10^{-3}
Angle bin	1°	1°
Doppler	$-9 \equiv -139.5$ Hz	$-9 \equiv -139.5$ Hz
Range bin	$290 \equiv 16.2$ miles	$295 \equiv 16.575$ miles

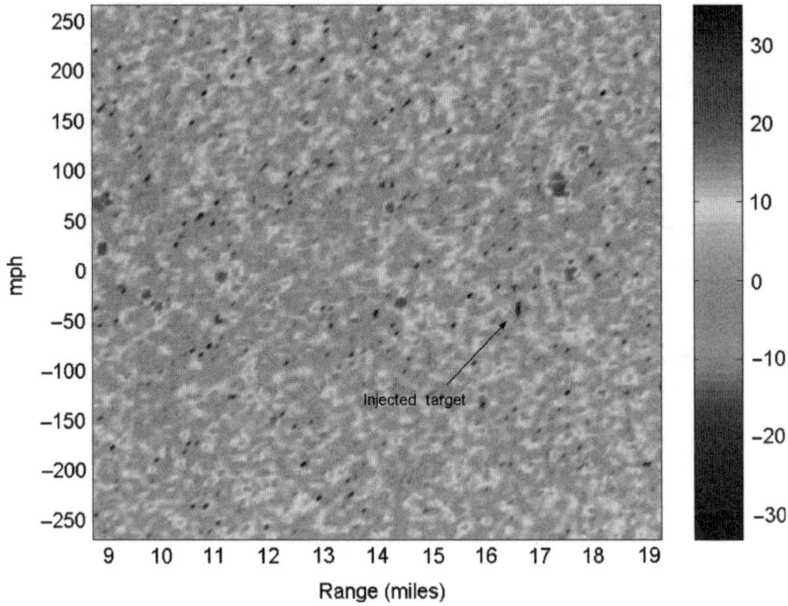

Figure 5.6 JDL processing ignoring array effects and nonhomogeneities. (See color insert.)

detected in almost all range and Doppler bins, including at extremely high velocities. If using the original JDL algorithm with measured data, therefore, one must deal with several false alarms. Also, whereas the second injected target is clearly visible, the first target is not detected at all. This inability to detect the target is because the second target is present in the secondary data while attempting to detect the first target at range bin 290. The presence of a target-like nonhomogeneity in the secondary data makes detection of a weak target practically impossible.

The KB processor, outlined in Fig. 5.4, matches the processing to the interference in that it uses JDL processing in the homogeneous range cells and hybrid processing in the nonhomogeneous cells. Figure 5.7 plots the AMF statistic obtained by using the KB processor. The improved discrimination, as compared to Fig. 5.6, between a few target signals and residual interference is clear. The first target is now clearly visible. This is possible because the NHD treats the second injected target as a nonhomogeneity and it is eliminated from the secondary data while processing the range cell corresponding to the first, weaker, injected target. The KB processor can, therefore, detect weak targets buried in nonhomogeneous interference.

The final step in determining the presence or absence of a target is to apply a threshold to the AMF statistic of Figs 5.6 and 5.7 to yield target declarations. Here, a target is declared at all points with an estimated MVDR statistic of greater than 40. Figures 5.8 and 5.9 plot the declared target locations as a function of Doppler and range. These locations are correlated with the map of Fig. 5.5. In Fig. 5.8, note the extremely high number of false alarms. Also, as in Fig. 5.6, the

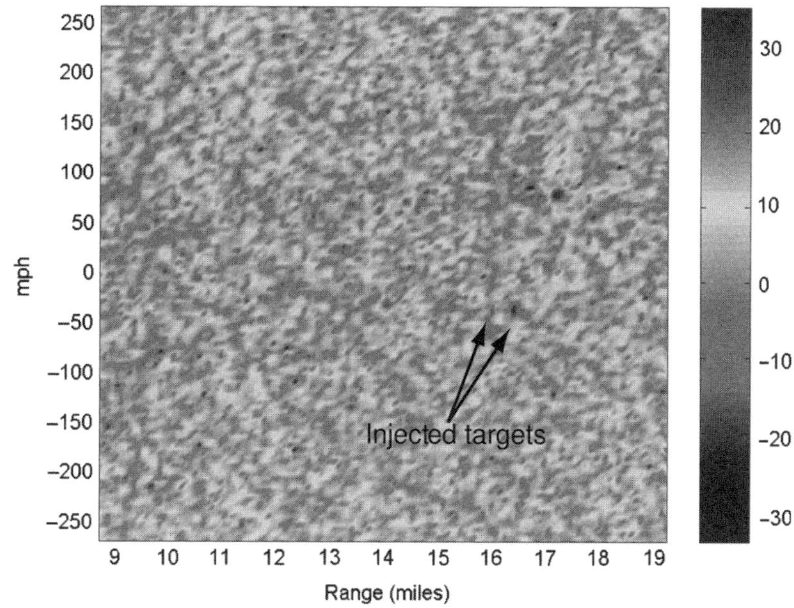

Figure 5.7 KB processor matching the STAP algorithm to the interference scenario.

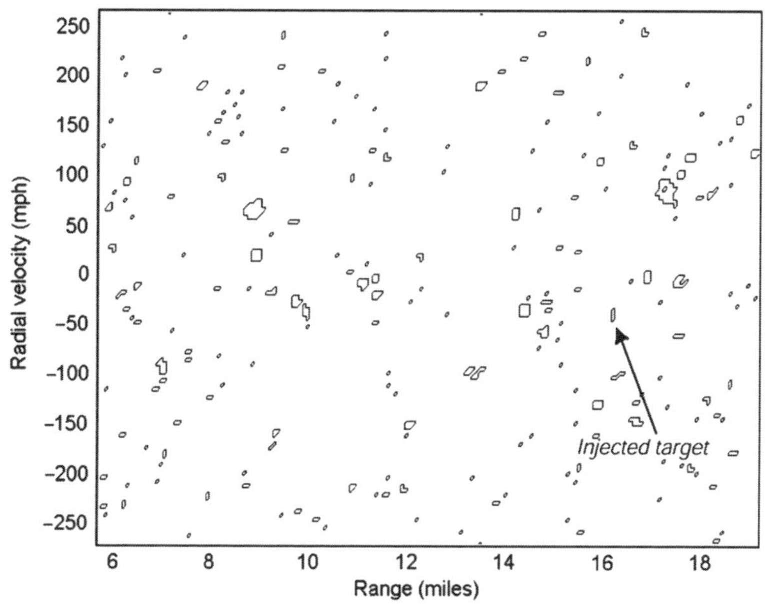

Figure 5.8 Target declarations using JDL, ignoring array effects and nonhomogeneities.

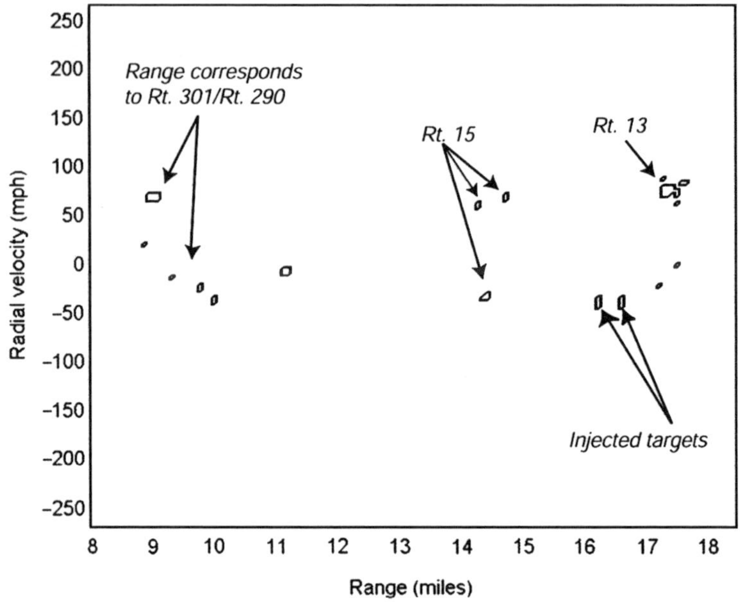

Figure 5.9 Target declarations using KB-STAP.

weak injected target is not detected. On the other hand, nearly all the target declarations by the KB processor, in Fig. 5.9, correlate directly with major highways in Maryland and Delaware illuminated by the radar mainbeam. Routes 290 and 301 in Maryland are closely spaced at a range of 9.0 and 9.8 miles. Accounting for the platform motion, the ground speed of the target(s) is approximately 50 mph.

The target detections at the far range shown in the plot are between 19.4 and 20.4 miles. The range to Route 9 varies between 19.1 and 21.1 miles within the transmit mainbeam. These far range detections therefore correspond to Route 9. The targets detected at these ranges are present in both Figs 5.8 and 5.9.

5.5.3 A Long-Term View

Several years of research has shown that KB approaches are essential for a practical implementation of STAP in airborne radars. The twin issues of data nonhomogeneity and adequate data support necessitate real-time analysis of the received data and the choice of an appropriate adaptive algorithm (with its associated parameters). As shown in Fig. 5.3, there are several knowledge sources that make the decision process more effective. Furthermore, as the references show, recent research has developed the many pieces of the overall knowledge-aided STAP puzzle.

The fundamental question that is yet to be addressed is feasibility. As must be clear, STAP by itself is a computationally complex process. Receiving, basebanding and processing multichannel signals in real-time places an enormous burden on available digital signal-processing technology. As it stands, KB-STAP will require several orders of magnitude gains in available computation capabilities. Furthermore, as is readily apparently, there is no single "best" solution and every choice involves a tradeoff. The feasibility of implementation of KB-STAP needs to be examined on a case-by-case basis. The computational requirements, storage, access, and communications overhead in addition to system considerations such as size, weight, power, and cost will dictate the implementation of KB-STAP for each application of interest. We predict that looking forward, it will be these factors that limit what will or will not be implemented.

5.6 CONCLUSIONS

This review has attempted to provide the reader with an intuitive and theoretical basis of space–time adaptive processing. The focus has been on the importance of STAP, and the fundamental issues that have guided research in this area. Two central problems arise in the application of STAP — the issue of computation load and the homogeneity of the sample support needed to train the adaptive filter. There have been several algorithms to address either of these issues, the key concepts of which have been presented here. However, most researchers would agree that there is no one best algorithm and the only practical approach is to use a *knowledge-based* scheme that best matches the signal processing to the interference scenario at hand. This matching could be in the choice of adaptive algorithm including its parameters, the scheme used to distinguish nonhomogeneities, and the training data used.

In Section 5.5.2 we presented an example of using a preliminary KB processor on measured data. The example illustrates the immense potential of KB approaches in detecting weak targets and reducing false-alarm rates. However, it must be emphasized that the algorithm used is extremely simple. In fact, the example emphasizes the vast amount of work remaining, such as that undertaken in the KASSPER program.

REFERENCES

[1] R.A. Monzingo, T.W. Miller. *Introduction to Adaptive Arrays*. New York: Wiley; 1980.

[2] J.R. Guerci. *Space–Time Adaptive Processing for Airborne Radar*. Boston: Artech House Publishers; 2003.

[3] R. Klemm. *Principles of Space–Time Adaptive Processing*. Bodmin, England: IEE Press; 2002.

[4] H.L. VanTrees. *Detection, Estimation and Modulation Theory Part-I*. New York: John Wiley & sons; 1968.

[5] J. Ward. Space–time adaptive processing for airborne radar. Technical Report 1015, MIT Lincoln Laboratory, December 1994.

[6] F.C. Robey, D.R. Fuhrmann, E.J. Kelly, R. Nitzberg. A CFAR adaptive matched filter detector. *IEEE Trans. Aerospace and Electronic Systems* 1992;AES-28:208–216.

[7] B. Widrow, P.E. Mantey, L.J. Griffiths, B.B. Goode. Adaptive antenna systems. *Proc. IEEE* 1967;55:2143–2161.

[8] S.P. Applebaum. Adaptive arrays. Technical report, Syracuse University Research Corporation, 1966.

[9] I.S. Reed, J.D. Mallett, L.E. Brennan. Rapid convergence rate in adaptive arrays. *IEEE Trans. Aerospace and Electronic Systems* 1974;AES-10:853–863.

[10] E.J. Kelly. An adaptive detection algorithm. *IEEE Trans. Aerospace and Electronic Systems* 1986;AES-22:115–127.

[11] S. Kraut, L.L. Scharf, L.T. McWhorter. Adaptive subspace detectors. *IEEE Trans. Signal Processing* 2001;49:1–16.

[12] J.S. Bergin, P.M. Techau. High-fidelity site-specific radar simulation: Kassper'02 workshop datacube. Technical Report ISL-SCRD-TR-02-105, Defense Advanced Research Projects Agency, 2002.

[13] M. Rangaswamy, J.H. Michels, B. Himed. Statistical analysis of the nonhomogeneity detector for STAP applications. *Digital Signal Processing* 2004;14:253–267.

[14] M. Rangaswamy. Statistical analysis of the non-homogeneity detector for non-Gaussian interference backgrounds. *IEEE Trans. Signal Processing* 2005;SP-53.

[15] M. Rangaswamy, F.C. Lin, K.R. Gerlach. Robust adaptive signal processing methods for heterogeneous radar clutter scenarios. *Signal Processing* 2004;84:1653–1665.

[16] J. Bergin, C. Teixeira, P. Techau, J.R. Guerci. STAP with Knowledge Aided Data Pre-Whitening. *Proc. IEEE Radar Conference*, Philadelphia, PA, 2004.

[17] S. Blunt, K.R. Gerlach, M. Rangaswamy. The Enhanced FRACTA Algorithm With Knowledge Aided Covariance Estimation. *Proc. Third IEEE Workshop on Sensor Array and Multichannel Processing*, Barcelona, Spain, 2004.

[18] W.L. Melvin, G.A. Showman, J.R. Guerci. A knowledge-aided GMTI detection architecture. *Proc. 2004 IEEE Radar Conference*, April 2004.

[19] D.J. Zywicki, W.L. Melvin, G.A. Showman, J.R. Guerci. STAP performance in site-specific clutter environments. *Proc. 2003 IEEE Aerospace Conference*, April 2004, p. 2005–2020.

[20] H. Wang, L. Cai. A localized adaptive mtd processor. *IEEE Trans. Aerospace and Electronic Systems* 1991;AES-27.

[21] J.R. Roman, M. Rangaswamy, D.W. Davis, Q. Zhang, B. Himed, J.H. Michels. Parametric adaptive matched filter for airborne radar applications. *IEEE Trans. Aerospace and Electronic Systems* 2000;36:677–692.

[22] J.S. Goldstein, L.L. Scharf, I.S. Reed. A multistage representation of the Wiener filter based on orthogonal projections. *IEEE Trans. Information Theory* 1998; IT-44:2943–2959.

[23] T.K. Sarkar, N. Sangruji. An adaptive nulling system for a narrowband signal with a look-direction constraint utilizing the conjugate gradient method. *IEEE Trans. Antennas and Propagation* 1989;37(7):940–944.

[24] R.S. Adve, T.B. Hale, M.C. Wicks. Joint domain localized adaptive processing in homogeneous and non-homogeneous environments. Part II: Nonhomogeneous environments. *IEE Proc. Radar Sonar and Navig.* 2000;147(2):66–73.

[25] R.S. Adve, T.B. Hale, M.C. Wicks. Joint domain localized adaptive processing in homogeneous and non-homogeneous environments. Part I: Homogeneous environments. *IEE Proc. Radar Sonar and Navig.* 2000;147(2):57–65.

[26] S.L. Marple. *Digital Spectral Analysis With Applications.* New Jersey: Prentice Hall; 1987.

[27] J.H. Michels, B. Himed, M. Rangaswamy. Robust STAP detection in a dense signal airborne radar environment. *Signal Processing* 2004;84:1625–1636.

[28] J.S. Goldstein, I.S. Reed, P.A. Zulch. Multistage partially adaptive STAP CFAR detection algorithm. *IEEE Trans. Aerospace and Electronic Systems* 1999;AES-35:645–661.

[29] M. Zoltowski. Conjugate gradient adaptive filtering with application to space–time processing for wireless digital communications. *Proc. Second IEEE Conference on Sensor Array and Multichannel Processing*, Key Bridge, VA, August 2002.

[30] I.P. Kirsteins, D.W. Tufts. Adaptive detection using a low rank approximation to a data matrix. *IEEE Trans. Aerospace and Electronic Systems* 1994;AES-30:55–67.

[31] K.R. Gerlach. Outlier resistant adaptive matched filtering. *IEEE Trans. Aerospace and Electronic Systems* 2002;38:885–901.

[32] W.L. Melvin, M.C. Wicks, P. Chen. Nonhomogeneity detection method and apparatus for improved adaptive signal processing. US patent 5706013, 1999.

[33] J.R. Guerci, J.S. Bergin. Principal components, covariance matrix tapers, and the subspace leakage problem. *IEEE Trans. Aerospace and Electronic Systems* 2002;38:152–162.

[34] S.A. Vorobyov, A.B. Gershman, Z.Q. Luo. Robust adaptive beamforming using worst-case performance optimization: a solution to the signal mismatch problem. *IEEE Trans. Signal Processing* 2003;51:313–324.

[35] Yong-Liang Wang, Jian-Wen Chen, Zheng Bao, Ying-Ning Peng. Robust space–time adaptive processing for airborne radar in nonhomogeneous clutter environments. *IEEE Trans. Aerospace and Electronic Systems* 2003;39:70–81.

[36] R.S. Adve, T.B. Hale, M.C. Wicks. Transform domain localized processing using measured steering vectors and non-homogeneity detection. *Proc. 1999 IEEE National Radar Conference*, April 1999, Boston, MA. p. 285–290.

[37] F. Gini, M. Greco. Covariance matrix estimation for CFAR detection in correlated heavy tailed clutter. *Signal Processing* 2002;82:1847–1859.

[38] R.S. Adve, T.B. Hale, M.C. Wicks, P.A Antonik. Ground moving target indication using knowledge based space–time adaptive processing. *Proc. 2000 IEEE International Radar Conference*, May 2000, Washington, DC.

[39] S.D. Blunt, K.R. Gerlach, M. Rangaswamy, A. Shackelford. STAP via knowledge-aided covariance estimation and the FRACTA meta-algorithm. In F. Gini, M. Rangaswamy, editors, Chapter 7, *Knowledge-Based Radar Detection, Tracking and Classification* New York; John Wiley & Sons, 2007.

[40] D. Sloper, D. Fenner, J. Arntz, E. Fogle. Multi-channel airborne radar measurement (MCARM), MCARM flight test. Contract F30602-92-C-0161, Westinghouse Electronic Systems, April 1996. Additional information available at http://sunrise.deepthought.rl.af.mil.

6

CFAR KNOWLEDGE-AIDED RADAR DETECTION AND ITS DEMONSTRATION USING MEASURED AIRBORNE DATA

Christopher T. Capraro, Gerard T. Capraro, Antonio De Maio, Alfonso Farina, and Michael C. Wicks

This chapter addresses the design and analysis of a knowledge-aided (KA) detector for airborne Space–Time Adaptive Processing (STAP) applications. The proposed processor is composed of a training data selector, which chooses secondary cells best representing the clutter statistics in the cell under test, and an adaptive processor for detection processing. The data selector is a hybrid algorithm, which pre-screens training data through the use of terrain information from the United States Geological Survey (USGS). Then, in the second stage, a data-driven selector attempts to eliminate residual nonhomogeneities.

The performance of this new approach is analyzed using measured airborne radar data, obtained from the Multi-Channel Airborne Radar Measurements (MCARM) program, and is compared with alternative STAP detectors proposed in the open literature.

6.1 INTRODUCTION

To mitigate the deleterious effects of clutter and jammer, modern radars have adopted adaptive processing techniques such as Constant False Alarm Rate (CFAR) detectors, adaptive arrays, and Space–Time Adaptive Processing (STAP). In the real world, however, suboptimal performance might occur (high false alarm rate and/or low detection probability) as a consequence of

Knowledge-Based Radar Detection, Tracking, and Classification. Edited by Fulvio Gini and Muralidhar Rangaswamy

- heterogeneous clutter (rapidly varying terrain, that is, mountainous areas with rapid elevation/reflectivity variation, rapid land cover variations, littorals);
- dense target backgrounds (moving clutter, military/civilian vehicles);
- large discretes and spiky clutter (urban clutter, power lines, towers, step mountainous terrain).

The aforementioned adaptive techniques are very restrictive because they require the environment to "remain stationary and homogeneous" during adaptation. Poor training data selection in such adaptive filters can produce bad output signals. A possible way to circumvent this drawback is the real-time exploitation of a priori knowledge concerning the environment surrounding the radar [1–8].[1] In fact, the environmental context is the key to efficient adaptation: sensors like humans might benefit from the context. Examples of a priori knowledge are Digital Terrain Elevation Models (DTEMs), previous look data, Geographic Information Systems (GISs), roadway maps (to highlight sectors of surveillance where moving cars or vehicles might be present), background of air/surface traffic, system calibration information, and so on. The ultimate goal is to make the radar an intelligent device, such that it is capable of developing cognition of the surrounding environment. Otherwise stated, the environment in which the radar system operates acts as a teacher and the radar can become more an expert with time by learning from the context. This is basically the concept of KA or cognitive radar, known to the radar community since the pioneering papers of Vannicola and colleagues [9, 10] and Haykin [11].

Recent advances in environmental measurements, DTEM, future information quality and accessibility, digital processing, mass and random-access memory technologies, have opened up many possibilities, unthinkable in the past, for radar systems to improve their on-line performance. New real-time processing techniques are required to take advantage of these opportunities to bring radar performance close to optimum under difficult operation conditions such as littorals that include mixed sea and variable terrain.

In a typical STAP algorithm [12] the disturbance covariance matrix is estimated using training data from range cells in close proximity to the cell under test with the assumption they have similar spectral properties. However, training data are often contaminated by power variations over range (in addition to radar range equation effects), clutter discretes, and other outliers. Moreover, the strength of the clutter also fluctuates with terrain type, elevation, ground cover, and the presence of man-made structures.

In these situations, training data may not be representative of the disturbance in the cell under test. As a consequence, covariance matrix estimates from heterogeneous environments with clutter discretes may result in significantly undernulled interference, which leads to degradation in detection performance as well as CFAR behavior.

1. The Knowledge-Aided Sensor Signal Processing and Expert Reasoning (KASSPER) program [4, 5] aims to improve the airborne radar performance by taking into account all the available a priori knowledge.

This is especially true in regions containing varying ground cover such as regions connecting land and sea [13–16].

It is thus of primary concern to design selection procedures that carefully choose the training set. Previous solutions to this challenging problem can be found in reference [17], where the Power Selected Training and the Power Selected De-emphasis, which use measurements of the interference environment to select training data, are described and analyzed using recorded radar data. In reference [18], the Nonhomogeneity Detector is introduced and assessed. In reference [15], it is further analyzed with emphasis on its use as a pre-processing step for STAP algorithms, while in reference [19], it is generalized to the case of small sample support introducing the concept of Innovation Inner Product IIP. In reference [20], approximate Maximum Likelihood ML procedures, also re-iterative, are proposed and analyzed. In reference [21], the problem of secondary data selection in non-Gaussian disturbance modeled as a Spherically Invariant Random Process SIRP is considered. Finally, in references [22] and [23], procedures relying upon the Generalized Likelihood Function GLF criterion are introduced. Among them we mention the Two-Step Data Selection Procedure (2S-DSP), which provides good performances with an acceptable computational complexity. All the quoted algorithms are usually referred to as data-adaptive procedures because the decision concerning which cells to be excised are driven by the data.

Nevertheless, data-independent methodologies such as the Sliding Window, Exponential Window, and Range Segmentation have also been proposed and analyzed (see references [24] and [25] for a survey) showing that they usually result in inadequately nulled clutter [17]. Finally the use of a KA data selection procedure and its application to post-Doppler radar processing is discussed in references [26–28]. The basic idea is to use digital terrain data to aid in choosing representative secondary data. The assumption is that the estimation of the covariance matrix will improve by choosing secondary data based upon how well its terrain classification compares with the cell under test. As such, STAP algorithms will cancel interference due to terrain clutter more effectively. Indeed there is a growing amount of terrain data publicly available at resolutions as small as 10 m. Agencies such as the National Imagery and Mapping Agency, NIMA recently renamed the National Geospatial-Intelligence Agency [29], NGA, and the USGS [30], offer digitized geospatial data containing terrain elevation, classification (urban, agricultural, forested, and so on), linear features (roads, power lines, railroads, and so on), and multispectral imagery. Several software products are also available to aid in converting and viewing the data.

In this chapter we still consider the problem of data selection and propose a hybrid processor composed of two blocks: the KA selector [26–28], which exploits the a priori knowledge of the terrain and the data-adaptive algorithm 2S-DSP devised in references [22] and [23]. The task of the first block is to perform a pre-screening of the data, attempting to enhance the ability of the 2S-DSP to filter other types of nonhomogeneities such as moving outliers.

The benefits of employing the hybrid selection strategy [31] as a pre-processing step of an adaptive detection algorithm are highlighted. More precisely, we devise a KA detector exploiting the aforementioned hybrid system as data selector and the Recursive Persymmetric Adaptive Normalized Matched Filter (RP-ANMF) [32] as the final stage performing the decision concerning the target presence. In the analysis stage, the performance of the new system is assessed on measured airborne data also in comparison with other adaptive detection algorithms available in the open literature.

The chapter is organized as follows. In Section 6.2 we deal with the formulation of the problem and the design issues. In Sections 6.3 and 6.4 we introduce respectively the GIS-aided and the data-adaptive selectors. In Section 6.5 we describe the RP-ANMF. In Section 6.6 we assess the performance of the proposed KA scheme on measured radar data, highlighting the benefits arising from the joint use of both KA and data-adaptive pre-processing. Finally, conclusions and directions for future research are given in Section 6.7.

6.2 PROBLEM FORMULATION AND DESIGN ISSUES

We refer to a single-bin post-Doppler STAP [33] where all the pulses of the Coherent Processing Interval (CPI) are filtered prior to adaptation. Doppler processing is performed separately on the signals from each array element. A different adaptive problem is then solved in each target Doppler bin, utilizing the signals from all elements (whose number is denoted by N). The basic concept in a post-Doppler processing is to suppress mainlobe clutter nonadaptively and to localize the competing sidelobe clutter in angle at the Doppler of interest, thereby reducing the required degrees of freedom and sample support. Although it has been shown in reference [33] that heavy Doppler tapering is needed and that multibin post-Doppler performance is theoretically better, the single-bin algorithm requires less sample support.

Denote by $\mathbf{r}$ the N-dimensional vector of the data from a given Doppler bin of the cell under test. Under the hypothesis H_0, namely target absence, the vector $\mathbf{r}$ contains clutter/interference plus noise only; that is,

$$\mathbf{r} = \mathbf{c} + \mathbf{n}, \tag{6.1}$$

where $\mathbf{c}$ and $\mathbf{n}$ denote respectively the clutter/interference and receiver noise vectors, which are assumed statistically independent. Under H_1, instead, $\mathbf{r}$ contains also a target component; that is,

$$\mathbf{r} = \alpha\mathbf{p} + \mathbf{c} + \mathbf{n}, \tag{6.2}$$

with α denoting the complex amplitude accounting for both the target as well as the channel propagation effects and $\mathbf{p}$ the filtered target space–time steering vector at the considered Doppler bin (assumed without loss of generality with unitary norm).

In many practical applications, it often happens that the statistics of the vector $\mathbf{c} + \mathbf{n}$, and in particular its covariance matrix $\mathbf{M}$, are unknown. It is usual, in order to estimate $\mathbf{M}$, to employ a set of secondary data $\mathbf{r}_1, \ldots, \mathbf{r}_H$, namely Doppler filtered vectors from range cells surrounding the one being tested and referring to the same

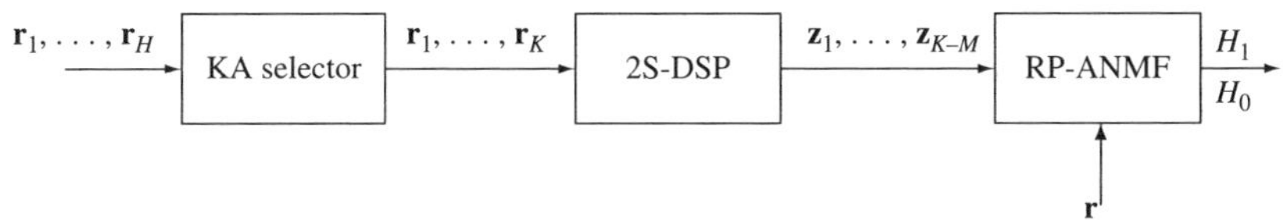

Figure 6.1 Block diagram of the KA-RP-ANMF detector.

Doppler bin of $\mathbf{r}$. Ideally the training data are assumed free of signal components, and with the same spectral properties of $\mathbf{r}$. However, although most STAP schemes have been designed employing the assumption that secondary data are independent and identically distributed vectors, experimental campaigns have demonstrated that such an assumption is not always verified [34]. More specifically, the size of the training set is often limited; namely, the large swaths of homogeneous clutter/interference necessary for estimating the covariance matrix may not be available. In addition, the analysis of several STAP algorithms, mostly conducted assuming homogeneity of the secondary data, has shown that nonhomogeneities magnify the loss between the adaptive implementation and optimum conditions [14, 35].

A possible method for circumventing the sample support shortcoming is to exploit data selection techniques that, as already pointed out, might be either knowledge-based (KB) or data-adaptive.

Following this guideline the first two blocks of the proposed KA system, shown in Fig. 6.1, try to select the most homogeneous training data and to eliminate secondary vectors containing strong static and/or dynamic outliers. The idea is to perform a hybrid KA and data-adaptive training selection [31]. Specifically, the first block implements a KA data screening resorting to the algorithm devised in reference [27], which utilizes terrain data from the USGS to aid the data-selection process. The second block is the 2S-DSP, a data-adaptive selection procedure devised in references [22] and [23], which attempts to remove dynamic outliers and other residual nonhomogeneities from the training set.

Finally the primary data and the secondary data passed by the screening process in both the selectors are processed using the RP-ANMF, an adaptive radar detector devised in reference [32], which performs the final decision concerning the target presence.

In the next sections we provide a more detailed description of the three aforementioned algorithms implemented by the proposed KA processor, referred to as the KA-RP-ANMF.

6.3 KA DATA SELECTOR

The basic assumption of the KA selection algorithm is that the dominant clutter competing with targets in the test cell is due to the patch of Earth (later referred to as the test clutter cell) within the same test ring, and corresponding to the target Doppler. As a consequence, the data selector chooses secondary range–Doppler cells that have the "same" terrain as the test clutter cell. To this end it exploits National Land Cover

Data (NLCD) [36] to classify the ground environment illuminated by the radar. The NLCD data were obtained from the USGS in a grid cell format with a spatial resolution of 30 m. The terrain is hierarchically grouped by 9 major classifications such as urban areas, barren land, and water, and subgrouped into 21 minor classifications such as high intensity residential urban areas and low intensity residential urban areas. The data were then converted to a nonprojected global geodetic coordinate system (latitude, longitude, and elevation) and stored in a relational database for flexible search and retrieval.

The position of four boundary points defining the area of each range–Doppler cell, for a given Doppler of interest, is calculated using the registration techniques described in reference [28] see also Appendix A. As these boundary points do not align with latitude and longitude like the terrain cells, a rectangle bounding them is computed that approximates the range–Doppler cell's area, as illustrated in Fig. 6.2.

Once a rectangle is calculated for a given range-Doppler cell, the database is queried to determine the terrain cell count for each type of land classification within its boundaries. The results are stored in a 21-element vector with each element corresponding to a land classification type. The vector is then normalized by dividing it by the total number of terrain cells contained in the bounding rectangle. This is necessary in order to account for the variation in area of the rectangles for different range–Doppler cells. The normalized vector for the lth range sample is represented by

$$\mathbf{t}_l = [t_{l,1,\ldots,}t_{l,21}]. \tag{6.3}$$

Once the terrain vectors for all of the range–Doppler cells at the Doppler of interest are determined, the vectors of potential secondary data cells are compared with the

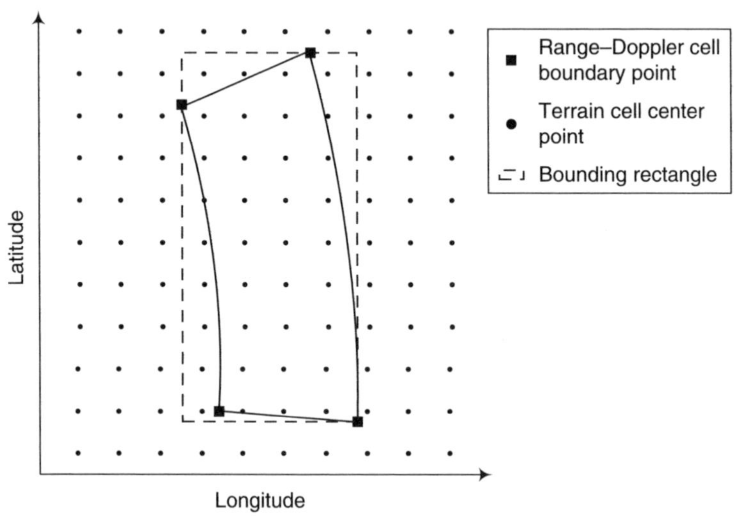

Figure 6.2 Bounding rectangle approximating the range–Doppler cell's area.

vector of the test clutter cell by computing the squared error between their elements. This provides a measurement or grade of how close the lth range–Doppler cell matched the test clutter cell and is defined by

$$\text{grade}_l = \sum_{i=1}^{21} (t_{c,i} - t_{l,i})^2 \tag{6.4}$$

where $t_{c,i}$ is the ith element of the test cell terrain vector and $t_{l,i}$ is the ith element of the potential secondary data terrain vector. Cells with lower grades are assumed to better match the test clutter cell. The grades are then sorted in decreasing order and the last K cells are chosen as secondary data.

6.4 2S-DSP DATA SELECTION PROCEDURE

Denote by $\mathbf{r}_1, \ldots, \mathbf{r}_K$ the secondary data provided by the KA selector and let M be the number of secondary cells that are to be further excised. The value of M might stem from KA criteria [2], which, depending on the type of operational environment and thus on the maximum number of expected residual outliers, suggest how many cells are to be excised. For instance, a possible source of a priori knowledge might be represented by maritime and/or ground traffic information. Otherwise stated, during hours of strong traffic one can set a certain value of M higher than the one chosen during hours of minimal traffic. In any case, one can always enhance the robustness of the system against the worst expected situation. Assume that

$$\begin{cases} \mathbf{r}_i = \mathbf{c}_i & \forall_i \in \Omega - \Omega_0 \\ \mathbf{r}_i = \mathbf{p}_i + \mathbf{c}_i & \forall_i \in \Omega_0 \end{cases} \tag{6.5}$$

where

- $\Omega = \{1, \ldots, K\}$ is a set of size K and $\Omega_0 = \{i_1, \ldots, i_M\}$ is a subset of Ω with distinct elements and of size M (see Fig. 6.3).
- $\mathbf{c}_1, \ldots, \mathbf{c}_K$, are independent, zero-mean, complex, circular, Gaussian, random vectors sharing the same covariance structure $\mathbf{M}$ assumed to be positive definite; that is,

  $$E[\mathbf{c}_i \mathbf{c}_i^\dagger] = \sigma_i^2 \mathbf{M}, \qquad i = 1, \ldots, K, \tag{6.6}$$

  where, in turn, $E[\cdot]$ denotes statistical expectation, $(\cdot)^\dagger$ conjugate transpose, and σ_i^2 the unknown, possibly random, power level of the ith random vector. It is supposed that $\sigma_i^2 \in \Xi = [\sigma^2, +\infty)$; namely, it cannot take values smaller than the noise floor at the receiver σ^2.
- $\boldsymbol{p}_i$'s are unknown, possibly random, N-dimensional complex vectors, representing outliers, namely discretes, nonhomogeneities, or multiple targets. $\forall i \in \Omega_0$,

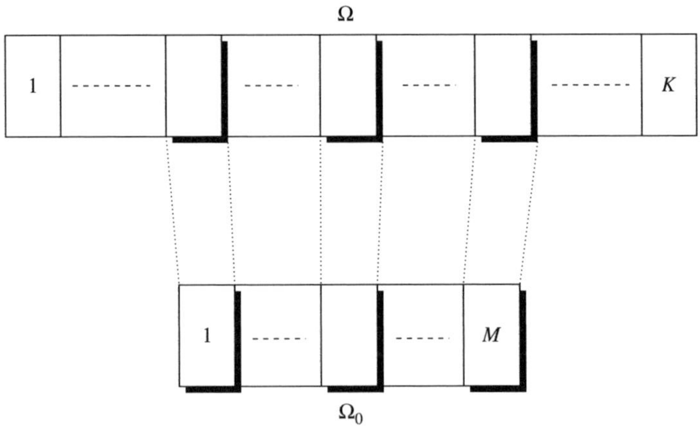

Figure 6.3 Diagram showing Ω and Ω_0.

they contaminate $\mathbf{c}_i$, producing vectors $\mathbf{r}_i$ with a covariance structure very different from $\mathbf{M}$. Otherwise stated, assuming that $\mathbf{p}_i$ is statistically independent of the $\mathbf{c}_i$, the covariance of $\mathbf{r}_i$ can be written as

$$E[\mathbf{r}_i\mathbf{r}_i^\dagger] = \sigma_i^2\mathbf{M} + E[\mathbf{p}_i\mathbf{p}_i^\dagger] = \sigma_i^2\mathbf{M}_i, \qquad \forall_i \in \Omega_0, \tag{6.7}$$

and the corresponding structure

$$\mathbf{M}_i = \mathbf{M} + \frac{E[\mathbf{p}_i\mathbf{p}_i^\dagger]}{\sigma_i^2} \tag{6.8}$$

may deeply deviate from $\mathbf{M}$.

The 2S-DSP algorithm, devised according to the two-step GLF criterion, is a data-adaptive selection technique capable of detecting the presence of outliers within the secondary data. Otherwise stated it chooses the most homogeneous $K-M$ secondary data, namely the vectors whose indexes belong to the set $\Omega - \Omega_0$.

The design of this selector is based on a two-step technique. In the former step we assume that $\mathbf{M}$ is known and derive the ML estimate of Ω_0 modeling $\sigma_1^2, \ldots, \sigma_K^2$, $\mathbf{p}_{i_1}, \ldots, \mathbf{p}_{i_M}$ as unknown parameters. Then, in the second step, a suitable estimate of $\mathbf{M}$ is substituted in place of its exact value.

Step 1. For known $\mathbf{M}$ the ML estimate of Ω_0 is the solution to the problem

$$\widehat{\Omega_0} = \arg\max_{\Omega_0}[\max_{\sigma_1^2,\ldots,\sigma_K^2} \max_{\mathbf{p}_{i_1},\ldots,\mathbf{p}_{i_M}} f(\mathbf{r}_1, \ldots, \mathbf{r}_K | \mathbf{M}, \sigma_1^2, \ldots, \sigma_K^2, \mathbf{p}_{i_1}, \ldots, \mathbf{p}_{i_M})]. \tag{6.9}$$

Maximizing over $\mathbf{p}_i$ $\forall i \in \Omega_0$, yields

$$\max_{\mathbf{p}_{i_1},\ldots,\mathbf{p}_{i_M}} f(\mathbf{r}_1,\ldots,\mathbf{r}_K|\mathbf{M}, \sigma_1^2,\ldots,\sigma_K^2, \mathbf{p}_{i_1},\ldots,\mathbf{p}_{i_M})$$

$$= \prod_{i\in\Omega_0} \frac{1}{\pi^N \sigma_i^{2N} \det(\mathbf{M})} \prod_{i\in\Omega-\Omega_0} \frac{1}{\pi^N \sigma_i^{2N} \det(\mathbf{M})} \times \exp\left[-\mathbf{r}_i^\dagger \frac{\mathbf{M}^{-1}}{\sigma_i^2} \mathbf{r}_i\right]. \tag{6.10}$$

Moreover the maximization, $\forall i \in \Omega_0$, of Equation 6.10 over $\sigma_i^2 \in \Xi$ leads to

$$\max_{(\sigma_{i_1}^2,\ldots,\sigma_{i_M}^2)\in\Xi^M} \max_{\mathbf{p}_{i_1},\ldots,\mathbf{p}_{i_M}} f(\mathbf{r}_1,\ldots,\mathbf{r}_K|\mathbf{M}, \sigma_1^2,\ldots,\sigma_K^2, \mathbf{p}_{i_1},\ldots,\mathbf{p}_{i_M})$$

$$= \frac{1}{\pi^{NM} \sigma^{2NM} \det^M(\mathbf{M})} \prod_{i\in\Omega-\Omega_0} \frac{1}{\pi^N \sigma_i^{2N} \det(\mathbf{M})} \exp\left[-\mathbf{r}_i^\dagger \frac{\mathbf{M}^{-1}}{\sigma_i^2} \mathbf{r}_i\right]. \tag{6.11}$$

It still remains to maximize Equation 6.11 over $\sigma_i^2 \in \Xi$, $\forall i \in \Omega - \Omega_0$. After some algebraic manipulations this last maximization yields

$$\frac{1}{\pi^{NM} \sigma^{2NM} \det^M(\mathbf{M})} \prod_{i\in\Omega-\Omega_0} \frac{1}{\pi^N \widehat{\sigma}_i^{2N} \det(\mathbf{M})} \exp\left[-\mathbf{r}_i^\dagger \frac{\mathbf{M}^{-1}}{\widehat{\sigma}_i^2} \mathbf{r}_i\right], \tag{6.12}$$

where

$$\widehat{\sigma}_i^2 = \max\left(\sigma^2, \frac{1}{N}\, \mathbf{r}_i^\dagger \mathbf{M}^{-1} \mathbf{r}_i\right). \tag{6.13}$$

As a consequence,

$$\widehat{\Omega_0} = \arg\max_{\Omega_0} \left\{ \prod_{i\in\Omega-\Omega_0} \frac{1}{\widehat{\sigma}_i^{2N}} \exp\left[-\mathbf{r}_i^\dagger \frac{\mathbf{M}^{-1}}{\widehat{\sigma}_i^2} \mathbf{r}_i\right] \right\}. \tag{6.14}$$

Step 2. The second step is tantamount to substituting a suitable estimate of $\mathbf{M}$ in place of its exact value. To this end we exploit the normalized covariance matrix estimator [37]; that is,

$$\widehat{\mathbf{M}} = \frac{N}{K} \sum_{i\in\Omega} \frac{\mathbf{r}_i \mathbf{r}_i^\dagger}{\|\mathbf{r}_i\|^2}, \tag{6.15}$$

where $\|\cdot\|$ denotes the Euclidean norm of a complex vector. As a consequence the devised procedure can be implemented as follows.

6.4.1 Two-Step Data Selection Procedure (2S-DSP)

1. Construct the normalized sample covariance matrix of Equation 6.15.
2. $\forall i \in \Omega$, construct the quadratic forms

$$\xi_i = \mathbf{r}_i^\dagger \widehat{\mathbf{M}}^{-1} \mathbf{r}_i, \tag{6.16}$$

   referred to as the Generalized Inner Product (GIP) with normalized sample covariance matrix, and sort them in decreasing order.
3. The set $\mathbf{\Psi}$, composed of the $K-M$ most homogeneous vectors, can be obtained excising the M vectors characterized by the highest values of ξ_i, from the initial set.

The computational complexity of the 2S-DSP is $\mathcal{O}(KN^2)$ floating point operations (flops).[2] In fact $\mathcal{O}(KN^2)$ flops are required by the evaluation of Equation 6.15 and K GIP with normalized covariance matrix must be calculated with an overall computational complexity of $\mathcal{O}(KN^2)$ flops.

A fundamental characteristic of the proposed algorithm is that it exploits a normalized covariance estimate (in which each outer product is normalized to the corresponding trace) in place of the usual sample covariance. It is thus desirable to introduce an intuitive explanation as to why exploiting the normalized sample covariance matrix might be more convenient than using the conventional one.

To this end we first recall that standard data selectors (nonhomogeneity detectors) such as the Non-reiterative Censored Sample Matrix Inversion (NCSMI) [20] and the Reiterative Censored Sample Matrix Inversion (RCSMI) [20] exploit as a selection metric the conventional GIP, which is equal to Equation 6.16 but for the presence of the conventional (non-normalized) sample covariance matrix in place of $\widehat{\mathbf{M}}$. Hence let us consider a scenario where a strong outlier is present in a certain data vector $\mathbf{r}_k$. This implies that the corresponding outer product $\mathbf{r}_k\, \mathbf{r}_k^\dagger$ dominates the other terms in the definition of the sample covariance matrix and, as a consequence, the inverse covariance matrix estimate places a deep null in the direction of the outlier. This nulling effect leads to a conventional GIP, in correspondence of $\mathbf{r}_k$, which might not be higher than the GIP evaluated from the other training data. This might preclude the possibility that the outlier is correctly detected.

On the contrary, normalizing each outer product, which is tantamount to exploiting $\frac{\mathbf{r}_k \mathbf{r}_k^\dagger}{\|\mathbf{r}_k\|^2}$ in place of $\mathbf{r}_k\, \mathbf{r}_k^\dagger$, it is possible to limit the dynamic range of the generic term in the definition of the covariance estimate, avoiding that an addend dominates the others. By doing so the null placed by the inverse normalized covariance estimate in correspondence of the outlier direction is less pronounced than the one introduced by the usual sample covariance matrix. It follows that the GIP with normalized sample covariance matrix, in correspondence of $\mathbf{r}_k$, is expected to be larger than those evaluated from the remaining training data, permitting a correct outlier detection.

2. Herein we use the usual Landau notation $\mathcal{O}(n)$; hence, an algorithm is $\mathcal{O}(n)$ if its implementation requires a number of flops proportional to n [38].

6.5 RP-ANMF DETECTOR

The RP-ANMF detector was devised in reference [32] and, under the design condition, it is capable of ensuring the CFAR property with respect to both the clutter powers as well as the clutter covariance structure [32, 40].[3] The detector is described as follows. Denote by $\mathbf{z}_1, \ldots, \mathbf{z}_{K-M}$, the vectors at the output of the 2S-DSP selector and construct

$$\mathbf{z}_{ek} = \frac{1}{\sqrt{2}}(\mathbf{z}_k + \mathbf{J}\mathbf{z}_k^*)$$

and

$$\mathbf{z}_{ok} = \frac{1}{\sqrt{2}}(\mathbf{z}_k - \mathbf{J}\mathbf{z}_k^*),$$

where $k = 1, \ldots, K - M$, where $\cdot^*$ denotes complex conjugate, and $\mathbf{J}$ is the permutation matrix, that is,

$$\mathbf{J} = \begin{bmatrix} 0 & 0 & \cdots & 0 & 1 \\ 0 & 0 & \cdots & 1 & 0 \\ \vdots & \vdots & \vdots & \vdots & \vdots \\ 1 & 0 & \cdots & 0 & 0 \end{bmatrix}.$$

Define the recursive persymmetric covariance structure estimator at the $(t+1)$ step as follows

$$\widehat{\mathbf{\Sigma}}^{t+1} = \frac{N}{K-M} \sum_{k=1}^{K-M} \frac{\mathbf{z}_{ek}\,\mathbf{z}_{ek}^{\dagger}}{\mathbf{z}_{ek}^{\dagger}\left(\widehat{\mathbf{\Sigma}}^{(t)}\right)^{-1}\mathbf{z}_{ek}}, \tag{6.17}$$

initializing the recursion with[4]

$$\widehat{\mathbf{\Sigma}}^{(0)} = \frac{1}{K-M} \sum_{k=1}^{K-M} \frac{\mathbf{z}_{ek}\mathbf{z}_{ek}^{\dagger}}{(\mathbf{T}\mathbf{z}_{ok}\mathbf{z}_{ok}^{\dagger}\mathbf{T}^{\dagger})_{\{i\}\{i\}}} \tag{6.18}$$

with $\mathbf{T} = (\mathbf{I} + \mathbf{J}) + j(\mathbf{I} - \mathbf{J})$, $j = \sqrt{-1}$, and $\mathbf{I}$ the identity matrix. The RP-ANMF is obtained substituting Equation 6.17 in place of $\mathbf{\Sigma}$ in the following decision rule,

$$\frac{|\mathbf{p}^{\dagger}\mathbf{\Sigma}^{-1}\mathbf{r}|^2}{(\mathbf{r}^{\dagger}\mathbf{\Sigma}^{-1}\mathbf{r})(\mathbf{p}^{\dagger}\mathbf{\Sigma}^{-1}\mathbf{p})} \underset{H_0}{\overset{H_1}{\gtrless}} T, \tag{6.19}$$

3. Please see reference [32] for a more detailed description of both the design hypotheses and the analytical derivations.

4. $\mathbf{A}_{\{i\}\{i\}}$, $i = 1, \ldots, N$, denotes the (i, i)th element of the $N \times N$ matrix $\mathbf{A}$.

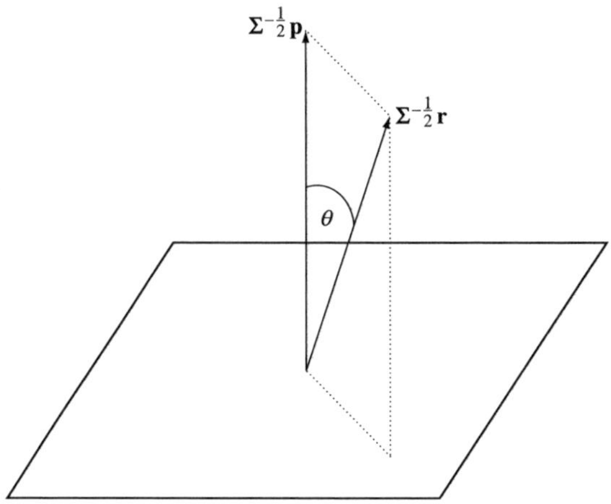

Figure 6.4 Geometric interpretation of the RP-ANMF of Equation 6.19.

where $\mathbf{r}$ is the primary data (namely the data vector where the presence of a useful target is sought), $|\cdot|$ is the modulus of a complex number, and T is the detection threshold.

An interesting geometric interpretation can be given to the left-hand side of Equation 6.19 according to the arguments in reference [39]. Precisely, the test statistic is the squared cosine of the angle θ between the vectors $\boldsymbol{\Sigma}^{-\frac{1}{2}}\mathbf{r}$ and $\boldsymbol{\Sigma}^{-\frac{1}{2}}\mathbf{p}$, namely the primary data and the useful signal whitened through an estimate of the covariance structure performed resorting to secondary data see Fig. 6.4. for a pictorial illustration.

The computational complexity required for the implementation of the RP-ANMF is $\mathcal{O}(N_{1,it}KN^2)$ flops where $N_{1,it}$ denotes the number of iterations required in order for the recursive covariance estimate to achieve convergence ($N_{1,it} = 3$ is sufficient [32]).

6.6 PERFORMANCE ANALYSIS

In this section we assess the performance of the proposed KA system and its comparison with three adaptive processors available in the open literature:

- The Modified Sample Matrix Inversion (MSMI) detector[5] [41], which does not exploit any selection algorithms (see Fig. 6.5); that is,

$$\frac{|\mathbf{r}^\dagger \mathbf{S}^{-1}\mathbf{p}|^2}{\mathbf{p}^\dagger \mathbf{S}^{-1}\mathbf{p}} \underset{H_0}{\overset{H_1}{\gtrless}} T \qquad \mathbf{S} = \frac{1}{K}\sum_{k=1}^{K} \mathbf{r}_k^\dagger \mathbf{r}_k. \tag{6.20}$$

5. This receiver is also known as the Adaptive Matched Filter (AMF) [42] or Wald Test [43].

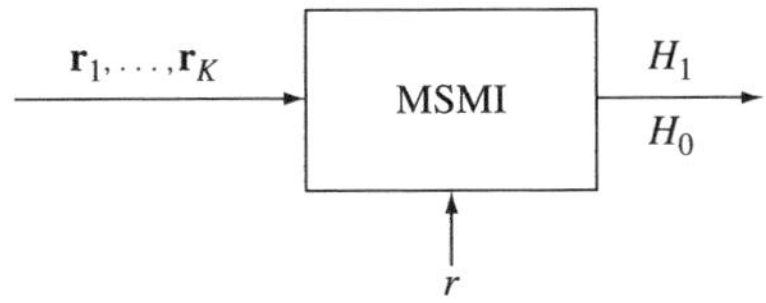

Figure 6.5 Block diagram of the MSMI detector.

In other words, the conventional sliding window (possibly with a number of guard cells) is used for training purposes.

- The Modified Sample Matrix Inversion (MSMI) detector with the KA data selector [28] (see Fig. 6.6).
- The 2S-DSP-RP-ANMF, namely the KA-RP-ANMF without the KA stage.

All the analysis is conducted on measured airborne radar data obtained from the AFRL Sensors Directorate's MCARM program.

The data sets consist of multichannel clutter data collected by an airborne platform with a side looking radar. The radar was configured with a 2×11 channel linear array including sum and delta analog beamformers. MCARM operated at L-Band, in low, medium and high pulse repetition frequency (PRF) modes. It had a range resolution of approximately 120 m. Each CPI consisted of 128 pulses and the clutter was typically unambiguous in Doppler. Northrop Grumman collected the data during flights over the Delmarva Peninsula and the East coast of the United States in the mid-1990s. The main parameters of the radar are summarized in Table 6.1 and the data set exploited in this paper refers to flight 5, acquisition 151.

As already pointed out, digital terrain data, exploited by the KA selector, have been obtained from the USGS to classify the ground environment illuminated by MCARM. NLCD have been chosen, which were collected in the 1990s at about the same time as the MCARM experiments. The terrain map corresponding to the considered data set is illustrated in Fig. 6.7, where different colors are used to denote different attributes of the scene. We also superimposed onto the image the position of the radar, the heading vector, and the main beam. The illuminated area is of Baltimore, Maryland (specifically the large red area in the center of the image is Baltimore). Finally we added, onto the map, approximate range rings for range bins 300 and 400, which correspond to our results. Notice that other available terrain data sets that provide elevation and linear feature information have not been used in this paper because the Delmarva Peninsula is relatively flat and the NLCD data already contains

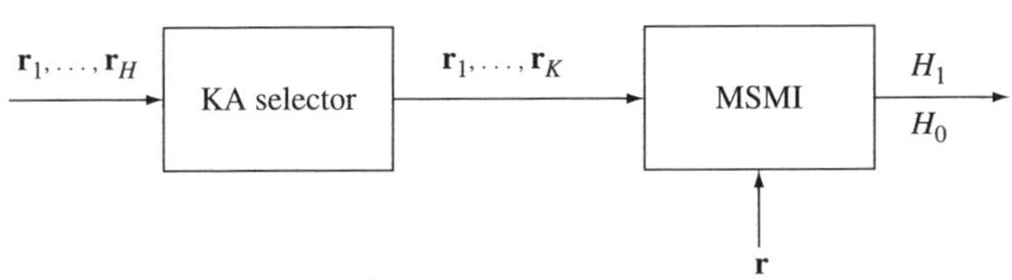

Figure 6.6 Block diagram of the KA-MSMI detector.

TABLE 6.1 Specifications of the MCARM dataset.

	Dataset 151
Number of array elements	22
Number of pulses per CPI	128
Number of usable range samples	500
Peak transmit power	1.5 kW
Instantaneous bandwith	800 KHz
Pulse repetition frequency (PRF)	1984 Hz
Interelement spacing	0.109 m
Radar frequency	1.24 GHz
Range resolution	120 m

some information about major roads, bridges, and railroads. However, these additional data sets should be considered, especially in mountainous environments or areas where more detailed information about linear features is required.

The first results we present concern the CFAR behavior of the three analyzed systems. We set the threshold of the receivers in order to ensure a probability of false alarm (P_{fa}) equal to 10^{-2} in the presence of homogeneous Gaussian white noise. (more precisely, 10^4 independent simulations of the data under the null hypothesis were run assuming homogeneous Gaussian white noise, and the decision statistics of the considered processors were computed and sorted. The thresholds were set from the $1 - P_{\text{fa}}$ quantile.) Then we evaluate the number of threshold crossings of the detection statistics in the presence of the measured radar data. In Fig. 6.8a–d the detection statistics are plotted versus the cell number for $K = 45$, $N = 22$, $M = 5$, angle bin 65 (corresponding to boresight), and Doppler bin 40. The plots clearly show that the proposed KA system presents a number of false alarms smaller than the number exhibited by the other analyzed processors. More precisely the new system shows zero false alarms out of 100 trials (0/100), which agrees with the nominal P_{fa} value. The KA-MSMI, the plain MSMI, and the 2S-DSP-RP-ANMF show respectively 3/100, 38/100, and 8/100 false alarms. (Notice that if the 2S-DSP is removed from the 2S-DSP-RP-ANMF, namely the plain RP-ANMF is used, a further increase in the number of false alarms can be observed.) The robust behavior of the new processor is confirmed by the plots of Fig. 6.9a–d, which assume $K = 49$ and the remaining parameters equal to those of the previous figures. In this case the number of false alarms shown by the proposed system, the KA-MSMI, the MSMI, and the 2S-DSP-RP-ANMF are respectively 1, 2, 35, and 12 out of 100 trials.

Further evidence, which generally agrees with the performance in Figs 6.8 and 6.9, is given by the plots of Fig. 6.10a–d, which assume $K = 45$, $N = 11$, angle bin 65, and Doppler bin 40. In this last case the number of false alarms exhibited by the proposed system, the KA-MSMI, the MSMI, and the 2S-DSP-RP-ANMF are respectively 1, 3, 4, and 0 out of 100 trials. Remarkably both the MSMI and the 2S-DSP-RP-ANMF exhibit a number of false alarms smaller than that in the previous figures.

In the last part of the chapter we provide some results concerning the capabilities of the analyzed algorithms to detect a target. To this end we inject a synthetic target

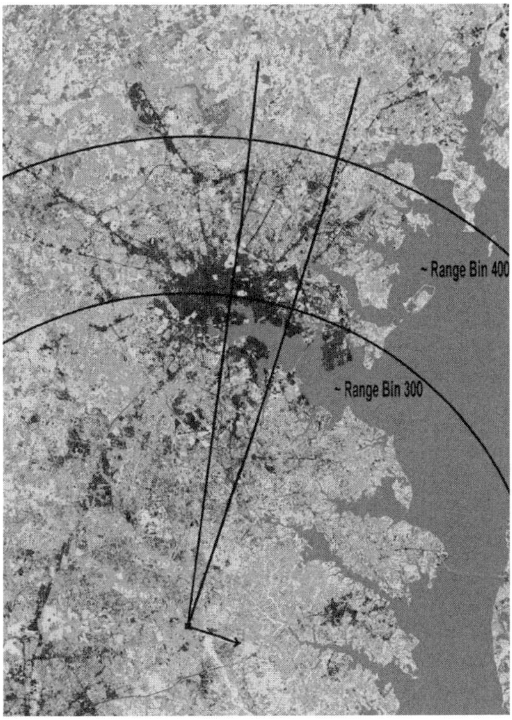

Color key	Class number and name
	11 • Open Water
	12 • Perennial Ice/Snow
	21 • Low Intensity Residential
	22 • High Intensity Residential
	23 • Commercial/Industrial/Transportation
	31 • Bare Rock/Sand/Clay
	32 • Quarries/Strip Mines/Gravel Pits
	33 • Transitional
	41 • Deciduous Forest
	42 • Evergreen Forest
	43 • Mixed Forest
	51 • Shrubland
	61 • Orchards/Vineyards
	71 • Grasslands/Herbaceous
	81 • Pasture/Hay
	82 • Row Crops
	83 • Small Grains
	84 • Fallow
	85 • Urban/Recreational Grasses
	91 • Woody Wetlands
	92 • Emergent Herbaceous Wetlands

Figure 6.7 Terrain map and its legend for MCARM flight 5, acquisition 151. (See color insert.)

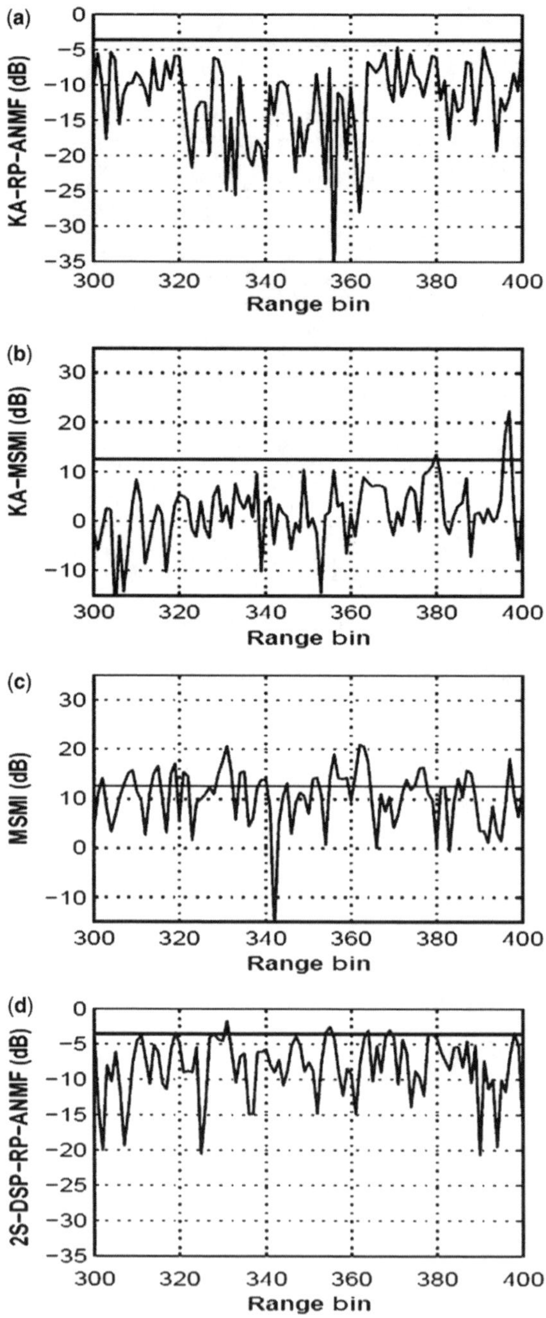

Figure 6.8 Decision statistic versus the cell number for $K = 45$, $N = 22$, $M = 5$, angle bin 65, and Doppler bin 40. (**a**) KA-RP-ANMF, (**b**) KA-MSMI, (**c**) MSMI, (**d**) 2S-DSP-RP-ANMF. Straight line: detection threshold for $P_{fa} = 10^{-2}$.

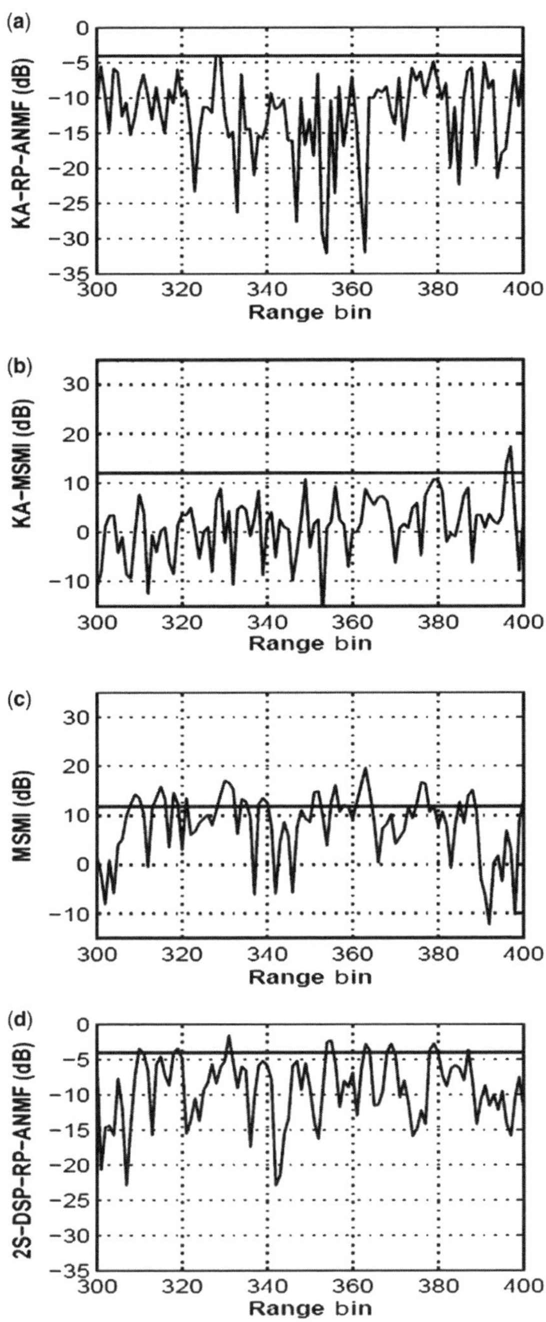

Figure 6.9 Decision statistic versus the cell number for $K = 49$, $N = 22$, $M = 5$, angle bin 65, and Doppler bin 40. (**a**) KA-RP-ANMF, (**b**) KA-MSMI, (**c**) MSMI, (**d**) 2S-DSP-RP-ANMF. Straight line: detection threshold for $P_{fa} = 10^{-2}$.

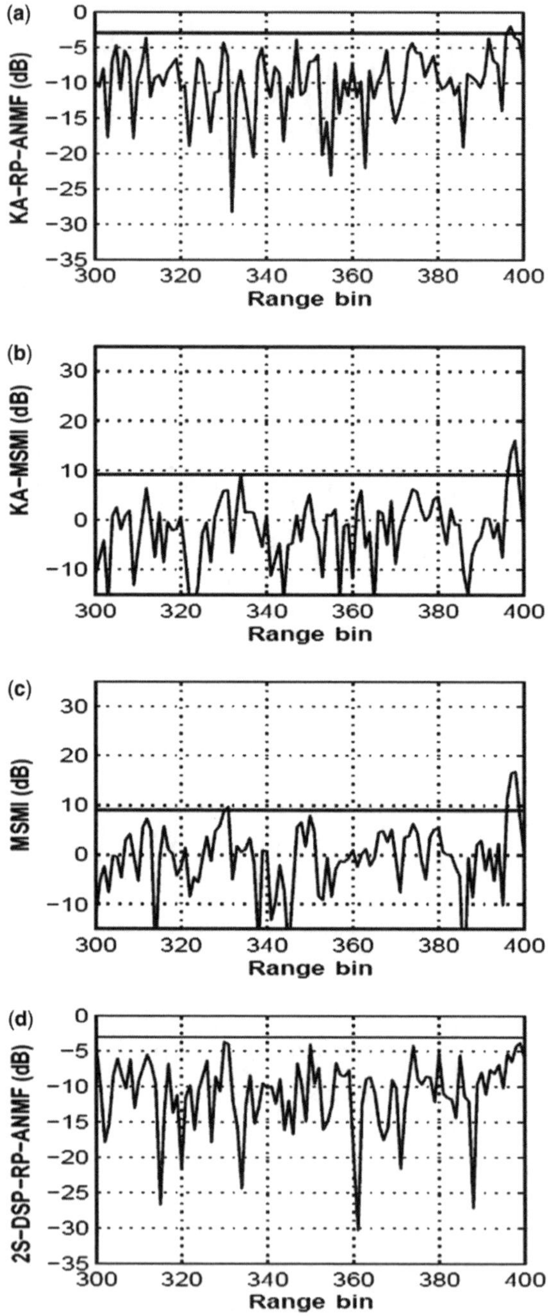

Figure 6.10 Decision statistic versus the cell number for $K = 45$, $N = 11$, $M = 5$, angle bin 65, and Doppler bin 40. (**a**) KA-RP-ANMF, (**b**) KA-MSMI, (**c**) MSMI, (**d**) 2S-DSP-RP-ANMF. Straight line: detection threshold for $P_{fa} = 10^{-2}$.

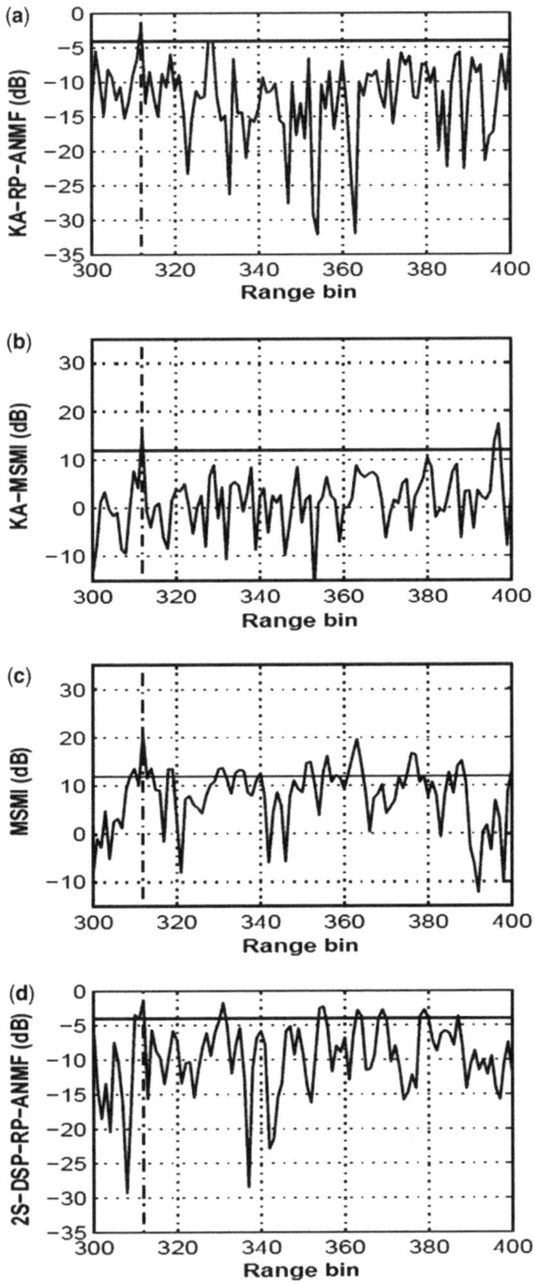

Figure 6.11 Decision statistic versus the cell number for $K = 49$, $N = 22$, $M = 5$, angle bin 65, Doppler bin 40, and target with SCR = 0dB (injected at range bin 312, angle bin 65, Doppler bin 40). (**a**) KA-RP-ANMF, (**b**) KA-MSMI, (**c**) MSMI, (**d**) 2S-DSP-RP-ANMF. Straight line: detection threshold for $P_{fa} = 10^{-2}$.

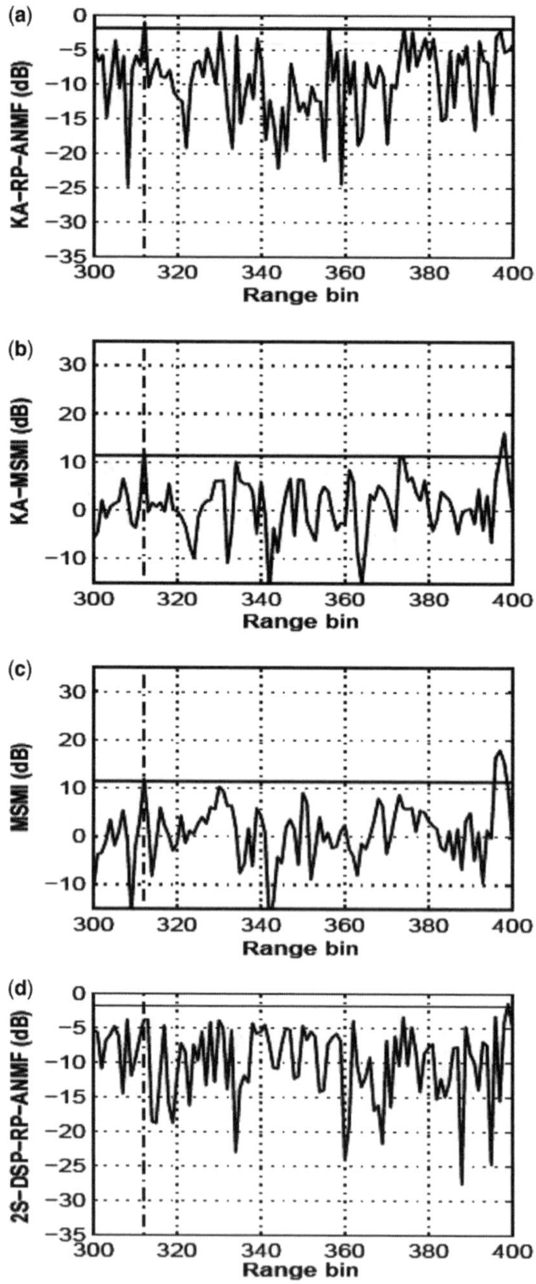

Figure 6.12 Decision statistic versus the cell number for $K = 27$, $N = 11$, $M = 5$, angle bin 65, Doppler bin 40, and target with SCR = 0 dB (injected at range bin 312, angle bin 65, Doppler bin 40). **(a)** KA-RP-ANMF, **(b)** KA-MSMI, **(c)** MSMI, **(d)** 2S-DSP-RP-ANMF. Straight line: detection threshold for $P_{fa} = 10^{-2}$.

in a given range cell and study whether the algorithms are capable of detecting it. The signal to clutter power ratio (SCR) is defined as

$$\text{SCR} = \frac{|\alpha|^2}{\overline{\sigma^2}}, \tag{6.21}$$

where $\overline{\sigma^2}$ is the average clutter power estimated from the measured data set.

In Figs 6.11a–d a target with SCR = 0 dB is injected in range bin 312, assuming $K = 49$, $N = 22$, $M = 5$, angle bin 65, and Doppler bin 40. The plots clearly indicate that all the analyzed detectors are capable of detecting the target. A similar behavior is obtained for SCR = −5 dB. Further results on the detection performance are illustrated in Figs 6.12a–d, which refer to the case $N = 11$ and the remaining simulation parameters equal to those of Fig. 6.11. Although the decision statistic of the proposed system, the KA-MSMI, and the MSMI exceeds the detection threshold in the target range location, the 2S-DSP-RP-ANMF is not capable of detecting the target.

Before concluding the analysis we explicitly point out that further results, obtained using additional data from the MCARM experiments, seem to confirm the performance demonstrated in this section.

6.7 CONCLUSIONS

In this chapter we have considered the design and the analysis of a KA radar processor for airborne STAP composed of a data selection stage and an adaptive detector. The selector is a hybrid KA and data-driven algorithm. Although the KA block exploits prior information concerning the terrain type illuminated by the radar, the data-driven block resorts to a statistical procedure to excise data containing outliers and nonhomogeneities. Finally the data that pass these screening procedure are fed, together with the data from the test cell, to the RP-ANMF, which performs the final decision as to the target presence.

At the analysis stage we have assessed the performance of the new system on measured airborne radar data from the AFRL MCARM program. Moreover, a careful comparison with other detectors available in the open literature has also been conducted. Remarkably, the new system presents a CFAR behavior stronger than that shown by the other analyzed systems without sacrificing detection performance.

Further study is needed to determine how well terrain classification data correlate with airborne radar clutter statistics. Other types of terrain data should be studied as well in order to explore their potential as an aid for STAP.

REFERENCES

[1] W.L. Melvin, M. Wicks, P. Antonik, Y. Salama, P. Li, H. Schuman. Knowledge-based space–time adaptive processing for airborne early warning radar. *IEEE AES Systems Magazine* 1998;13:37–42.

[2] R. Adve, P. Antonik, W. Baldygo, C. Capraro, G. Capraro, T. Hale, R. Schneible, M. Wicks. Knowledge-base application to ground moving target detection, AFRL-SN-TR-2001-185, in-house technical report, September 2001.

[3] A. Farina, G. Capraro, H. Griffiths, M. Wicks. Knowledge-based radar signal & data processing. Research and Technology Organisation (RTO) Lecture Series 233, NATO Rome, Italy, November 2003.

[4] "KASSPER: Knowledge-Aided Sensor Signal Processing and Expert Reasoning," Workshop on Knowledge-Aided Sensor Signal Processing and Expert Reasoning (KASSPER), Research Laboratory, DARPA, USA, April 2004. Available at http://www.darpa.mil/STO/space/kassper.html.

[5] J.R. Guerci. "Knowledge-Aided Sensor Signal Processing and Expert Reasoning," Workshop on Knowledge-Aided Sensor Signal Processing and Expert Reasoning (KASSPER), Washington, DC, USA, April 2002. Available at http://www.darpa.mil/STO/space/kassper.html.

[6] D.J. Zywicki, W.L. Melvin, G.A. Showman, J.R. Guerci. STAP Performance in Site-Specific Clutter Environments. *Proc. 2003 IEEE Aerospace Conference*, March 2003, p. 1–16.

[7] F. Gini, editor. Knowledge-based systems for adaptive radar: Detection, tracking, and classification. *IEEE Signal Processing Magazine* 2006;23:14–76.

[8] J.R. Guerci, W.L. Melvin, editors. Special section on Knowledge-Aided Sensor Signal and Data Processing. *IEEE Trans. Aerospace and Electronic Systems* 2006;42:983–1120.

[9] V.C. Vannicola, J.A. Mineo. Applications of knowledge-based systems to surveillance. *Proc. 1988 IEEE National Radar Conference*, April 1988, 157–164.

[10] V.C. Vannicola, L.K. Slaski, G.J. Genello. Knowledge-based resource allocation for multifunction radars. *Proc. 1993 SPIE Conference on Signal and Data Processing of Small Targets*, April 1993, p. 410–425.

[11] S. Haykin. Radar vision. *Proc. 1991 Second International Specialist Seminar on the Design and Application of Parallel Digital Processors*, April 1991, p. 75–78.

[12] E.J. Kelly. An adaptive detection algorithm. *IEEE Trans. Aerospace and Electronic Systems* 1986;22:115–127.

[13] R. Nitzberg. An effect of range heterogeneous clutter on adaptive Doppler filters. *IEEE Trans. Aerospace and Electronic Systems* 1990;26:475–480.

[14] W.L. Melvin. Space–time adaptive radar performance in heterogeneous clutter. *IEEE Trans. Aerospace and Electronic Systems* 2000;36:621–633.

[15] M. Rangaswamy, J.H. Michels, B. Himed. Statistical analysis of the nonhomogeneity detector for STAP applications. *Digital Signal Processing* 2004;14:253–267.

[16] W.L. Melvin, G.A. Showman. An approach to knowledge-aided covariance estimation. *IEEE Trans. Aerospace and Electronic Systems* 2006;42:1021–1042.

[17] D.J. Rabideau, A.O. Steinhardt. Improved adaptive clutter cancellation through data-adaptive training. *IEEE Trans. Aerospace and Electronic Systems* 1999;35:879–891.

[18] P. Chen, W.L. Melvin, M.C. Wicks. Screening among multivariate normal data. *Journal of Multivariate Analysis* 1999;69:10–29.

[19] J.H. Michels, B. Himed, M. Rangaswamy. Robust STAP detection in a dense signal airborne radar environment. *Signal Processing* 2004;84:1625–1636.

[20] K. Gerlach. Outlier resistant adaptive matched filter. *IEEE Trans. Aerospace and Electronic Systems* 2002;38:885–901.

[21] M. Rangaswamy. Statistical analysis of the nonhomogeneity detector for non-Gaussian interference backgrounds. *IEEE Trans. Signal Processing* 2005;53:2101–2111.

[22] E. Conte, A. De Maio, A. Farina, G. Foglia. Training data selection for adaptive radar detection. *Proc. Radar 2004*, Toulouse, October 2004, p. 18–22.

[23] E. Conte, A. De Maio, A. Farina, G. Foglia. Design and analysis of a knowledge-based radar detector for Doppler processing. *IEEE Trans. Aerospace and Electronic Systems* 2006;42:1058–1079.

[24] N.L. Owsley. Sonar array processing. In S. Haykin, editor. *Array Signal Processing* Englewood Cliffs, NJ: Prentice Hall; 1985.

[25] G.K. Borsari, A.O. Steinhardt. Cost-efficient strategies for space–time adaptive processing algorithms. *Proc. 1996 29th Asilomar Conference on Signals, Systems and Computers*, October 1996, p. 650–654.

[26] D.D. Weiner, G.T. Capraro, M.C. Wicks. An approach for utilizing known terrain and land feature data in estimation of the clutter covariance matrix. *Proc. 1998 IEEE Radar Conference*, May 1998, p. 381–386.

[27] C.T. Capraro, G.T. Capraro, D.D. Wiener, M.C. Wicks. Knowledge-based map space–time adaptive processing (KBMapSTAP). *Proc. 2001 International Conference on Imaging Science, Systems, and Technology*, Las Vegas, NV, June 2001, p. 533–538.

[28] C.T. Capraro, G.T. Capraro, D.D. Wiener, M.C. Wicks, W.J. Baldygo. Improved STAP performance using knowledge-aided secondary data selection. *Proc. 2004 IEEE National Radar Conference*, Philadelphia, PA, April 2004, p. 361–365.

[29] Available at http://www.nga.mil.

[30] Available at http://www.usgs.gov.

[31] C.T. Capraro, J.T. Capraro, A. De Maio, A. Farina, M. Wicks. Demonstration of knowledge aided STAP using measured airborne data. *IEE Proc. Radar, Sonar and Navigation* 2006;153:487–494.

[32] E. Conte, A. De Maio. Mitigation techniques for non-Gaussian sea clutter. *IEEE Journal on Oceanic Engineering* 2004;29:284–302.

[33] J. Ward. Space–time adaptive processing for airborne radar. Technical Report No. 1015, Lincoln Laboratories, 13 December, 1994.

[34] B. Himed, W.L. Melvin. Analyzing space–time adaptive processors using measured data. *Proc. 1998 31th Asilomar Conference on Signals, Systems & Computers*, October 1998, p. 930–935.

[35] C.D. Richmond. Performance of a class of adaptive detection algorithms in nonhomogeneous environments. *IEEE Trans. Signal Processing* 2000;48:1248–1262.

[36] National Land Cover Data (NLCD). Available at http://landcover.usgs.gov/

[37] E. Conte, M. Lops, G. Ricci. Adaptive detection schemes in compound-Gaussian clutter. *IEEE Trans. Aerospace and Electronic Systems* 1998;34:1058–1069.

[38] G.H. Golub, C.F. Van Loan. *Matrix Computations*. The Johns Hopkins University Press, Baltimore, MA, 1989.

[39] S. Kraut, L.L. Scharf, L.T. McWhorter. Adaptive subspace detectors. *IEEE Trans. Signal Processing* 2001;49:1–16.

[40] A. De Maio, G. Foglia, E. Conte, A. Farina. CFAR behavior of adaptive detectors: An experimental analysis. *IEEE Trans. Aerospace and Electronic Systems* 2005;41:233–251.

[41] L. Cai, H. Wang. On the adaptive filtering with the CFAR feature and its performance sensitivity to non-Gaussian interference. *Proc. 1990 24th Annual Conference on Information Sciences and Systems*, Princeton, NJ, March 1990, p. 558–563.

[42] F.C. Robey, D.R. Fuhrmann, E.J. Kelly, R. Nitzberg. A CFAR adaptive matched filter detector. *IEEE Trans. Aerospace and Electronic Systems* 1992;28:208–216.

[43] A. De Maio. A new derivation of the adaptive matched filter. *IEEE Signal Processing Letters* 2004;11:792–793.

[44] W.H. Press, S.A. Teukolsky, W.T. Vetterling, B.P. Flannery. *Numerical Recipes in C: The Art of Scientific Computing*. Cambridge University Press, Newyork, NY, 1992.

APPENDIX 6A

REGISTRATION GEOMETRY

Registration of the radar with the Earth was performed to determine the terrain illuminated by the radar during a CPI. A system of three nonlinear equations was developed to calculate the position of a point on the Earth given a slant range, a Doppler frequency, and an oblate spheroid (elliptical) model of the Earth. Figure 6A.1 illustrates the geometry of the problem. An Earth-Centered Earth-Fixed (ECEF) geographical coordinate system was used. The point $P_r(x_r; y_r; z_r)$ represents the position of the radar, and the point on the Earth to be determined is designated as $P_e(x; y; z)$. Also shown in Fig. 6A.1 is the slant range R_s to point P_e and the iso-Doppler of interest (represented as a dashed line). Assuming the radar is flying slower than the maximum unambiguous Doppler velocity, the intersection of the slant range with the iso-Doppler contour and the Earth's surface occurs at two points, P_e and a mirror point on the iso-Doppler. However, as the radar data was gathered by a side-looking radar, we need only determine the location of one of the two points depending on the orientation of the radar platform. The first equation is related to

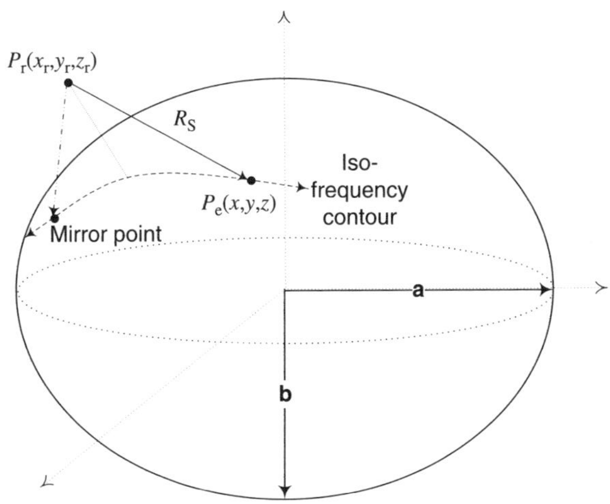

Figure 6.A.1 Registration geometry.

the slant range and is simply the Euclidian distance between points P_e and P_r. The functional form of the equation is given as

$$F_1(x; y; z) = (x - x_r)^2 + (y - y_r)^2 + (z - z_r)^2 - R_s^2 = 0. \tag{6A.1}$$

The second equation represents the iso-Doppler contour on the Earth and takes the form

$$F_2(x; y; z) = (x - x_r)v_{rx} + (y - y_r)v_{ry} + (z - z_r)v_{rz} - \frac{\lambda f_d}{2} R_s = 0, \tag{6A.2}$$

where v_{rx}, v_{ry}, and v_{rz} are the components of the radar's velocity vector, f_d is the Doppler frequency, and λ is the wavelength of the radar.

The last equation models the Earth's surface as an oblate spheroid and is defined as

$$F_3(x; y; z) = \frac{x^2}{a^2} + \frac{y^2}{a^2} + \frac{z^2}{b^2} - 1 = 0, \tag{6A.3}$$

where a and b are the semi-major and semi-minor radii of the Earth, respectively. Values for these parameters were obtained from the World Geodetic System 1984 (WGS84) because they were used by the USGS to define the coordinates of the terrain data. However, depending upon where registration is to be performed on the Earth, there may be a more accurate local datum available. In order to find solutions for x, y, and z, an iterative Newton–Raphson method [44] was used until the method converged to a solution within a certain tolerance. The initial point of the iteration was calculated from a spherical Earth model and was chosen to be near the point of interest, P_e. This helped the Newton–Raphson method converge to P_e instead of its mirror point. A check was done to ensure the result was on the correct side of the radar platform.

7

STAP VIA KNOWLEDGE-AIDED COVARIANCE ESTIMATION AND THE FRACTA META-ALGORITHM[1]

Shannon D. Blunt, Karl Gerlach, Muralidhar Rangaswamy, and Aaron K. Shackelford

In this chapter, we address knowledge-aided processing as it pertains to airborne/space-based radar for the Ground Moving Target Indication (GMTI) problem. Specifically, knowledge-aided processing is used as a means to supplement the performance of the FRACTA meta-algorithm for Space–Time Adaptive Processing (STAP). STAP necessitates accurate estimation of the unknown input covariance matrix, which is estimated via multiple sensory channels and used in the adaptive Weiner Filter solution in order to maximize the signal-to-interference ratio by cancelling the interference from clutter and jamming, thus enabling the detection of moving targets. However, real-world effects such as data heterogeneity and training data contamination (e.g. by other targets of interest) greatly limit the performance of standard STAP processors. An effective STAP processor must be robust to these effects, facilitate rapid convergence of the input covariance matrix estimate with a minimal number of training samples, and allow for implementation with a relatively low computational cost.

1. The material presented in this chapter is expanded from the collection of papers in references 1, 2, 4, and 5, © 2004, 2005, 2006 IEEE. Reprinted, with permission, from *IEEE Transactions on Aerospace & Electronic Systems*.

Knowledge-Based Radar Detection, Tracking, and Classification. Edited by Fulvio Gini and Muralidhar Rangaswamy

To accomplish these goals, approximate prior knowledge of the physical configuration and operation of the radar is used to enhance the performance of the FRACTA meta-algorithm. FRACTA is a holistic STAP approach that employs multiple metrics to systematically extract potential targets and excise false alarms. However, due to the heterogeneous nature of the interference from clutter and the internal culling process of FRACTA, insufficient training data may be available with which to adequately suppress interference. A simple bald-Earth knowledge-aided model for the clutter is used to compensate for the training data deficiency and is shown to significantly improve target detection performance in extreme cases of "sample starvation." Also, a reduced-dimension formulation for knowledge-aided processing is presented and used in conjunction with a reduced-dimension version of FRACTA to yield a robust STAP processor with reduced computational cost.

7.1 INTRODUCTION

For airborne/space-based radar detection, STAP is used to find a desired target return in the presence of additive interference from Doppler/angle spread ground clutter and man-made jamming. This detection is accomplished through cancellation of the interference, which necessitates second-order statistical knowledge via a covariance matrix formed from the input channels. Accurate estimation of the unknown input covariance matrix of the input channels is required to achieve good performance. Presuming clairvoyant knowledge of the covariance matrix, the optimal linear weighting of the input could be achieved such that the output signal-to-interference ratio is maximized. However, because precise knowledge is generally unavailable, adaptive techniques are required to estimate effectively the input covariance matrix. If utilized correctly, prior knowledge may be exploited to improve covariance matrix estimation and, as a result, overall STAP performance. In this chapter we shall address the benefit provided by model-based prior knowledge when used to supplement the FRACTA meta-algorithm [1–5], a multistage/multimetric approach that is robust to data heterogeneity.

In the STAP framework, the input channels corresponding to a particular range cell comprise the return signal from a set of consecutive pulses (the temporal component) for each of the antenna array elements (the spatial component). The collection of all space–time input channels at a given range is a vector "snapshot," which is used to determine if a moving target resides at the given range. To suppress the interference present in a given snapshot, the covariance matrix of the interference must first be determined. In order to estimate the interference covariance matrix, a group of snapshots in near proximity (in space and/or time) must be found that share the same interference properties as the given snapshot. Normally, for adaptive radar applications, the sample covariance matrix is estimated using training data from range cells close to the range cell under test (CUT). In the derivation of the maximum likelihood estimate of the covariance matrix, it is normally assumed that the CUT sample vector (the data vector to which the adaptive weight is applied) and all of the training data vectors are independent and identically distributed (i.i.d.). Such sample data are characterized as being homogeneous. In practice, however, the sample data is generally heterogeneous in nature, thereby fundamentally invalidating the underlying assumption of the maximum likelihood estimate. As a result, straightforward statistical solutions often perform poorly, thus necessitating a more robust solution that addresses the underlying causes of heterogeneity in the sample data.

For the airborne/space-based radar application, the cause of data heterogeneity can be roughly separated into two classes: [1] the variability of ground clutter and [2] data outliers. The former may result from (perhaps significant) variation in local terrain, especially in areas such as littoral environments and mountainous regions. The presence of man-made structures such as buildings and roads also contributes to ground clutter variability. In regard to the latter, data outliers may arise for a variety of reasons (some of which, such as sidelobe clutter discretes, may dually be denoted as ground clutter variability, although the underlying mechanism is generally unimportant). Outliers may be caused by temporally sporadic electromagnetic interference (EMI), malfunctioning input sensor channels, or an intentional blinking interference source. However, for airborne/space-based radar, the most deleterious form of outlier is one that possesses a spatio-temporal steering vector that is highly correlated with the desired signal. For example, if one is trying to detect adaptively an individual target in the presence of a formation of targets (e.g. an airborne formation, or a convoy traveling along a road), the other target returns located in the range cells near the desired target will have approximately the same steering vector. Thus the snapshots in nearby ranges that would typically be used as training data will produce a weight vector that lies in the null space of both the interference and the desired signal. This form of "data contamination" effectively nulls the desired signal in the CUT along with the ambient interference, thereby precluding detection.

The problem of data contamination is particularly difficult for the airborne/space-based radar scenario. If the desired signal is present in a given snapshot, it will generally possess a much lower amplitude than the clutter, which is the reason why interference cancellation is necessary. However, in order to select uncontaminated training data, it is first necessary to have identified all of the surrounding range cells whose corresponding snapshots possess the desired signal. To address this seeming impasse, nonhomogeneity detectors have been developed that attempt to identify outliers from a given collection of data samples. These detectors typically use either the Generalized Inner Product (GIP) metric or the Adaptive Power Residue (APR) metric. The result is a sorting of a given set of training data snapshots into two subsets. Depending upon the efficacy of the sorting procedure, one subset is purported to contain uncontaminated training data and the other may contain potential targets.

The remainder of this chapter details the development of a holistic approach to airborne/space-based radar for ground moving target indication (GMTI) that systematically addresses the two general classes of heterogeneous data. Section 7.2 describes the FRACTA meta-algorithm [1–3], which, being comprised of multiple stages/metrics, provides robustness against data contamination. One facet of the FRACTA meta-algorithm is the censoring from the training data of snapshots that are too similar to the desired signal. The practical aspects of censoring are discussed in Section 7.3. The result of this censoring, in conjunction with heterogeneity due to ground clutter variability, may potentially yield a "sample starvation" effect whereby insufficient data samples are available for accurate estimation of the input covariance matrix. To compensate, a model-based approach is described in Section 7.4 that uses current radar/platform measurements to generate approximate prior knowledge of the input covariance matrix based on the bald-Earth model, which is denoted as knowledge-aided covariance estimation (KACE). This prior knowledge is used to supplement the outlier-resistant property of the FRACTA meta-algorithm to provide further robustness against data heterogeneity [4]. Finally, Section 7.5 discusses partially adaptive (i.e. reduced dimension) implementation of FRACTA-KACE as a means to further compensate for sample starvation as well as to reduce the computational cost [5].

7.2 THE FRACTA META-ALGORITHM

FRACTA [1] is referred to as a meta-algorithm because it is composed of several individual algorithms/metrics. Individually, the original components consist of the Fast Maximum Likelihood (FML) algorithm, Reiterative Censoring (RC), the Adaptive Power Residue (APR) metric for outlier censoring, Concurrent Block Processing (CBP), the Two-Weight Method (TWM), and the Adaptive Coherence Estimation (ACE) metric. A subsequent enhanced version of FRACTA [2, 3] provides an automated censoring stopping mechanism, an alternative data blocking approach for APR censoring, and a fast reiterative censoring procedure to replace the computationally intensive FML algorithm. In the following, the individual components are examined with regard to the role they play as part of the overall methodology.

7.2.1 The General STAP Model

Consider a radar antenna system that consists of an N-element uniform linear array. The radar antenna system provides N radio frequency (RF) input signals or channels. Time-delayed inputs of these N inputs are to be combined via linear weighting to form an output such that an output performance metric (such as signal-to-noise (S/N) power ratio) is optimized. If the linear weighting is derived adaptively, such an implementation is called a STAP processor (depicted in Fig. 7.1). Assume that for each of these RF channels, the radar front-end carries out amplification, filtering, reduction to baseband, and analog-to-digital (A/D) conversion. The output of each A/D is a data stream of in-phase and quadrature-phase (I,Q) output pairs. The I and Q components represent the real and imaginary parts, respectively, of the complex-valued data stream.

The radar waveform is assumed to be a burst of M identical pulses with pulse repetition interval (PRI) equal to T. Target detection is based upon the returns from this burst. The input data in the respective channels are sampled to form range-gate samples for each pulse. For a given range gate, we form an MN-length sample

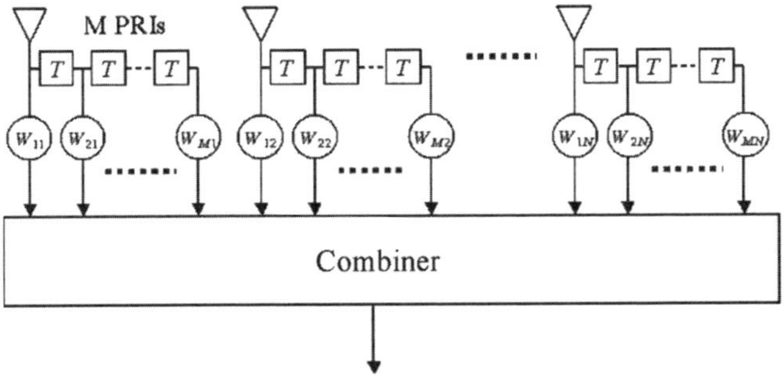

Figure 7.1 Generic STAP for airborne/space-based radar.

vector (called a snapshot) by stacking in succession the N-length data vectors associated with each of the antenna channels for each of the M pulses. Signal presence is sought in one range gate (called the CUT) at a time. A collection of length-MN data vectors (called the training data) are used to estimate an adaptive matched filter (AMF) weight vector (also of length MN). This adaptive weight vector is applied to the CUT data vector, which may or may not be one of the training data vectors. In the derivation of the optimal adaptive weighting vector, the training data and CUT vectors are assumed to have the same interference covariance matrix. However, in practice this is not necessarily an accurate assumption.

The input interference present in the main and auxiliary channels consists of three statistically independent components: thermal noise (system noise and external thermal noise), clutter, and jamming. Let $\mathbf{z}$ represent one of the MN-length data vectors. Assuming that there is no desired signal present, then

$$\mathbf{z} = \mathbf{z}_{\mathrm{T}} + \mathbf{z}_{\mathrm{C}} + \mathbf{z}_{\mathrm{J}}, \tag{7.1}$$

where $\mathbf{z}_{\mathrm{T}}$, $\mathbf{z}_{\mathrm{C}}$, and $\mathbf{z}_{\mathrm{J}}$ represent the thermal noise, clutter, and jamming components, respectively. Let $\mathbf{R}_{\mathrm{T}}$, $\mathbf{R}_{\mathrm{C}}$, and $\mathbf{R}_{\mathrm{J}}$ represent the $MN \times MN$ covariance matrices associated with the thermal noise, clutter, and jamming components, respectively, where each is a positive semi-definite Hermitian (PSDH) matrix. If $\mathbf{R}$ is the covariance matrix of $\mathbf{z}$ defined as $\mathbf{R} = E\{\mathbf{z}\mathbf{z}^H\}$, where $E\{\cdot\}$ denotes the expected value and $(\cdot)^H$ denotes the complex conjugate transpose operation, then because of the mutual statistical independence of the thermal noise, clutter, and jamming components,

$$\mathbf{R} = \mathbf{R}_{\mathrm{T}} + \mathbf{R}_{\mathrm{C}} + \mathbf{R}_{\mathrm{J}}. \tag{7.2}$$

We can assume without loss of generality that the thermal noise on each of the MN elements of $\mathbf{z}_{\mathrm{T}}$ is statistically independent and with power equal to one. Thus $\mathbf{R}_{\mathrm{T}} = \mathbf{I}_M \otimes \mathbf{I}_N$ where $\mathbf{I}_h$ denotes the hth-order identity matrix and $\otimes$ is the Kronecker matrix product [6]. Furthermore, it is commonly assumed that the jamming components are statistically independent from pulse to pulse (i.e. barrage jamming); thus it can be shown that $\mathbf{R}_{\mathrm{J}} = \mathbf{I}_M \otimes \mathbf{R}_{\mathrm{J}}$, where $\mathbf{R}_{\mathrm{J}}$, is the spatial $N \times N$ jamming covariance matrix associated with the elements of the linear array.

A data vector may contain a desired signal vector denoted by $a\mathbf{s}$ where a is an unknown complex amplitude and $\mathbf{s}$ is an MN-length column vector related to the desired signal. For the input data structure previously described, the spatio-temporal steering vector $\mathbf{s}$ takes the form

$$\mathbf{s} = \mathbf{s}_{\mathrm{D}} \otimes \mathbf{s}_{\mathrm{S}}, \tag{7.3}$$

where $\mathbf{s}_{\mathrm{D}}$ is an M-length temporal steering vector associated with the desired signal's Doppler shift and $\mathbf{s}_{\mathrm{s}}$ is an N-length spatial steering vector associated the desired signal's spatial phase shift across the antenna array. For a moving target with Doppler phase shift per pulse equal to ϕ, the temporal steering vector is $\mathbf{s}_{\mathrm{D}} = \left[1\; e^{j\phi}\; e^{j2\phi} \ldots\; e^{j(M-1)\phi}\right]^T$, where the superscript T denotes the transpose operation. For a target with a spatial phase shift across the array of θ, the spatial steering vector is defined as $\mathbf{s}_{\mathrm{s}} = \left[1\; e^{j\theta}\; e^{j2\theta} \ldots\; e^{j(N-1)\theta}\right]^T$.

Given $\mathbf{s}$ and the $MN \times MN$ interference covariance matrix $\mathbf{R}$, it is well known [7] that conjugate weighting of the MN-length CUT data vector that maximizes the output signal-to-interference power ratio (SIR) is given by

$$\mathbf{w} = \mathbf{R}^{-1}\mathbf{s}. \tag{7.4}$$

For adaptive problems, $\mathbf{R}$ is generally not known, although we may have *a priori* information regarding the structure of $\mathbf{R}$. For example, it was previously stated that the jamming component of $\mathbf{R}$ has the form $\mathbf{I}_M \otimes \mathbf{R}_{\mathrm{J}}$ where $\mathbf{R}_{\mathrm{J}}$ is the unknown $N \times N$ spatial jamming covariance matrix. Furthermore, for radar systems operating at microwave frequencies, the thermal noise is generally dominated by the internal noise power, which can be readily measured, and thus the thermal noise covariance matrix is known. Hence, it can be assumed that the unknown interference covariance matrix has the form of a known diagonal matrix (assumed to be the identity matrix) plus an unknown positive semi-definite Hermitian (PSDH) matrix. This particular form allows the covariance matrix to be estimated using fast convergence techniques such as FML [8], loaded sample matrix inversion (LSMI) [9], and eigen-projection [10, 11]. Convergence "time" is generally measured as the number of training snapshots necessary to attain an output signal-to-noise power ratio performance that is within 3 dB of optimal.

In general, the covariance matrix $\mathbf{R}$ will have to be estimated from a number of training data input vectors. Let K be the number of training data vectors. Normally, for a given range CUT, the training data vectors used to calculate the adaptive weight are chosen to be associated with range indices that are in close proximity to the CUT range index. The motivation for choosing these indices is that it increases the likelihood that the training data and CUT will have approximately the same interference covariance matrix.

7.2.2 FRACTA Description

The FRACTA meta-algorithm comprises multiple algorithms and metrics that, when employed in conjunction, provide a holistic STAP processor that is robust to data

Table 7.1 Nomenclature.

ACE	Adaptive coherence estimate
AMF	Adaptive matched filter
APR	Adaptive power residue
CBP	Concurrent block processing
CA-CFAR	Cell-averaging constant false alarm rate
CPI	Coherent processing interval
CTD	Censored training data
CUT	Cell under test
DS-ACE	Doppler-sensitive ACE
FML	Fast maximum likelihood algorithm
FRC	Fast reiterative censoring
GC	Global censoring
GCM	Generalized clutter model
GIP	Generalized inner product
GMTI	Ground moving target indicator
ICM	Internal clutter motion
ITD	Initial training data
KACE	Knowledge-aided covariance estimation
KASSPER	Knowledge-aided sensor signal processing and expert reasoning
LSMI	Loaded sample matrix inverse
PRF	Pulse repetition frequency
PRI	Pulse repetition interval
PSDH	Positive semi-definite Hermitian
RC	Reiterative censoring
SMI	Sample matrix inverse
STAP	Space–time adaptive processing
SWP	Sliding window processing
TD	Training data
TWM	Two-weight method
UTD	Uncensored training data

heterogeneity resulting from ground clutter variability and outliers. The reader is referred to Table 7.1 for a list of acronyms of the algorithms, metrics, and structures used throughout. FRACTA uses a three-stage detection process: [1] a censoring stage to identify potential targets and mitigate training data contamination [2] a constant false alarm rate (CFAR) detector, and [3] a coherence test to screen out detections that are not closely aligned with the desired steering vector. We first consider the reiterative censoring stage.

7.2.2.1 Reiterative Censoring As stated above, censoring performs the twofold function of both identifying potential targets and excising snapshots from the training data (TD) that may be closely correlated with the desired signal and would thus result in data contamination. For a given CUT and desired steering vector **s**, the associated set of training data vectors is denoted as the initial training data (ITD) and is comprised of K input snapshots of length MN. The set of K snapshots in the ITD is to be segregated into two mutually exclusive subsets denoted as the censored training data (CTD) and the uncensored training data (UTD). The CTD is composed of some K_C snapshots and the UTD is composed of K_U snapshots, whereby $K = K_C + K_U$. Thus CTD ∪ UTD = ITD and CTD ∩ UTD = Ø. Let $\mathbf{z}_k$, $k = 1, 2, \ldots, K$ denote the $MN \times 1$ vectors of the initial TD and $\tilde{\mathbf{R}}$ be the estimate of the covariance matrix, which is derived from the initial TD using one of the fast convergence techniques such as FML, LSMI, or eigen-projection. Note that due to its demonstrated convergence performance properties and robustness to thermal noise mis-estimation [8] FRACTA originally used the FML algorithm (hence the 'F' in FRACTA), which is based on the assumption that the unknown interference covariance matrix comprises a known diagonal matrix (related to the noise component) and a positive semi-definite Hermitian matrix (related to the clutter and jamming components). However, FML was later replaced with LSMI, which requires significantly less computation for only a marginal reduction in performance.

To censor data vectors from the ITD, the FRACTA meta-algorithm uses the adaptive power residue (APR) metric. The APR metric can be defined as [1]

$$\text{APR:} \quad |\mathbf{s}^H \tilde{\mathbf{R}}^{-1} \mathbf{z}_k|^2, \quad k = 1, 2, \ldots, k. \tag{7.5}$$

Note that the APR is a coherent metric, as it measures the similarity of a given data vector $\mathbf{z}_k$ with the desired steering vector **s**. As such, for the objective of removing data vectors from the ITD that are "close" to the desired steering vector (so as to identify potential targets and eliminate the most deleterious cause of data contamination) the APR metric has been shown to be superior to another popular censoring metric denoted as the generalized inner product (GIP), which is noncoherent [1].

Although numerous censoring strategies exist, the simplest (and perhaps the most powerful) approach is as follows. Prior to censoring we have $K_U = K$ (UTD = ITD) and $K_C = 0$ (CTD = Ø). Based upon the covariance matrix estimate $\tilde{\mathbf{R}}$ derived from the UTD vectors, let m_k ($k = 1, 2, \ldots, K_U$) denote the set of resulting APR values for the K_U vectors in the UTD. Transfer the TD vector associated with $\max_k m_k$ from UTD to CTD and subsequently reduce K_U by one and increase K_C by one. The covariance matrix estimate $\tilde{\mathbf{R}}$ is then re-computed from the remaining UTD vectors, and the censoring process is repeated. In this manner, censoring can be performed either until some prescribed number of TD vectors has been censored or until a censoring stopping criterion is met (an approach to achieve the latter will be discussed in Section 7.3).

The FRACTA meta-algorithm uses the reiterative censoring process as an initial detection stage, so the CUT is included in the initial training data set. In contrast to sliding window processing (SWP), in which a new covariance matrix is estimated for each CUT, censoring enables use of concurrent block processing (CBP), whereby, for a given set of K snapshots of ITD, the interference covariance matrix estimated after the final pass of censoring is applied as in Equation 7.4 to all of the snapshots in the CTD that are concurrent CUTs.

To enable the censored targets to stand out further from the quiescent noise level, two adaptive weight vectors are computed; one each for the CTD and the UTD. The snapshots in the CTD are considered potential targets, as they have passed the initial screening process that is accomplished through censoring. Thus, the CTD weight vector $\mathbf{w}_C$ is computed using the UTD covariance matrix estimate $\tilde{\mathbf{R}}_U$ (determined from the snapshots of UTD) as

$$\mathbf{w}_C = \tilde{\mathbf{R}}_U^{-1}\mathbf{s}, \tag{7.6}$$

because it is assumed that all snapshots that resemble the desired signal have been removed. In so doing, the "self-cancellation" effect due to data contamination can be avoided. The UTD weight vector $\mathbf{w}_U$ is computed using the ITD covariance matrix $\tilde{\mathbf{R}}_I$ (determined from the snapshots of ITD) as

$$\mathbf{w}_U = \tilde{\mathbf{R}}_I^{-1}\mathbf{s}, \tag{7.7}$$

because the UTD is assumed to contain no snapshots with targets and the ITD has more samples than the UTD, thus providing a more accurate estimate of the covariance matrix and avoiding the problem of "sample starvation." Also, because the CTD contains all the potential targets, the problem of "self-cancellation" is likewise avoided when applying the weight vector $\mathbf{w}_U$ to the UTD. The weight vectors $\mathbf{w}_C$ and $\mathbf{w}_U$ are then applied to the respective censored and uncensored snapshots that constitute the set of CUTs. This procedure is denoted as the two-weight method (TWM) for concurrent block processing (CBP) and is found to make a significant impact on the performance of a cell-averaging constant false alarm rate (CA-CFAR) detector.

7.2.2.2 CFAR Detector In the traditional one-weight sense for SWP, a CA-CFAR detector is applied to each individual CUT by estimating a detection threshold based upon an acceptable false alarm rate and the local quiescent noise power. The local quiescent noise power is estimated by averaging the linearly weighted outputs of the nearby surrounding range cells. In this manner, a unique CA-CFAR detection threshold is computed for each range cell.

In contrast, for the TWM and CBP of FRACTA, the local quiescent noise power is estimated by averaging the weighted outputs from the UTD. As such, a single threshold is determined for all the local CUTs. The set of CUT values to be compared with the CA-CFAR threshold are the linearly weighted outputs from the CTD set.

The use of TWM results in targets standing out dramatically from the suppressed noise and interference, thus greatly improving the detectability of small target returns. Care must be taken, however, because censored cells that do not contain targets can stand out somewhat as well. For this reason, in addition to a CA-CFAR detector, the ACE metric is also used as a measure of target detection.

7.2.2.3 ACE Detector As a further measure of detectability, the ACE metric is used. For a given range index k, the ACE metric is defined as [12]

$$\mathrm{ACE}(k) = \frac{|\mathbf{s}^H \tilde{\mathbf{R}}_k^{-1} \mathbf{z}_k|^2}{(\mathbf{s}^H \tilde{\mathbf{R}}_k^{-1} \mathbf{s})(\mathbf{z}^H \tilde{\mathbf{R}}_k^{-1} \mathbf{z}_k)}, \tag{7.8}$$

where $\tilde{\mathbf{R}}_k$ is the estimated covariance matrix for the kth range index (according to the TWM). The ACE test statistic is often used in cascade with the CFAR test statistic in order to determine target detectability. The ACE screens out detections that are not closely aligned with the desired steering vector $\mathbf{s}$, such as undernulled clutter or targets that reside in the antenna/Doppler filter sidelobes. In essence, it is a signal correlation parameter that is normalized between 0 and 1.

The use of the ACE test statistic is highly motivated by the two-weight CBP adaptive censoring algorithm previously discussed. It was found that the APR of the range indices associated with the censored range indices (the ones that use the CTD to calculate the adaptive weight) was noticeably higher than the uncensored range indices (the ones that use the ITD to calculate the adaptive weight). Thus, in some cases, a typical CA-CFAR algorithm that averages the power of the cells about a given CUT might detect censored range indices that do not contain targets of interest (i.e. are not closely aligned with the desired steering vector). Hence, a secondary test such as ACE is necessary in order to screen these detections. Note that the CA-CFAR and ACE thresholds will have to be set accordingly in order to achieve a given probability of false alarm. In Section 7.4, a Doppler-sensitive approach to setting the ACE threshold is discussed.

A flowchart of the FRACTA meta-algorithm is depicted in Fig. 7.2. FRACTA operates as follows. APR censoring is applied to a set of ITD snapshots $\mathbf{z}_1, \mathbf{z}_2, \ldots, \mathbf{z}_K$,

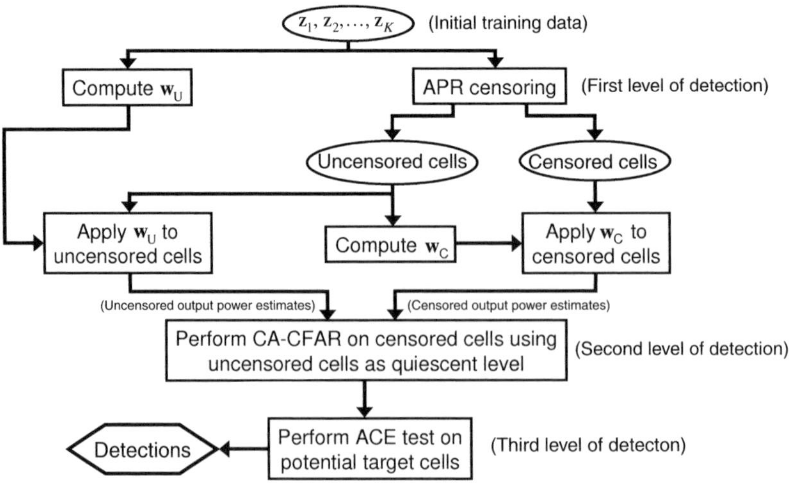

Figure 7.2 Operation of FRACTA meta-algorithm.

such that the ITD is divided into CTD and UTD sets. The UTD is used to compute the adaptive weight vector $\mathbf{w}_C$, which is applied to the CTD to generate the censored output power residues. The ITD is used to compute the adaptive weight vector $\mathbf{w}_U$, which is applied to the UTD to generate the uncensored output power residues. The censored output power residues are considered as CUTs for the local CA-CFAR, which uses the uncensored output power residues to estimate the quiescent local noise power level. The cells passing the CA-CFAR detector are also tested with an ACE detector that compares the ACE test statistic of the respective candidate cells with some predetermined threshold. Cells that pass the ACE detector are declared detections.

7.3 PRACTICAL ASPECTS OF CENSORING

The censoring stage of FRACTA plays a pivotal role in the overall detection performance, as it is tasked with ensuring that all potential targets are excised from the ITD, both to avoid "self-cancellation" problems due to data contamination and to provide a "down-selected" set of range cells that may be considered as CUTs. We now consider some of the practical aspects of censoring related to the distribution/density of targets in range, the criterion for stopping the censoring process, and the computational cost of censoring [2].

7.3.1 Global Censoring

As previously discussed, a single pass of censoring involving the removal of a snapshot from the current set of UTD is essentially an adaptive matched filter (AMF) operation in which the covariance matrix is estimated from the remaining UTD vectors, a weight vector is computed as in Equation 7.4, and the resulting APR values for the UTD are calculated as in Equation 7.5. The APR values are ranked, the largest of which is censored (i.e. moved from the UTD to the CTD), and the process is repeated.

Based on the premise of concurrent block processing (CBP), the data block of snapshots forming the ITD is the same as that used to subsequently compute the actual AMFs $\mathbf{w}_C$ and $\mathbf{w}_U$ and thereafter perform CA-CFAR and ACE detection. This general approach can be denoted as "local censoring." Although it is typically desired that the AMF maximally suppress interference in order to enable good target detection performance, censoring only requires that the interference be suppressed sufficiently to allow cells containing targets to possess a larger APR and thus be censored. This key difference in the application of the AMF to some degree relaxes the strict requirement of stationarity of the interference for the purpose of censoring, because it is not necessary to fully suppress the interference at this stage. Therefore, the size of the data block $\tilde{K}$ for censoring can be considerably larger than that specified for AMF determination. This fact is useful because, although in general it can be assumed that the targets constitute a small portion of the total number of cells, targets may occur in dense clusters (e.g. convoys), which may cause them to potentially dominate the covariance matrix estimate $\tilde{\mathbf{R}}$ if the data block size is not sufficiently large. When the targets dominate the covariance matrix estimate, the target cells will experience relatively low APR values and hence will not be censored. By enlarging the data block for censoring, the relative proportion of cells in the data block is reduced, thereby improving the likelihood that the target cells are censored. The concept of using a censoring block size that is considerably larger than that used for AMF computation is denoted as global censoring (GC). For example, in reference 2 it is shown for the KASSPER I simulated datacube [13] that global censoring provides for a 33% increase in the number of detected targets relative to local censoring at the same false alarm rate.

Besides improving censoring capability for dense target clusters, global censoring also reduces computational complexity. This is due to the fact that censoring large blocks of data does not require that the data blocks overlap as is the case when local censoring is performed, which may necessitate as much as 50% overlap to ensure adequate sample support [1]. The lack of overlapping data blocks eliminates redundant censoring of data vectors, thus reducing computation.

7.3.2 Censoring Stopping Criterion

It was discussed previously that, for a given desired steering vector and data block, some pre-determined number of snapshots are censored. This approach may be employed if some average number of snapshots expected to be censored is known *a priori*. However, for any particular data block it is likely that either too many snapshots will be censored (thus unnecessarily reducing the sample support) or too few of the snapshots will be censored (thus resulting in missed detections). Furthermore, for ground moving target indicator (GMTI) scenarios it is often the case that most, if not all, of the targets are located very close to the clutter ridge. Hence, performing censoring in Doppler bands far from the clutter ridge may be wasted effort. However, the Doppler region far from the clutter ridge cannot simply be ignored or targets may be missed. To remedy this, as well as to minimize computation, it is desired to halt the censoring process for a given block of training data at a given Doppler when there are no more detectable targets left to censor. Besides the computational benefit, this also reduces the false alarm rate by reducing the number of nontarget cells that are considered candidate target detections to be passed on to the CA-CFAR detector and possibly even the ACE detector.

One approach found to be effective is to halt censoring upon the identification of a probe vector that is appended to the set of training data. The probe vector takes the form

$$\text{probe: } \alpha_{\rm p}\mathbf{s}, \tag{7.9}$$

where $\alpha_{\rm p}$ is a pre-determined magnitude that is set such that the probe vector is nominally detectable (say 10–15 dB above the noise floor). The APR of the probe is found to be $|\alpha_{\rm p}\mathbf{s}^H\tilde{\mathbf{R}}^{-1}\mathbf{s}|^2$, and whenever this value exceeds the APRs of all the remaining data vectors in the set of UTD, censoring is halted. This censoring stopping rule is a somewhat conservative stopping criterion, because the probe does not contain noise or interference that would increase its APR, especially near the clutter ridge. To that end, it is still useful to maintain an upper limit on the number of cells allowed to be censored for a given data block.

7.3.3 Fast Reiterative Censoring

Finally, consider the computation cost incurred by the repeated estimation and subsequent inversion of an $MN \times MN$ covariance matrix that is required for censoring. This cost can be minimized by using the loaded sample matrix inversion (LSMI) algorithm, which predicates the use of a fast rank-1 matrix update via Woodbury's identity [14]. The LSMI algorithm to estimate the covariance matrix is defined as

$$\tilde{\mathbf{R}} = \frac{1}{K}\sum_{k=1}^{K} \mathbf{z}_k \mathbf{z}_k^H + \rho \mathbf{I}_{MN}, \tag{7.10}$$

where ρ is a scalar loading factor and $\mathbf{I}_{MN}$ is the order MN identity matrix. The LSMI algorithm has been shown [15] to yield signal-to-interference ratio performance only slightly below that of FML, while requiring a substantially lower computational cost.

Most importantly, the use of LSMI enables the use of Woodbury's identity so that the full $MN \times MN$ matrix inverse at each iteration of censoring can be replaced by a rank-1 adjustment to the inverted covariance matrix estimate from the previous iteration after a snapshot has been censored. The LSMI estimate from Equation 7.10 is used as the initial covariance matrix estimate so that $\tilde{\mathbf{R}}_0 = \tilde{\mathbf{R}}$. For $\mathbf{z}_k$ the mth data vector being censored, the rank-1 matrix $\mathbf{z}_k \mathbf{z}_k^{\mathrm{H}}$ is subtracted from the current covariance matrix estimate $\tilde{\mathbf{R}}_{(m-1)}$, which is then renormalized by $(K_{(m-1)} - 1)/K_{(m-1)}$, where $K_{(m-1)}$ is the previous number of uncensored samples and $K_0 = K$ from Equation 7.10. Therefore, the rank-1 adjustment to the inverted covariance matrix estimate is accomplished as

$$\tilde{\mathbf{R}}_m^{-1} = \left(\frac{K_{(m-1)} - 1}{K_{(m-1)}}\right)\left(\tilde{\mathbf{R}}_{(m-1)}^{-1} + \frac{\left(\tilde{\mathbf{R}}_{(m-1)}^{-1}\mathbf{z}_k\right)\left(\tilde{\mathbf{R}}_{(m-1)}^{-1}\mathbf{z}_k\right)^H}{K_{(m-1)} - \mathbf{z}_k^H \tilde{\mathbf{R}}_{(m-1)}^{-1}\mathbf{z}_k}\right). \tag{7.11}$$

Compared to the iterative application of FML, which uses a singular value decomposition and requires approximately $2(MN)^2K + 4MNK^2$ computational operations at each iteration, the fast reiterative censoring (FRC) method has a computational complexity of only approximately $3(MN)^2$ at each iteration given the inverse of the covariance matrix estimate computed from the ITD.

7.3.4 FRACTA Performance

To demonstrate the performance of FRACTA, it is applied to the KASSPER (knowledge-aided sensor signal processing & expert reasoning) challenge data cube 1 [13]. The operating parameters for KASSPER 1 are given in Table 7.2.

Table 7.2 Operating parameters for KASSPER 1 data cube.

Parameter	Value
Carrier frequency	1240 MHz
Bandwidth	10 MHz
Number of pulses (M)	32
Number of array channels (N)	11
Minimum range	35 km
Maximum range	50 km
Number of range cells	1000
Pulse repetition frequency	1984 Hz
β parameter	0.923
Peak power	15 kW
Platform speed	100 m/s
Crab angle	3°

The KASSPER 1 data cube comprises simulated high-fidelity airborne radar data for the heterogeneous terrain near Olancha, CA, which consists of mountains and deserts. The incident radar return power as a function of range and Doppler is depicted in Fig. 7.3, where the clutter ridge is easily visible. The average clutter power incident upon a single antenna element is found to be 37 dB above the noise and has a dynamic range of roughly 28 dB. The peaks and troughs in range are believed to be the results of shadowing due to mountains. The simulated airborne radar is flying at 3000 m altitude at 100 m/s traveling due East (270° measured from true North) with a 3° crab angle. The radar is operating at 1240 MHz with a peak

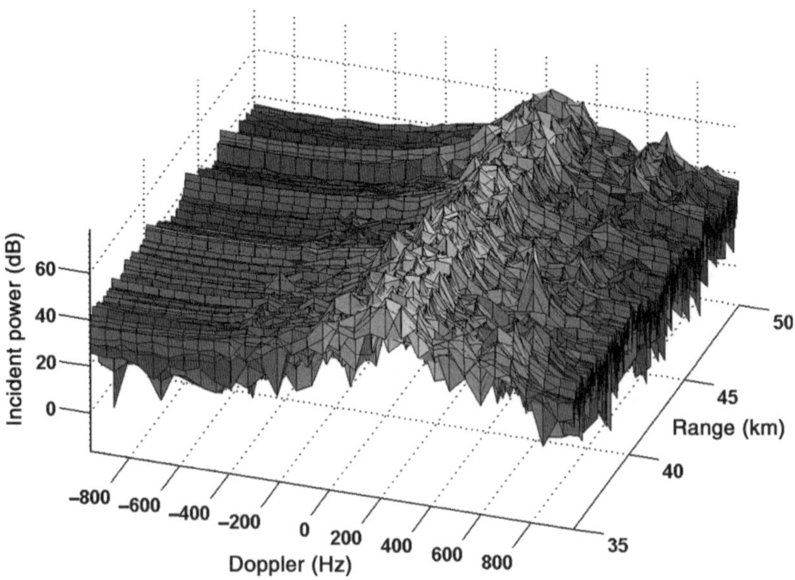

Figure 7.3 Incident power versus range and Doppler for KASSPER 1.

power of 15 kW. The 11 (virtual) antenna array elements are spaced slightly less than a half-wavelength apart at 0.1092 m (0.9028 half-wavelength spacing), and the antenna boresight is pointed at 177° with a 5° depression angle. The pulse repetition frequency (PRF) is 1984 Hz and the CPI contained 32 pulses. There are 1000 range cells of generated data, covering 35 km to 50 km. Finally, the KASSPER 1 data cube is found to possess a sample covariance matrix with roughly 50 dominant eigenvalues so that according to the $2J$ rule established in reference 7, the required sample support is at least 100 samples. Note that clairvoyant covariance matrices are provided for each range bin and are used to determine a benchmark for performance.

Due to range/Doppler sidelobes resulting from pulse compression and Doppler filtering, it is common for a target to spread into nearby range-Doppler cells. For this reason, it is standard procedure for a radar to cluster target detections such that a detection in a given range-Doppler cell is associated with a target that lies in a contiguous range cell or Doppler band. For the results presented here, clustering (or target detection compactification) is performed on the basis of ± 1 range cells and ± 1 Doppler bands.

For the purpose of comparison, the standard SWP approach, which utilizes no censoring, is applied to the KASSPER 1 data cube. For each CUT, three guard cells are employed on either side, with 40 cells surrounding the CUT and guard cells used in a CA-CFAR detector. The 120 cells surrounding the CUT exclusive of the guard cells are used to estimate the covariance matrix by LSMI. The estimated covariance matrix is applied to the respective space–time steering vector associated with each of the 32 Doppler frequency bands to form an AMF Doppler filter bank for each range cell that results in the output power residue (i.e. the APR metric) and the ACE metric depicted in Figs. 7.4 and 7.5, respectively. When applying a

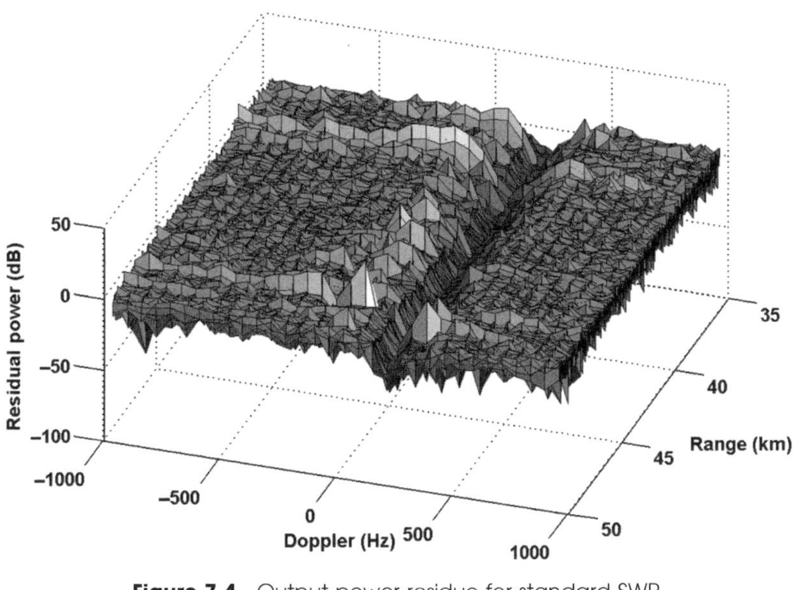

Figure 7.4 Output power residue for standard SWP.

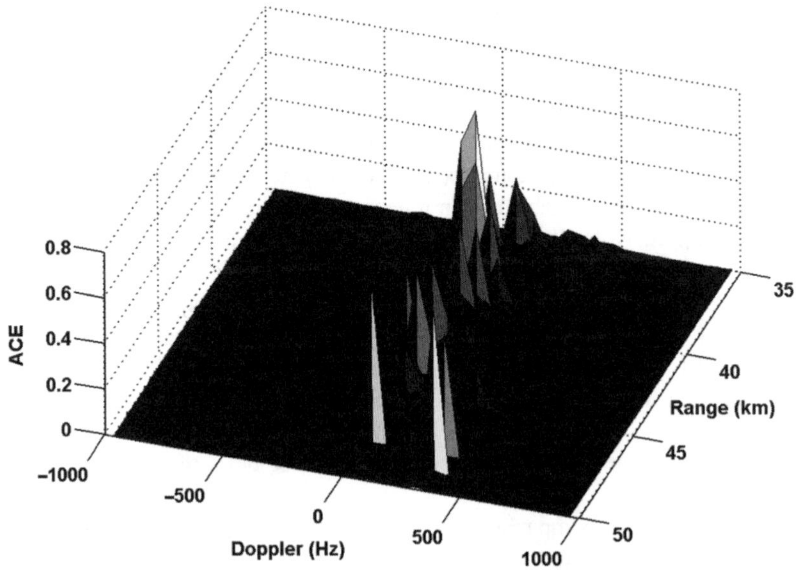

Figure 7.5 ACE for standard SWP.

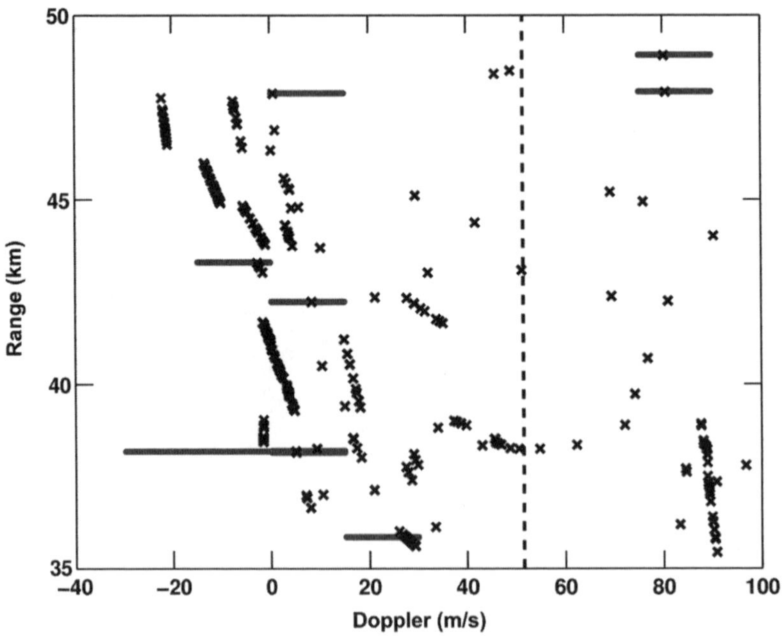

Figure 7.6 Detection map for standard SWP using CA-CFAR detector.

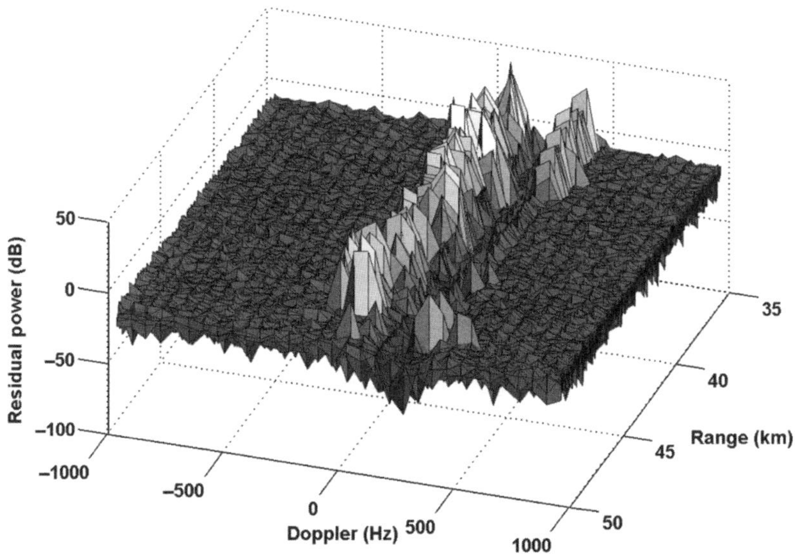

Figure 7.7 Output power residue for FRACTA.

local CA-CFAR detector to the output power residues for the SWP approach, only 8 of the 268 targets present are detected for a single false alarm. A detection map for SWP is shown in Fig. 7.6 in terms of range and Doppler where the true targets are represented by a black × and the detected range-Doppler cells by a gray bar.

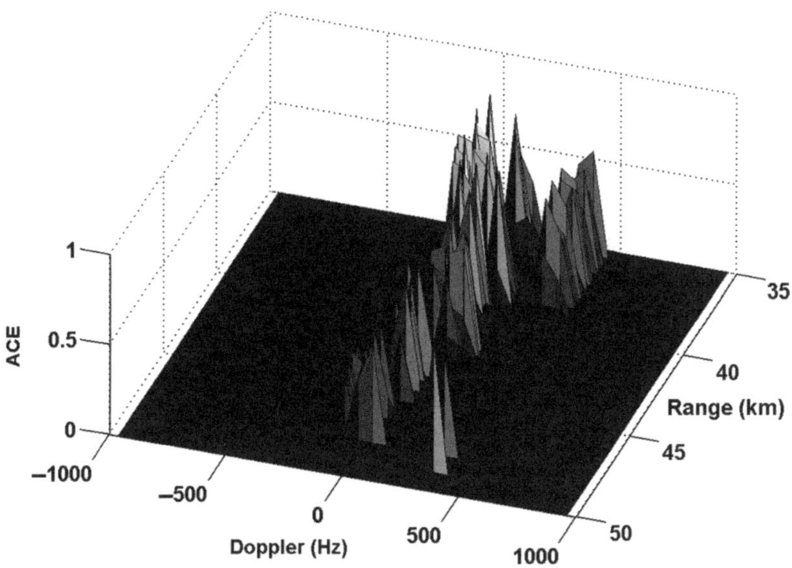

Figure 7.8 ACE for FRACTA.

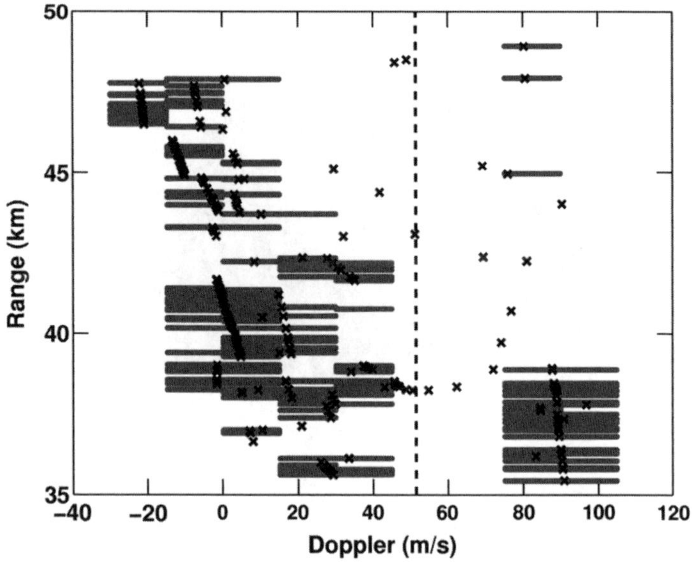

Figure 7.9 Detection map for FRACTA.

The spectral width of the detected range-Doppler cells is the Doppler filter bandwidth. The dashed vertical line near the center is the peak of the clutter ridge. The estimated false alarm probability is $P_f = 1/(32 \times 1000) = 3.125 \times 10^{-5}$. When a CA-CFAR followed by an ACE detector is used for SWP, 65 of the 268 targets are detected for a single false alarm.

In comparison, FRACTA is applied to the KASSPER 1 data cube, in which global censoring is applied to the entire set of 1000 range cells using a censoring stopping rule for each Doppler bin with a probe signal that, after integration gain, is 10 dB above the noise with the maximum number of censored cells per Doppler set to 100. For computing the AMF weight vectors, based on the CBP approach, blocks of 120 contiguous snapshots are used for covariance estimation, with the center 50 cells comprising the CUTs to which the AMF weight vectors will be applied. The output power residue (the APR metric) for FRACTA is illustrated in Fig. 7.7 along with the ACE in Fig. 7.8. FRACTA does well at locating slow-moving targets very close to the peak of the clutter ridge. Of the 32 Doppler frequency bands, only 9 contain true targets and they are all close to the clutter ridge. Using the censoring stopping rule, a total of only 4 cells were censored in all the Doppler bands not containing targets, so that essentially two-thirds of the computation requirement for censoring was eliminated. Figure 7.9 presents the true targets (black ×) along with the cells detected by FRACTA (gray bar) in terms of Doppler and range. For a single false alarm, FRACTA detects 192 of the 268 targets present. This is the same number of targets detected when using clairvoyant covariance matrices to compute the adaptive weights followed by an ACE detector.

7.4 KNOWLEDGE-AIDED FRACTA

The FRACTA meta-algorithm is robust to heterogeneous data, especially in regard to the deleterious effects of data contamination due to signals closely aligned to the desired steering vector being present in the training data. However, in general, there is only a finite amount of training data available. As such, the process of censoring may result in "sample starvation," when insufficient snapshots remain after censoring to provide an accurate covariance matrix estimate. In this section we shall consider the use of approximate prior knowledge to supplement the training data and thereby offset the deleterious effects of sample starvation. In addition, a Doppler-sensitive version of the ACE detector will be discussed that utilizes some prior knowledge of the general STAP configuration.

7.4.1 Knowledge-Aided Covariance Estimation

To compensate for the performance loss associated with low sample support, prior knowledge of the interference environment can be exploited to generate a structured covariance matrix to supplement the estimated data covariance matrix that was previously discussed. Knowledge-aided covariance estimation (KACE) uses partial approximate knowledge of the clutter covariance matrix based upon the simplified general clutter model (GCM) [9, 16]. The spatial input channels are assumed to be highly matched (or calibrated). Prior knowledge is assumed for the number of antenna elements and pulses in the coherent processing interval (CPI), the radar β parameter [17] (the number of half inter-element spacings traversed by the platform between successive pulses, which measures the slope of the clutter ridge in azimuth-Doppler), the crab angle, the clutter power, the model for the intrinsic clutter motion (i.e. Gaussian [18], Billingsley [19], etc.) and its associated parameters, the element-spacing to wavelength ratio, an antenna transmit pattern (approximate), and the look direction azimuth and depression angles. All of these are either system design parameters or are readily measurable (but not necessarily accurate). In general, at a given nominal range from the radar, ideally the KACE covariance matrix takes the form [20]

$$\tilde{\mathbf{R}}_{\mathrm{KACE}} = \sum_{l=1}^{N_c} \xi_\ell (\Gamma_\ell \circ \boldsymbol{b}_\ell \boldsymbol{b}_\ell^H) \otimes (\mathbf{a}_\ell \mathbf{a}_\ell^H) \tag{7.12}$$

where N_c is the number of independent clutter patches evenly distributed in azimuth, ξ_ℓ, Γ_ℓ, $\mathbf{b}_\ell$, and $\mathbf{a}_\ell$ are the power, intrinsic clutter covariance matrix (ICM), $M \times 1$ temporal steering vector, and $N \times 1$ spatial steering vector, respectively, of the ℓth clutter patch, which are functions of the *a priori* parameters listed previously. The operators $\circ$ and $\otimes$ are the Hadamard and Kronecker matrix product operators, respectively. The N-length temporal and M-length spatial steering vectors are defined as

$$\mathbf{a}_\ell(n) = \exp\left(j(n-1)2\pi \frac{d}{\lambda_0} \cos\theta_\ell \sin\phi_\ell \right), \quad \ell = 1, 2, \ldots, N_\text{c}, \quad n = 1, 2, \ldots, N, \tag{7.13}$$

and

$$\mathbf{b}_\ell(m) = \exp\left(j(m-1)2\pi \frac{d}{\lambda_0} \beta \cos\theta_\ell \sin(\phi_\ell + \phi_a) \right), \quad \ell = 1, 2, \ldots, N_\text{c}, \quad m = 1, 2, \ldots, M, \tag{7.14}$$

where d is the distance between adjacent antenna elements, λ_0 is the wavelength of the center frequency of the transmitted waveform, (ϕ_ℓ, θ_ℓ) are the true azimuth and depression angles of the ℓth clutter patch measured from the reference of the antenna platform, ϕ_a is the misalignment angle (or crab angle), and $\beta = 2vT/d$, in which v is the platform velocity. Following the development in references 16 and 20, the intrinsic clutter covariance matrix is modeled by a constant $M \times M$ matrix Γ using the Gaussian clutter spectral model [18], where

$$\Gamma_{q,r} = \rho_\text{c}^{(q-r)^2}, \quad q, r = 1, 2, \ldots, M, \tag{7.15}$$

where ρ_c is the clutter correlation parameter and is necessarily close to 1. Finally, known array channel mismatch effects as a function of ϕ_ℓ and θ_ℓ can be included in $\mathbf{b}_\ell(m)$, although these effects are not included in the formulation of $\tilde{\mathbf{R}}_\text{KACE}$ in Equation 7.12.

For the bald-Earth model the clutter patch powers are assumed to be uniform in azimuth and are set according to the average measured clutter power over range. Hence, $\tilde{\mathbf{R}}_\text{KACE}$ is scaled such that each of its diagonal elements is the average of the diagonal elements of the data-estimated covariance matrix $\tilde{\mathbf{R}}$. Of course, superior performance could be expected if the individual patch powers for each particular range cell were available via some *a priori* knowledge database. However, this would necessitate the calculation of $\tilde{\mathbf{R}}_\text{KACE}$ for each individual range cell, which is quite computationally burdensome.

KACE is used in the computation of the AMF weight vectors as [16]

$$\mathbf{w} = (\tilde{\mathbf{R}} + \tilde{\mathbf{R}}_\text{KACE})^{-1}\mathbf{s}, \tag{7.16}$$

such that the effective covariance matrix comprises clutter covariance information both measured directly from the environment in the form of the data vector samples and estimated *a priori* information via KACE. In particular, KACE is used to supplement FRACTA under the condition of low sample support. Thus $\tilde{\mathbf{R}}_\text{KACE}$ acts as a colored loading term similar to that used in references 21 and 22, and provides an approximation to the information contained in the data-estimated clutter covariance matrix $\tilde{\mathbf{R}}$ thereby enabling sample support lower than the usual $2J$ samples (for J sources of interference) required by standard rank reduction techniques [8, 9, 23–26]. Note that the KACE covariance matrix in Equation 7.16 generally has a loading factor associated with it. For this work, the loading factor is unity.

The *a priori* clutter covariance matrix $\tilde{\mathbf{R}}_{\mathrm{KACE}}$ resulting from the bald-Earth approximation and Gaussian clutter spectral model for the ICM is by definition slightly mismatched to the true clutter covariance matrix $\tilde{\mathbf{R}}$. In addition, radar platform parameters such as crab angle and platform velocity are estimates and thus may contain errors. However, as discussed in reference 16, there are still salient features of the true clutter covariance matrix that are incorporated into $\tilde{\mathbf{R}}_{\mathrm{KACE}}$. Furthermore, the part of $\tilde{\mathbf{R}}$ that is not known or properly incorporated into $\tilde{\mathbf{R}}_{\mathrm{KACE}}$ is contained partially in the data-estimated covariance matrix $\tilde{\mathbf{R}}$. The use of $\tilde{\mathbf{R}}_{\mathrm{KACE}}$ thus provides a means of (at least approximately) estimating the true covariance matrix $\tilde{\mathbf{R}}$ when too few samples are available due to clutter heterogeneity.

7.4.2 Doppler-Sensitive ACE Detector

The ACE detector, which is the last detection stage of FRACTA, has been found to be effective at discriminating true targets from large returns associated with space–time sidelobes, thereby minimizing false alarms [1, 12]. In general, the ACE values close to the clutter ridge in Doppler tend to be larger than those farther away, due to the possible presence of undernulled clutter. As the sample support for AMF weight vector estimation decreases, the amount of undernulled clutter subsequently increases. Hence, it would be prudent to use an ACE threshold that is Doppler dependent and increases according to the proximity to the clutter ridge in order to improve the robustness to undernulled clutter.

It is known that for side-mounted antenna arrays and small crab angle, the Doppler frequency of the clutter ridge for the spatial direction of interest is approximately [17]

$$f_{\mathrm{c}}(\phi_{\mathrm{look}}, \theta_{\mathrm{look}}) = \frac{2v}{\lambda_0} \cos\theta_{\mathrm{look}} \sin\phi_{\mathrm{look}}, \tag{7.17}$$

where ϕ_{look} and θ_{look} are the azimuth and depression angles for the spatial look direction, respectively. At a given spatial look direction, when the clutter ridge Doppler frequency is higher than half the pulse repetition frequency (PRF), such as may occur with low PRF radar, the clutter ridge is aliased to the Doppler frequency

$$f_{\mathrm{c,alias}} = \left[\left(f_{\mathrm{c}} + \frac{\mathrm{PRF}}{2}\right) \bmod \mathrm{PRF}\right] \frac{\mathrm{PRF}}{2}, \tag{7.18}$$

where $x \bmod y$ represents the "x modulo y" operation.

For each individual Doppler bin, the ACE threshold should be set according to its proximity to the clutter ridge (taking into account the potential wrap-around in Doppler frequency) and the desired false alarm probability. Hence, a large ACE threshold should be set for Dopplers close to the clutter ridge, and the Dopplers further from the clutter ridge are allowed to have a relatively low ACE threshold, thereby maintaining sensitivity. This methodology is denoted the Doppler-sensitive ACE (DS-ACE) detector [27].

The DS-ACE methodology can be implemented in a data-dependent manner by taking a CFAR-like approach (i.e. setting the respective ACE thresholds according to the surrounding quiescent ACE levels). It is assumed that local snapshots (in terms of range) that may potentially contain a target have been previously censored such that the set of K ITD snapshots has been separated into a set of K_{C} CTD snapshots and a set of K_{U} UTD snapshots. As with Equations 7.6 and 7.7, two covariance matrix estimates are computed; $\tilde{\mathbf{R}}_{\mathrm{U}}$, the UTD covariance matrix estimate, comprising only the uncensored local snapshots, and $\tilde{\mathbf{R}}_{\mathrm{I}}$, ITD covariance matrix estimate, composed of both censored and uncensored local snapshots. The UTD covariance matrix $\tilde{\mathbf{R}}_{\mathrm{U}}$ is employed in Equation 7.8 to compute the ACE values for the block of $k_{\mathrm{C}} = 1, 2, \ldots, K_{\mathrm{C}}$ local censored snapshots denoted as $\mathbf{z}_{\mathrm{CTD},k_{\mathrm{C}}}$ for each of the $m = 1, 2, \ldots, M$ Doppler frequency bands using the associated steering vector $\mathbf{s}_m$ as

$$\mathrm{ACE}_{\mathrm{CTD}}(m, k_{\mathrm{C}}) = \frac{\left|\mathbf{s}_m^H \tilde{\mathbf{R}}_{\mathrm{U}}^{-1} \mathbf{z}_{\mathrm{CTD},k_{\mathrm{C}}}\right|^2}{\left(\mathbf{s}_m^H \tilde{\mathbf{R}}_{\mathrm{U}}^{-1} \mathbf{s}_m\right)\left(\mathbf{z}_{\mathrm{CTD},k_{\mathrm{C}}}^H \tilde{\mathbf{R}}_{\mathrm{U}}^{-1} \mathbf{z}_{\mathrm{CTD},k_{\mathrm{C}}}\right)}. \tag{7.19}$$

It is this value that is to be compared with the ACE threshold to determine if a target exists at a particular range cell and Doppler.

To determine the level of the ACE threshold, the quiescent ACE level is computed that corresponds to the H_0 hypothesis where no target is present. This is accomplished by determining the ACE values for the set of $k_{\mathrm{U}} = 1, 2, \ldots, K_{\mathrm{U}}$ local uncensored snapshots denoted as $\mathbf{z}_{\mathrm{UTD},k_{\mathrm{U}}}$ for each of the $m = 1, 2, \ldots, M$ Doppler frequency bands as

$$\mathrm{ACE}_{\mathrm{UTD}}(m, k_{\mathrm{U}}) = \frac{\left|\mathbf{s}_m^H \tilde{\mathbf{R}}_{\mathrm{I}}^{-1} \mathbf{z}_{\mathrm{UTD},k_{\mathrm{C}}}\right|^2}{\left(\mathbf{s}_m^H \tilde{\mathbf{R}}_{\mathrm{I}}^{-1} \mathbf{s}_m\right)\left(\mathbf{z}_{\mathrm{UTD},k_{\mathrm{C}}}^H \tilde{\mathbf{R}}_{\mathrm{I}}^{-1} \mathbf{z}_{\mathrm{UTD},k_{\mathrm{C}}}\right)}. \tag{7.20}$$

To set the local relative quiescent level for the M Doppler bins of the ACE detector, the ACE values from the local UTD snapshots from Equation 7.20 are averaged for each Doppler, which yields the M-length vector

$$\boldsymbol{\gamma} = \begin{bmatrix} \frac{1}{K_{\mathrm{U}}} \sum_{k_{\mathrm{U}}=1}^{K_{\mathrm{U}}} \mathrm{ACE}_{\mathrm{UTD}}(1, k_{\mathrm{U}}) \\ \vdots \\ \frac{1}{K_{\mathrm{U}}} \sum_{k_{\mathrm{U}}=1}^{K_{\mathrm{U}}} \mathrm{ACE}_{\mathrm{UTD}}(M, k_{\mathrm{U}}) \end{bmatrix}, \tag{7.21}$$

in which each element of the vector will lie between 0 and 1. As an example, Fig. 7.10 illustrates the ACE threshold values computed for CPI 22 of the KASSPER 2 data set, in which the peak of the clutter ridge in Doppler from Equations 7.17 and 7.18 is located at -414 Hz. In order to obtain the desired level of false alarm, $\boldsymbol{\gamma}$ is scaled to yield the DS-ACE threshold vector $\boldsymbol{\tau} = \alpha\boldsymbol{\gamma}$ where α is a constant scale factor (across the Doppler bins) used to control the false alarm

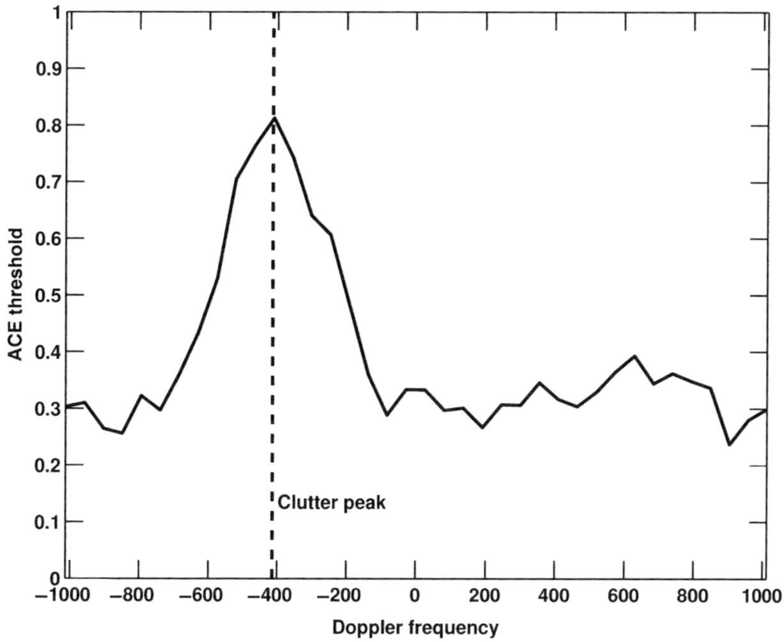

Figure 7.10 DS-ACE computed threshold for KASSPER 2 data cube.

probability and is set such that the smallest value of $\boldsymbol{\tau}$ is some minimum desired threshold level $\tau_{\min}$. Note that, to some degree, a nonadaptive approach similar to the DS-ACE is already employed in most pulse-Doppler radars, whereby returns in Doppler bins near the clutter ridge peak are simply excluded from detection. The benefit of DS-ACE therefore lies in the fact that it adaptively controls the ACE threshold at the "shoulders" of the clutter ridge outside the excluded Doppler bands, where targets are detectable yet undernulled clutter may still exist.

7.4.3 Performance of Knowledge-Aided FRACTA

As an example of the performance benefit from utilizing prior knowledge, FRACTA is now applied to the KASSPER 1 data cube and the 22nd CPI of the KASSPER 2 data set [28] (for which clairvoyant knowledge was provided).

Based upon the number of dominant eigenvalues of its corresponding covariance matrix (discussed in Section 7.3.4), the KASSPER 1 data cube requires a sample support of at least 100 in order to accurately estimate the covariance matrix, at which point FRACTA detects 192 of the 268 targets present for a single false alarm, the same as clairvoyant. However, based on the detection results for a single false alarm depicted in Fig. 7.11, when KACE is used to supplement the AMF covariance matrix, the sample support can be as low as 30 and near-clairvoyant target detection performance is maintained. Also, note that FRACTA both with and without KACE significantly outperforms the standard SWP method.

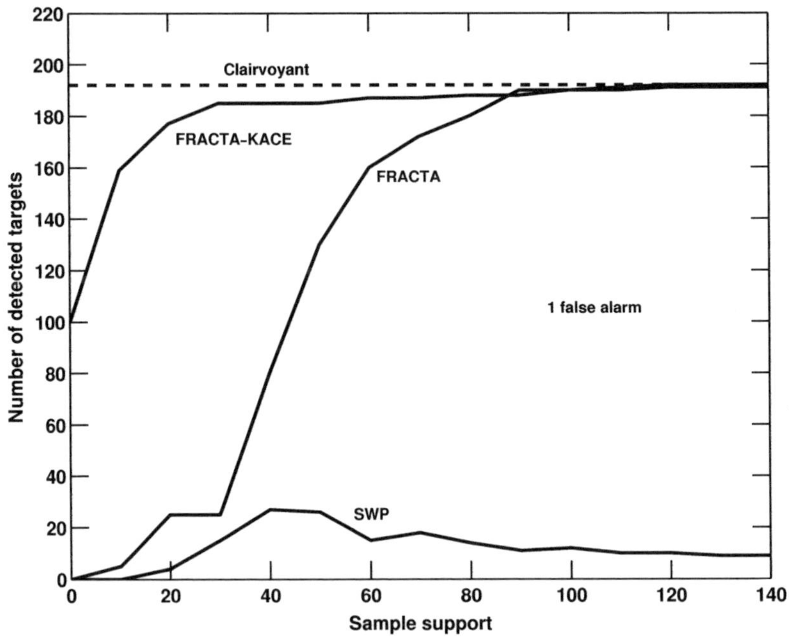

Figure 7.11 FRACTA with/without KACE applied to KASSPER 1.

The detection performance of FRACTA with and without KACE for a single false alarm when the DS-ACE detector is used is illustrated in Fig. 7.12. Although FRACTA does not quite reach clairvoyant performance (because the threshold is now estimated from the data), it does achieve a higher number of target detections for low sample support (steady state reached at $\sim$60 samples) relative to when DS-ACE was not used (Fig. 7.11). Also, supplementing FRACTA with KACE results in detection performance that is close to clairvoyant for a sample support of only 20. KACE improves performance by providing some of the deficient clutter covariance information when sample support is low. Also, the DS-ACE detector adds additional robustness to FRACTA, because it reduces the probability of false alarms being detected near the clutter ridge where undernulled clutter may exist, especially for low sample support. For a more stressing scenario where the need for both robust processing and prior knowledge is even more evident, we now consider the KASSPER 2 data set.

Relative to the KASSPER 1 data cube, the KASSPER 2 data has the added difficulty of the radar being operated at a higher frequency with approximately the same PRF, which gives rise to significant aliasing of the clutter (recall that the Doppler offset frequency is proportional to the transmitted frequency). As discussed previously, clustering of target detection is performed on the basis of ± 1 cells in range and ± 1 bands in Doppler. The incident power as a function of range-Doppler is depicted in Fig. 7.13 in which, due to Doppler aliasing, a single clutter ridge cannot be discerned. The average power incident on a single antenna element

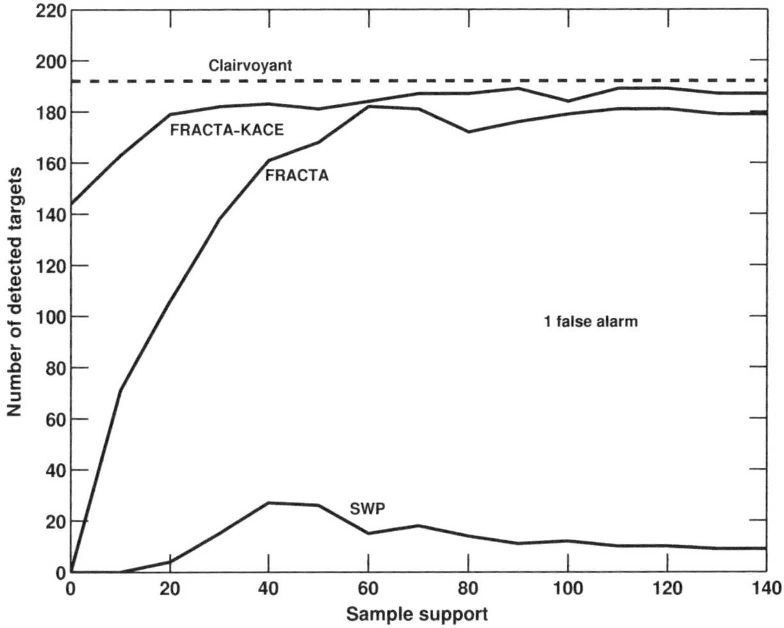

Figure 7.12 FRACTA using DS-ACE with/without KACE applied to KASSPER 1.

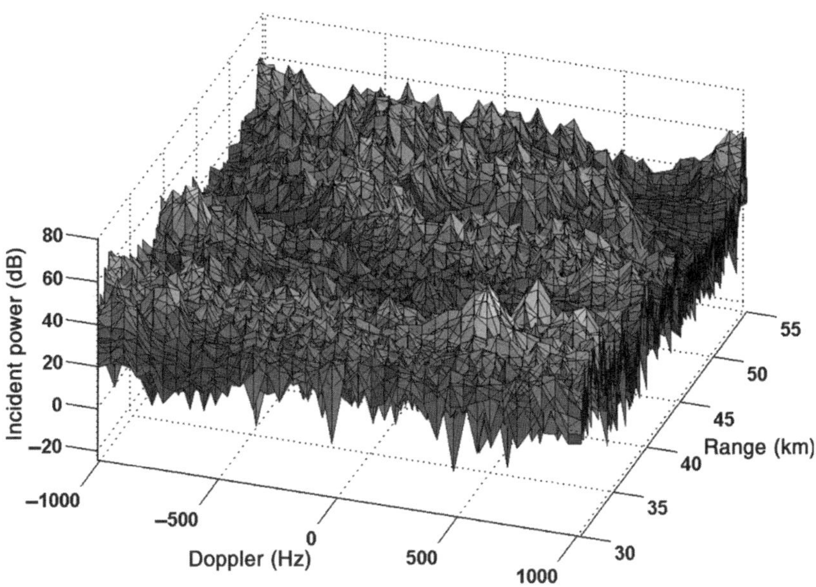

Figure 7.13 Incident power versus range and Doppler for KASSPER 2.

Table 7.3 Operating parameters for CPI 22 of KASSPER 2 data set.

Parameter	Value
Carrier frequency	10 GHz
Bandwidth	10 MHz
Number of pulses (M)	38
Number of array channels (N)	12
Minimum range	30 km
Maximum range	55 km
Number of range cells	1667
Pulse repetition frequency	2081 Hz
β parameter	9.616
Peak power	10 kW
Platform speed	150 m/s
Crab angle	3°

is 29 dB above the noise and has a dynamic range of roughly 50 dB. The simulated airborne radar is flying at 7000 m altitude, traveling at an average velocity of 150 m/s due East (270° measured from true North) with a 3° average crab angle. The 12 input channels are obtained after sub-array beamforming on 96° antenna array elements in which the antenna elements were spaced exactly one half-wavelength apart, with the antenna boresight pointed at 182° and a 5° depression angle. The operating parameters for KASSPER 2 are given in Table 7.3.

The sample covariance matrix is found to possess approximately 150 dominant eigenvalues, thus necessitating at least 300 samples to estimate accurately. The high clutter dimensionality is to some degree a result of Doppler aliasing, and is expected to cause sample starvation problems as well as yield substantially undernulled clutter after the application of the AMF weight vector. It is common practice for pulse-Doppler radar to censor a number of the Doppler bins surrounding the expected location of the clutter ridge peak (which can be estimated using Equations 7.17 and 7.18 for a given space–time steering vector), as the false alarms generated in those Doppler bins greatly outweigh the possibility of detecting targets. To that end, the returns in the Doppler bin corresponding to the peak of the clutter ridge and three Doppler bins on either side of the peak are not considered for detection. The DS-ACE detector therefore mitigates the undernulled clutter around the "shoulders" of the clutter ridge.

Using clairvoyant knowledge of the clutter covariance matrices yields 82 detected targets of the 127 targets present for a single false alarm. However, 21 of these targets lie within a 3 range cell cluster almost exactly at the peak of the clutter ridge. Therefore, eliminating the 21 targets at the clutter ridge peak because they are apparently not moving, the clairvoyant case yields 61 target detections for a single false alarm. As depicted in Fig. 7.14, FRACTA without the use of prior knowledge detects 47 targets for a single false alarm when the sample support for each data block reaches 250. When KACE is employed to supplement FRACTA, the number of

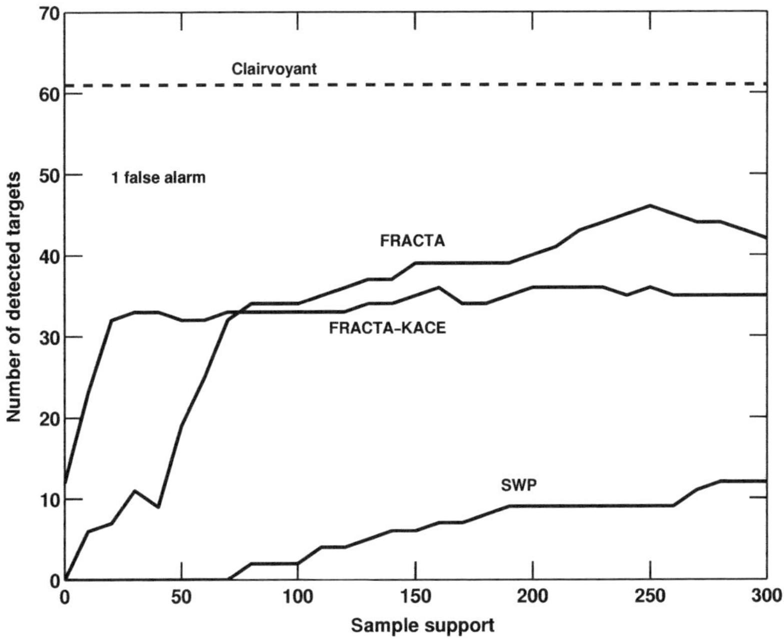

Figure 7.14 FRACTA with/without KACE applied to KASSPER 2.

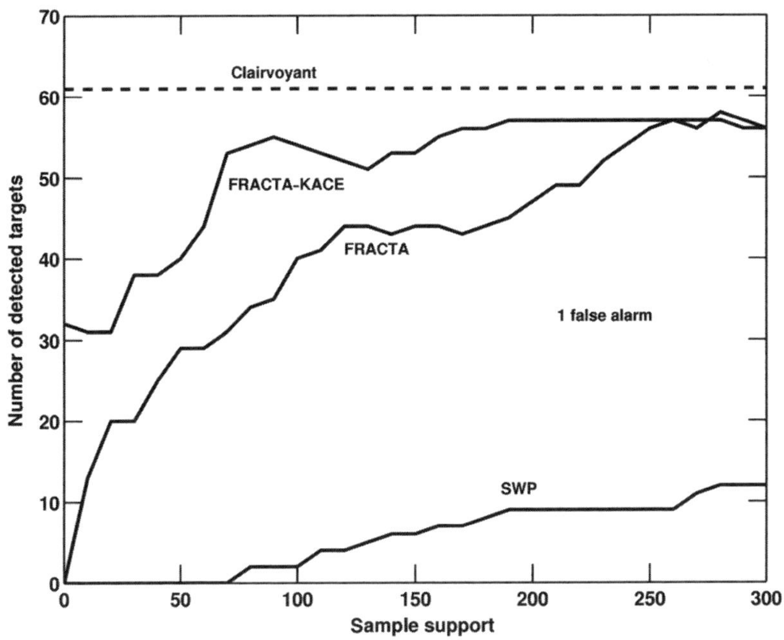

Figure 7.15 FRACTA using DS-ACE with/without KACE applied to KASSPER 2.

detections drops to 36 for the same sample support. However, KACE provides as many as 25 more target detections when the sample support is low (less than 50).

As illustrated in Fig. 7.15, using DS-ACE with FRACTA boosts the number of detections for a single false alarm to 55 when the sample support reaches 250. When using KACE as well, the same level of detection performance is attained for a sample support of 80. Hence, the use of prior knowledge enables FRACTA to achieve near-clairvoyant performance (outside of the clutter peak), even for very low sample support.

7.5 PARTIALLY ADAPTIVE FRACTA

To provide the adaptivity required for interference suppression in an airborne/space-based radar environment, fully adaptive STAP must often operate in a high dimensional space, which gives rise to two major limitations. First, the computational load required to solve the high dimensional adaptive problem is quite intense. In addition, the number of training data samples required for an accurate estimate of the interference covariance matrix can become impractical when dimensionality is high. If the sample matrix inverse (SMI) is used for weight computation, the number of training data samples required for performance within 3 dB of optimum is $K = MN$ snapshots [29]. If MN is large, K can exceed the number of available snapshots. Techniques such as LSMI [9] and FML [8] only require twice the number of dominant eigenvalues of the sample covariance matrix. For a side-looking uniform linear array, the number of dominant eigenvalues in the sample covariance matrix related to clutter can be estimated using the Brennan–Staudaher rule [30] as

$$r_c \approx [N + (M - 1)\beta], \tag{7.22}$$

where β is the number of half inter-element spacing traversed by the platform during one PRI, as discussed in Section 7.4.1. If barrage noise jamming is present, the number of dominant eigenvalues in the covariance matrix due to jamming is [17]

$$r_j = MJ, \tag{7.23}$$

where J is the number of jammers. Thus for techniques such as LSMI and FML, the number of training samples required for near-optimal performance is $K = 2(r_c + r_j)$, which can be significantly less than the $2MN$ samples required for the full sample matrix inverse. However, in many situations K is still quite large, and hence it is therefore desirable to explore dimensionality reduction techniques that maintain the performance of fully adaptive STAP.

Within the context of FRACTA, the dimension of the input data is reduced by projecting the input data vectors to a lower dimensional subspace via a linear transformation prior to adaptive processing, resulting in a partially adaptive implementation of FRACTA. We consider transforms that are defined *a priori* and are independent of the data. Reduced-rank STAP approaches utilizing data-dependent transformations [31] are not considered. The data transformation typically takes the form of either spatial or temporal filtering [17]. In this section, two well-known temporal filtering transformations (where the data is passed through a bank of Doppler filters prior to adaptation) are considered for use with FRACTA. PRI-Staggered [32] and Adjacent-Bin [33]. In addition, the inclusion of *a priori* clutter knowledge according to Section 7.4.1. is addressed for the reduced-dimension setting. For completeness, the following briefly summarizes the pertinent information from Ward [17] on reduced-dimension STAP and multiwindow post-Doppler STAP.

7.5.1 Reduced-Dimension STAP

Due to the computational complexity of fully adaptive STAP and the sample support required to provide an accurate estimate of the full $MN \times MN$ covariance matrix, reduced-dimensionality techniques are advantageous. These techniques separate the large problem into multiple smaller problems, which are more manageable. The dimensionality of the input data is reduced prior to adaptation by projecting the MN-dimensional input snapshots into a lower P-dimensional subspace via a linear transformation, where $P < MN$. The $P \times 1$ transformed snapshot $\hat{\mathbf{z}}$ is found by

$$\hat{\mathbf{z}} = \mathbf{T}^H \mathbf{z}, \tag{7.24}$$

where $\mathbf{T}$ is the $MN \times P$ transformation matrix and $\mathbf{z}$ is an $MN \times 1$ input snapshot. The $P \times P$ covariance matrix of the transformed data is

$$\hat{\mathbf{R}} = E\{\hat{\mathbf{z}}\hat{\mathbf{z}}^H\} = \mathbf{T}^H \mathbf{R} \mathbf{T}. \tag{7.25}$$

The transformed $P \times 1$ weight vector is therefore given by

$$\hat{\mathbf{w}} = \hat{\mathbf{R}}^{-1} \hat{\mathbf{s}}, \tag{7.26}$$

where $\hat{\mathbf{s}}$ is the transformed space–time steering vector

$$\hat{\mathbf{s}} = \mathbf{T}^H \mathbf{s}. \tag{7.27}$$

The transformation $\mathbf{T}$ should be designed such that $P \ll MN$ is as small as possible while maintaining performance close to that of the fully adaptive case. Although a variety of reduced-dimension partially adaptive STAP techniques exist (see references 17 and 31), for the sake of brevity we shall consider implementations of the multiwindow post-Doppler formulation [17]. Multiwindow post-Doppler techniques are chosen for integration into the FRACTA framework because they allow for significant dimensionality reduction while maintaining the flexibility to handle unknown jamming environments.

7.5.2 Multiwindow Post-Doppler STAP

Significant dimensionality reduction can be realized by performing Doppler filtering separately on each array element prior to adaptation. A separate spatially adaptive problem is then solved for each individual Doppler bin. This approach, known as factored post-Doppler STAP [17], reduces the MN-dimensional fully adaptive problem into M separate N-dimensional adaptive problems, resulting in a significant reduction in computational complexity. Factored post-Doppler STAP is adaptive solely in the spatial domain; interference rejection in the temporal domain is performed nonadaptively by the Doppler filter banks. Full adaptivity in the spatial domain endows post-Doppler STAP techniques with the ability to provide robust performance in unknown jamming environments. However, due to the lack of temporal adaptivity, the performance of factored post-Doppler STAP is significantly degraded from that of fully adaptive STAP in stressing clutter/jamming scenarios. However, this limitation can be overcome by utilizing a small degree of temporal adaptivity.

Temporal adaptivity is incorporated into factored post-Doppler STAP by combining adaptively weighted output of multiple Doppler filters from each array element. Each Doppler filter can be viewed as a different windowing of the time-delayed pulses; hence this approach is denoted as multiwindow post-Doppler STAP [17]. The general architecture for multiwindow post-Doppler STAP is shown in Fig. 7.16. The MN-dimensional fully adaptive problem is reduced to solving M separate ND-dimensional adaptive problems, where D is the number of Dopplers included from each element. For the mth Doppler bin, the outputs of D Doppler filters for each of the N array elements are adaptively combined. This process is repeated for each of the M Doppler bins. The transformation matrix for the mth Doppler bin is given by

$$\mathbf{T}_m = \mathbf{F}_m \otimes \mathbf{I}_N, \tag{7.28}$$

where $\mathbf{I}_N$ is the $N \times N$ identity matrix and $\mathbf{F}_m$ is an $M \times D$ Doppler filter matrix whose columns form a bank of FIR filters for the mth Doppler bin. The $ND \times 1$ transformed

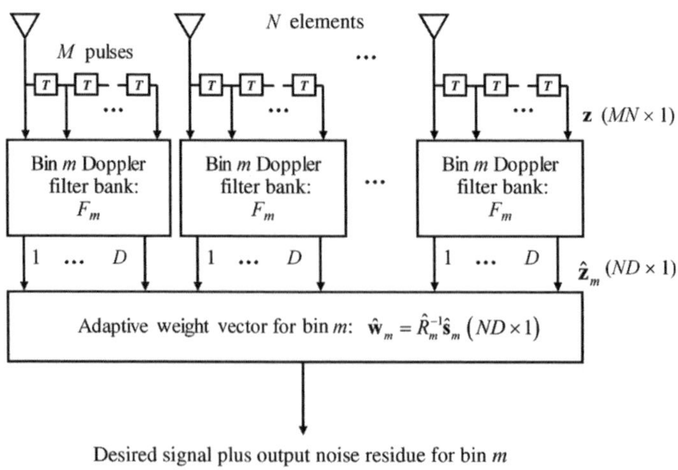

Figure 7.16 Multiwindow post-Doppler STAP for a single Doppler bin.

space–time snapshot for the mth Doppler bin is therefore

$$\hat{\mathbf{z}} = (\mathbf{F}_m \otimes \mathbf{I}_N)^H \mathbf{z}. \tag{7.29}$$

The adaptive weight vector for the mth Doppler bin is given by

$$\hat{\mathbf{w}}_m = \hat{\mathbf{R}}_m^{-1} \hat{\mathbf{s}}_m, \tag{7.30}$$

where $\hat{\mathbf{s}}_m$ is the $ND \times 1$ transformed steering vector and $\hat{\mathbf{R}}_m$ is the $ND \times ND$ transformed covariance matrix defined as

$$\hat{\mathbf{R}}_m = E\{\hat{\mathbf{z}}_m \hat{\mathbf{z}}_m^H\} = (\mathbf{F}_m \otimes \mathbf{I}_N)^H \mathbf{R} (\mathbf{F}_m \otimes \mathbf{I}_N). \tag{7.31}$$

Of course, in practice $\mathbf{R}$ is replaced by the estimated covariance matrix $\tilde{\mathbf{R}}$ as discussed in Section 7.2.2. After transformation, the number of dominant eigenvalues in the covariance matrix due to jamming is now [17]

$$\hat{r}_j = DJ. \tag{7.32}$$

If the Doppler filter matrix $\mathbf{F}_m$ is designed properly (see reference 32), the number of dominant eigenvalues in the covariance matrix due to clutter for a side-looking uniform linear array is given by the post-Doppler version of the Brennan–Staudaher rule (derived by Ward and Steinhardt [32]) as

$$\hat{r}_c = N + (D - 1)\beta. \tag{7.33}$$

For $D \ll M$, the partially adaptive $\hat{r}_c$ is much less than the fully adaptive $\hat{r}_c$ given by Equation 7.22. Hence, many fewer snapshots are required for good performance for the partially adaptive STAP than the fully adaptive. For use with FRACTA, two implementations of the Doppler filter matrix $\mathbf{F}_m$ are considered: PRI-staggered post-Doppler (developed by Brennan and Staudaher [30]) and adjacent-bin post-Doppler (developed by DiPietro [33]). These are briefly summarized in the following.

7.5.2.1 PRI-Staggered Post-Doppler STAP The mth bin Doppler filter matrix $\mathbf{F}_m$ for PRI-staggered post-Doppler consists of a bank of D identical-length $\hat{M}$ Doppler filters tuned to the frequency of bin m, each of which process a length $\hat{M}$ sub-CPI, where $\hat{M} = M - D + 1$ [32]. For example, if the CPI consists of $M = 30$ pulses and $D = 2$ Doppler filters, then the first Doppler filter would process pulses $1, \ldots, 29$ and the second Doppler filter would process pulses $2, \ldots, 30$. Let $\hat{\mathbf{U}} = [\hat{\mathbf{u}}_1 \hat{\mathbf{u}}_2 \cdots \hat{\mathbf{u}}_M]$ be the $\hat{M} \times M$ matrix formed from the first $\hat{M}$ rows of the $M \times M$ discrete Fourier transform (DFT) matrix and $\hat{\mathbf{t}}_f$ be an $\hat{M} \times 1$ Doppler filter taper. The $\hat{M} \times 1$ Doppler filter $\hat{\mathbf{f}}_m$ for the mth Doppler bin is therefore

$$\hat{\mathbf{f}}_m = \hat{\mathbf{t}}_f \circ \hat{\mathbf{u}}_m \tag{7.34}$$

where $\circ$ is the Hadamard product operator. The mth Doppler filter matrix $\mathbf{F}_m$ is thus the $M \times D$ Toeplitz matrix

$$\mathbf{F}_m = \begin{bmatrix} \hat{\mathbf{f}}_m & 0 & \cdots & 0 \\ 0 & \hat{\mathbf{f}}_m & & \vdots \\ \vdots & & \ddots & 0 \\ 0 & \cdots & 0 & \hat{\mathbf{f}}_m \end{bmatrix}. \tag{7.35}$$

7.5.2.2 Adjacent-Bin Post-Doppler STAP

As opposed to the previous implementation where the filtered outputs of D sub-CPIs within the same Doppler bin are combined for each element, Adjacent-bin post-Doppler adaptively combines the outputs of neighboring Doppler bins for each element [33]. As such, the full M-length CPI is processed. Let $\hat{\mathbf{U}} = [\hat{\mathbf{u}}_1 \hat{\mathbf{u}}_2 \cdots \mathbf{u}_M]$ be the full $M \times M$ DFT matrix. The mth bin M-length Doppler filter is

$$\mathbf{f}_m = \mathbf{t}_{\mathrm{f}} \circ \mathbf{u}_m, \tag{7.36}$$

where $\mathbf{t}_{\mathrm{f}}$ is an $M \times 1$ Doppler filter taper. The $M \times D$ Doppler filter matrix $\mathbf{F}_m$ is formed by selecting the D Doppler filters with indices $m - (d - 1)/2, \ldots, m + (D - 1)/2$ and thus has the form

$$\mathbf{F}_m = [\mathbf{f}_{m-(D-1)/2} \cdots \mathbf{f}_m \cdots \mathbf{f}_{m+(D-1)/2}], \tag{7.37}$$

where D is assumed to be odd. The adjacent filters are allowed to wrap around the edges of the Doppler space.

7.5.3 Multiwindow Post-Doppler FRACTA

Combining the multiwindow post-Doppler dimensionality reduction techniques with FRACTA is straightforward. Because a different transformation is performed for each Doppler bin, FRACTA must be applied separately to each of the M sets of ND-dimensional data. For each Doppler bin, the input snapshots are first transformed using Equation 7.29, where $\mathbf{F}_m$ is given by either Equation 7.35 or 7.37, depending upon whether the PRI-staggered or adjacent-bin implementation is being utilized. Censoring is performed globally, as described in Section 7.3.1, on the transformed snapshots for each of the M Dopplers. The input data covariance matrix for the mth Doppler bin is estimated using LSMI as

$$\tilde{\mathbf{R}}_m = \frac{1}{K} \sum_{k=1}^{K} \hat{\mathbf{z}}_k \hat{\mathbf{z}}_k^H + \rho (\mathbf{F}_m \otimes \mathbf{I}_N)^H \mathbf{I}_{ND} (\mathbf{F}_m \otimes \mathbf{I}_N), \tag{7.38}$$

where ρ is a scalar loading factor. Fast reiterative censoring (FRC) is utilized, followed by the remaining detection stages of a CA-CFAR detector and an ACE detector for each of the M Doppler bins.

7.5.4 Multiwindow Post-Doppler FRACTA + KACE

As with full-dimension STAP, approximate *a priori* knowledge of the clutter covariance matrix (KACE as discussed in Section 7.4.7) can also be utilized for reduced-dimension STAP. To operate in the multiwindow post-Doppler setting, the data-estimated covariance matrix $\tilde{\mathbf{R}}$ is linearly transformed as in Equation 7.31 into $\hat{\mathbf{R}}_m$ for each of the M Doppler bins, thus resulting in a reduced-dimension $ND \times ND$ transformed covariance matrix. Recall that the data-estimated covariance matrix and the KACE covariance matrix each provide some information regarding the second-order statistical structure of the interference and are combined as in Equation 7.16 in order to estimate the AMF weight vector. For the data-estimated and KACE covariance matrices to be combined for the reduced-dimension setting, the KACE covariance matrix must therefore undergo the same linear transformation as $\hat{\mathbf{R}}$. The transformed KACE matrix is thus determined as

$$\hat{\mathbf{R}}_{\text{KACE},m} = (\mathbf{F}_m \otimes \mathbf{I}_N)^H \tilde{\mathbf{R}}_{\text{KACE}} (\mathbf{F}_m \otimes \mathbf{I}_N), \tag{7.39}$$

utilizing the same $\mathbf{F}_m$ as employed for the data-estimated covariance matrix in Equation 7.31. In the same manner as in Equation 7.16, the mth transformed versions of the data-estimated covariance matrix $\hat{\mathbf{R}}_m$ and the KACE covariance matrix $\hat{\mathbf{R}}_{\text{KACE},m}$ are combined to yield the knowledge-aided reduced-dimension filter

$$\hat{\mathbf{w}}_m = (\tilde{\mathbf{R}}_m + \tilde{\mathbf{R}}_{\text{KACE},m})^{-1} \mathbf{s}_m. \tag{7.40}$$

Note that, within the context of FRACTA, regardless of whether dimensionality reduction and/or KACE are employed, each Doppler bin for a given data block will result in two weight vectors, one for the censored snapshots and one for the uncensored snapshots according to Section 7.2.2.

7.5.5 Performance of Partially Adaptive FRACTA + KACE

The performance of the partially adaptive FRACTA meta-algorithm with and without prior knowledge is evaluated through the use of the KASSPER I data cube (see Section 7.3.4). The PRI-staggered approach was found to perform best when $D = 6$ Doppler filters were utilized from each element, whereas the adjacent-bin dimensionality reduction approach performs best when $D = 7$. The number of target detections for a single false alarm when FRACTA, with and without KACE, is implemented using the PRI-staggered approach with $D = 6$ and the adjacent-bin approach with $D = 7$ are shown in Figs. 7.17 and 7.18, respectively. Without the use of KACE, both partially adaptive implementations of FRACTA achieve near-clairvoyant performance (192 detections for a single false alarm) once the sample support reaches approximately 40. This result is to be expected, as the number of dominant eigenvalues in the clutter covariance matrix predicted by Equation 7.33 is 16 for the PRI-staggered case and 17 for the adjacent-bin case, necessitating sample support greater than 32. Note that for full-dimension FRACTA (without KACE), at least 90 samples are required to achieve near-clairvoyant detection performance.

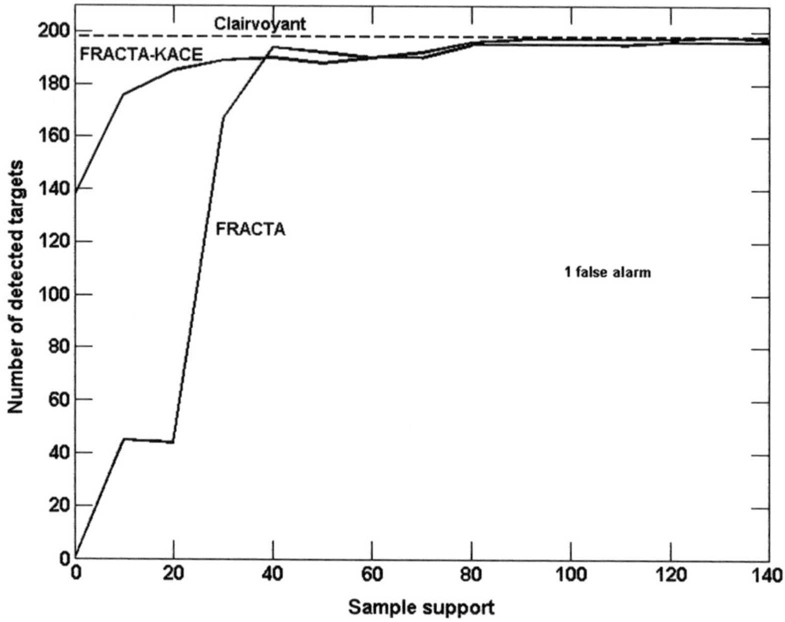

Figure 7.17 PRI-staggered ($D = 6$) FRACTA with/without KACE applied to KASSPER 1.

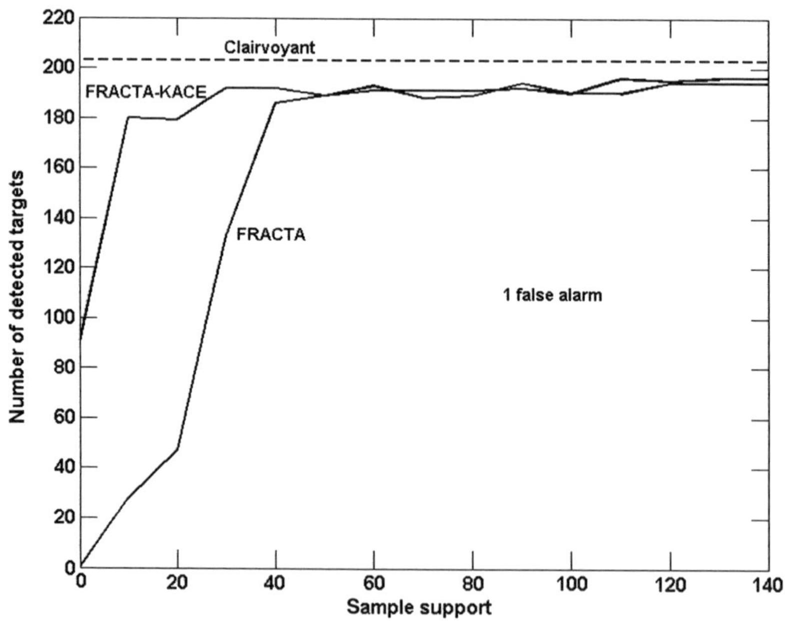

Figure 7.18 Adjacent-bin ($D = 7$) FRACTA with/without KACE applied to KASSPER 1.

When KACE is employed to supplement the covariance estimate, both reduced-dimension FRACTA implementations detect in excess of 175 targets (for a single false alarm) with a sample support of only 10. With zero sample support (i.e. nonadaptive) the PRI-staggered FRACTA implementation with KACE detects approximately 140 targets. The full-dimension and adjacent-bin implementations of FRACTA detect approximately 100 targets when using only KACE to provide the covariance estimate.

7.6 CONCLUSIONS

The FRACTA meta-algorithm is a powerful and robust tool in the arsenal of STAP techniques. It utilizes three stages of detection, which, individually, systematically identify potential targets while eliminating data contamination (censoring), detect targets within the clutter-suppressed environment (cell-averaging CFAR), and eliminate false alarms that may arise due to undernulled clutter and/or space–time filter sidelobes (ACE detector). Approximate prior knowledge in the form of knowledge-aided covariance estimation (KACE) further improves the robustness of FRACTA by supplementing interference covariance estimation in scenarios with insufficient sample support, which would otherwise lead to "sample starvation" problems. In addition, from the perspective of practical implementation, reduced-dimension FRACTA also combats sample starvation effects with the inclusion of reduced-dimension KACE, providing even further performance gains. Taken as a whole, knowledge-aided FRACTA in a reduced-dimension framework provides an effective processing mechanism for practical STAP implementation.

REFERENCES

[1] K. Gerlach, S.D. Blunt, M.L. Picciolo. Robust adaptive matched filtering using the FRACTA algorithm. *IEEE Trans. on Aerospace & Electronic Systems* 2004;40:929–945.

[2] S.D. Blunt, K. Gerlach. Efficient robust AMF using the FRACTA algorithm. *IEEE Trans. Aerospace & Electronic Systems* 2005;41:537–548.

[3] K. Gerlach, S.D. Blunt, inventors. The United States of America as represented by the Secretary of the Navy, assignee. Radar processor system and method. US patent 7,193,558. 20 March 2007.

[4] S.D. Blunt, K. Gerlach, M. Rangaswamy. STAP using knowledge-aided covariance estimation and the FRACTA algorithm. *IEEE Trans. Aerospace & Electronic Systems* 2006; 42:1043–1057.

[5] A.K. Shackelford, K. Gerlach, S.D. Blunt. Partially adaptive STAP using the FRACTA algorithm. *IEEE Trans. Aerospace & Electronic Systems* 2007 (in press).

[6] R. Horn, C. Johnson. *Matrix Analysis*. New York, NY: Cambridge University Press; 1983.

[7] L.E. Brennan, I.S. Reed. Theory of adaptive radar. *IEEE Trans. Aerospace & Electronics Systems* 1973;9:237–252.

[8] M.J. Steiner, K. Gerlach. Fast converging adaptive canceller for a structured covariance matrix. *IEEE Trans. Aerospace & Electronics Systems* 2000;36:1115–1126.

[9] B.D. Carlson. Covariance matrix estimation errors and diagonal loading in adaptive arrays. *IEEE Trans. Aerospace & Electronics Systems* 1988;24:397–401.

[10] A.M. Haimovich, Y. Bar-Ness. An eigenanalysis interference canceller. *IEEE Trans. Signal Processing* 1991;39:76–84.

[11] J.-L. Yu, C.-C. Yeh. Generalized eigenspace-based beamformers. *IEEE Trans. Signal Processing* 1995; 43:2453–2461.

[12] S. Kraut, L.L. Scharf, L.T. McWhorter. Adaptive subspace detectors. *IEEE Trans. Signal Processing* 2001;49:1–16.

[13] J.S. Bergin, P.M. Techau. High-fidelity site-specific radar simulation: KASSPER '02 workshop datacube. ISL Technical Note ISL-SCRD-TR-02-105, Vienna, VA, May 2002.

[14] T.K. Moon, W.C. Stirling. *Mathematical Methods and Algorithms for Signal Processing*. Upper Saddle River, NJ: Prentice-Hall; 2000. p. 258–261.

[15] K. Gerlach. Outlier resistant adaptive matched filters. *IEEE Trans. Aerospace & Electronic Systems* 2002;38:885–901.

[16] K. Gerlach, M.L. Picciolo. SINR improvement in airborne/spaceborne STAP using a priori platform knowledge. *Proc. IEEE Sensor Array & Multichannel Processing Conference*, August 2002, p. 548–552.

[17] J. Ward. Space–time adaptive processing for airborne radar. Technical Report ESC-TR-94-109, MIT Lincoln Lab, December 1994.

[18] D.K. Barton, editor. *Radar Clutter* (Radars vol. 5). Norwood, MA: Artech House; 1975.

[19] J.B. Billingsley. *Low-Angle Radar Land Clutter: Measurements and Empirical Models*. Norwich, NY: William Andrew Publishing; 2002.

[20] K. Gerlach, M.L. Picciolo. Airborne/spacebased radar STAP using a structured covariance matrix. *IEEE Trans. Aerospace & Electronic Systems* 2003;39:269–281.

[21] J.S. Bergin, C.M. Teixeira, P.M. Techau, J.R. Guerci. STAP with knowledge-aided data prewhitening. *Proc. IEEE Radar Conference*, April 2004, p. 26–29.

[22] J.D. Hiemstra. Colored diagonal loading. *Proc. IEEE Radar Conference*, April, 2002, p. 386–390.

[23] I.P. Kirsteins, D.W. Tufts. Adaptive detection using low rank approximations to a data matrix. *IEEE Trans. on Aerospace & Electronic Systems* 1994;30:55–67.

[24] A. Haimovich. The eigencanceler: adaptive radar by eigenanalysis methods. *IEEE Trans. Aerospace & Electronic Systems* 1996;32:532–542.

[25] J.S. Goldstein, I.S. Reed, L.L. Scharf. A multistage representation of the Wiener filter based on orthogonal projections. *IEEE Trans. Information Theory* 1998;44:2943–2959.

[26] J.R. Guerci, J.S. Goldstein, I.S. Reed. Optimal and adaptive reduced-rank STAP. *IEEE Trans. Aerospace & Electronic Systems* 2000;36:647–663.

[27] S.D. Blunt, K. Gerlach, inventors. The United States of America as represented by the Secretary of the Navy, assignee. Doppler-sensitive adaptive coherence estimate detector methods. US patent 7,212,150. 1 May 2007.

[28] J.S. Bergin, P.M. Techau, G. Chaney. High-fidelity site-specific radar simulation: KASSPER data set 2. ISL Technical note ISL-SCRD-TR-02-106, Vienna, VA, October 2002.

[29] I.S. Reed, J. Mallet, L. Brennan. Rapid convergence rate in adaptive arrays. *IEEE Trans. Aerospace & Electronic Systems* 1974;10:853–863.

[30] L.E. Brennan, F.M. Staudaher. Subclutter visibility demonstration. Technical Report RL-TR-92-21, Adaptive Sensors Incorporated, March 1992.

[31] W.L. Melvin. A STAP overview. *IEEE Aerospace & Electronic Systems Magazine* 2004; 19:19–35.

[32] J. Ward, A.O. Steinhardt. Multi-window post-Doppler space–time adaptive processing. *Proc. IEEE Workshop on Statistical Signal and Array Processing* 1994;461–464.

[33] R.C. DiPietro. Extended factored space–time processing for airborne radar. *Proc. 26th Asilomar Conference on Signals, Systems, and Computers*, October 1992, p. 425–430.

8

KNOWLEDGE-BASED RADAR TRACKING

Alessio Benavoli, Luigi Chisci, Alfonso Farina, Sandro Immediata, and Luca Timmoneri

This chapter addresses how to efficiently exploit the "knowledge base" (KB), e.g. environmental maps and characteristics of the targets, in order to gain improved performance in the tracking of multiple targets via radar measurements.

8.1 INTRODUCTION

In this chapter, attention is devoted to multitarget tracking in a surveillance region characterized by highly nonhomogeneous clutter. The objective is the development of a knowledge-based tracker (KBT) that, exploiting environmental maps as well as information on the targets' characteristics, is able to

1. reduce the number of false tracks;
2. speed up and make more efficient the initiation of true (target) tracks;
3. improve the life of true tracks.

The KBT achieves these goals by combining the information coming from the radar sensor (e.g. echoes from targets and clutter) with the contextual information (e.g. geographic, meteorological, clutter and road maps) as well as target characteristics. The architecture of the tracking system must take into account the scenario where the radar system is operating, which is characterized by the presence of multiple targets with different kinematic behaviors. To this end, the extended

Knowledge-Based Radar Detection, Tracking, and Classification. Edited by Fulvio Gini and Muralidhar Rangaswamy

Kalman filter (EKF) allows the design of a filter for each type of target that can be present in the environment, while the selection of the appropriate filter for each target can be managed by the Interacting Multiple Model (IMM) strategy [1].

Another key problem in radar tracking is data association, i.e. the assignment of plots to tracks. There exists a wide range of techniques, including NN (Nearest Neighbor), PDA (Probabilistic Data Association), JPDA (Joint Probabilistic Data Association), and MHT (Multiple Hypothesis Tracking). Although the NN algorithm is still widely used and attractive for its low computational requirements, a global optimum approach (e.g. JPDA) is certainly preferable in a multitarget scenario. As global optimum approaches are too computationally expensive, a sub-optimal technique known as NNCJPDA (NN Cheap JPDA) (2, ch. 1) is adopted in this work.

Another task to be accomplished by the tracking system is the initiation/termination of tracks, which has been implemented using the well-known M/N logic [3].

A KBT [4, 5] is realized by integrating the filtering, association and track formation procedures with the available knowledge base (KB). In particular, the KBT proposed in this chapter will accomplish this integration by

- exploiting the geographical maps in order to get a measurement, either discrete or continuous, of local clutter density for each radar cell;
- tuning some key parameters of the filtering, association and track formation algorithms based on the clutter density measurement;
- introducing appropriate strategies for the management of the critical zones, that is, zones with high clutter density;
- exploiting the targets' characteristics for target identification and the identified target type in the filtering and data association.

As a further contribution, this work will analyze the potential benefits arising from the exploitation of the target amplitude provided by the radar sensor. It is well known that the amplitude of the echo may help in discrimination among useful target returns and reflections of the environment surrounding the radar. In particular, the advantage of using the amplitude information in data association will be evaluated.

The rest of the chapter is organized as follows. Section 8.2 reviews the basic building blocks of a modern multitarget tracking system, and presents the architecture of the tracker, including the design of the filter bank and the selection of the data association logic. Sections 8.3–8.5 deal with KB tracking; specifically Section 8.3 describes processing of geographical information, Section 8.4 addresses KB target identification, and Section 8.5 discusses how to exploit amplitude information. Then, Section 8.6 presents the results obtained on simulated and live data relative to a naval surveillance radar. Finally, Section 8.7 concludes the chapter.

8.2 ARCHITECTURE OF THE TRACKING FILTER

Target tracking from radar observations is a difficult task, because of false alarms and the simultaneous presence of multiple targets with *detection probability* $P_d < 1$. At each scan, the radar provides a set $Z(k) = \{\mathbf{z}_i(k) = [r_i(k), \theta_i(k)]' : i = 1, 2, \ldots, m(k)\}$ of position measurements in polar coordinates (range and azimuth) with respect to the radar position. These measurements are employed by the tracking system to perform three main tasks [6]:

- *filtering*, that is, the update of the target state using measurements and a model of the target motion;
- *data association*, that is, the association of measurements to already established tracks;
- *track initiation*, that is, the detection of the presence of new targets in the surveillance area.

8.2.1 Filtering

The filtering task consists of estimating the kinematic variables (e.g. positions, velocities) of the target using measurements and a model of the target motion. The basic tool used for this is the Kalman filter (KF), whenever the model is linear, or its linearization around the current estimate, known as the extended Kalman filter (EKF), when the model is nonlinear. Unfortunately, it is well known [7] that a single-model filter is inadequate for tracking targets with fast maneuvering capabilities (e.g. helicopters or military aircraft). To this end, multiple model (MM) filters have been proposed in order to provide a greater flexibility in modeling different behaviors of the target. In fact, MM algorithms use a bank of filters, each of which is based on a specific model tailored to a possible target behavior (e.g. straight line motion, left turn, right turn). There exists a large variety of MM algorithms [8] depending on the *model set selection* as well as on the type of interaction between the filters; e.g. SMM = Static Multiple Model [8], GPB = Generalized Pseudo-Bayesian [9, 10], IMM = Interacting Multiple Model [1], VS-IMM = Variable Structure IMM [8]. In order to cope with frequent and sudden maneuvers of the target, the so-called Interacting Multiple Model (IMM) approach has been adopted in this work. This choice has been motivated by the observation, after extensive computer simulations, that IMM provides an optimal trade-off between on-line computational burden and tracking performance.

Let $\mathcal{M} = \{\mathcal{M}_1, \mathcal{M}_2, \ldots, \mathcal{M}_\mu\}$ denote the model set where each model $\mathcal{M}_i$ is of the form

$$\mathcal{M}_i: \begin{cases} \mathbf{x}(k+1) = \mathbf{f}_i(\mathbf{x}(k)) + \mathbf{w}_i(k) \\ \mathbf{z}(k) = \mathbf{h}(\mathbf{x}(k)) + \mathbf{v}(k), \end{cases} \tag{8.1}$$

where $\mathbf{x}$ is the target state, $\mathbf{z}$ is the target measurement, and $\mathbf{w}_i$ and $\mathbf{v}$ are the process and the measurement noise, respectively. In the IMM filter, the transition mechanism between the various models $\mathcal{M}_i$ is described by a *homogeneous Markov chain*, wherein the probabilities of transition from model $\mathcal{M}_i$ to model $\mathcal{M}_j$ are assumed constant and equal to a priori given values π_{ij}. For the sake of completeness, a full cycle of the IMM tracking filter is summarized in Table 8.1 (the reader can refer to

TABLE 8.1 One cycle of the IMM filter.

1. Re-initialization (for $j = 1, 2, \ldots, \mu$)

Predicted model probability:
$p_j(k|k-1) = \sum_{i=1}^{\mu} \pi_{ij} p_i(k-1)$

Mixing probabilities:
$p_{i|j}(k-1) = \frac{\pi_{ij}\, p_i(k-1)}{p_j(k|k-1)}$

Mixing estimate:
$\bar{x}_j(k-1|k-1) = \sum_{i=1}^{\mu} p_{i|j}(k-1)\, \hat{x}_i(k-1|k-1)$

Mixing covariance:
$\bar{P}_j(k-1|k-1) = \sum_{i=1}^{\mu} p_{i|j}(k-1)[P_i(k-1|k-1) +$
$(\bar{x}_j(k-1|k-1) - \hat{x}_i(k-1|k-1))\,(\bar{x}_j(k-1|k-1) - \hat{x}_i(k-1|k-1))']$

2. Update of the filter bank (for $j = 1, 2, \ldots, \mu$)

Model linearization:
$F_j(k) = \left[\frac{\partial f(\cdot)}{\partial x}\right]_{x=\bar{x}_j(k-1|k-1)}$ $\quad H_j(k) = \left[\frac{\partial h(\cdot)}{\partial x}\right]_{x=\hat{x}_j(k|k-1)}$

Predicted state:
$\hat{x}_j(k|k-1) = f_j(k-1, \bar{x}_j(k-1|k-1))$

Predicted covariance:
$\hat{P}_j(k|k-1) = F_j(k-1)\, \bar{P}_j(k-1|k-1)F_j'(k-1) + Q_j(k-1)$

Measurement residual:
$\tilde{z}_j(k) = z(k) - h(k, \hat{x}_j(k|k-1))$

Residual covariance:
$S_j(k) = H_j(k)\, P_j(k|k-1)H_j'(k) + R_j(k)$

Filter gain:
$K_j(k) = P_j(k|k-1)H_j'(k)S_j^{-1}(k))$

Update state:
$\hat{x}_j(k|k) = \hat{x}_j(k|k-1) + K_j(k)\tilde{z}_j(k)$

Update covariance:
$P_j(k|k) = P_j(k|k-1) - K_j(k)S_j(k)K_j'(k)$

3. Model probability update (for $j = 1, 2, \ldots, \mu$)

Model likelihood:
$L_j(k) = \mathcal{N}(\tilde{z}_j(k); 0, S_j(k))$

Model probability:
$p_j(k) = \frac{p_j(k|k-1)\, L_j(k)}{\sum_{i=1}^{\mu} p_i(k|k-1)\, L_i(k)}$

4. Estimate fusion

Overall estimate:
$\hat{x}(k|k) = \sum_{i=1}^{\mu} p_i(k)\, \hat{x}_i(k|k)$

Overall covariance:
$P(k/k) = \sum_{i=1}^{\mu} p_i(k)\, [P_i(k/k) + (\hat{x}(k|k) - \hat{x}_i(k|k))(\hat{x}(k|k) - \hat{x}_i(k|k))']$

reference 8 for details). The degrees of freedom in the IMM filter are the a priori model transition probabilities π_{ij} and, most importantly, the model set $\mathcal{M}$. In particular, the choice of the model set is of vital importance and must trade off computational load (i.e. low μ), on one hand, and target modeling flexibility (i.e. high μ) on the other. A variety of models has been proposed in the literature (e.g. the Singer model [11], constant velocity model, coordinated turn models [7]) to describe different kinematic behaviors of the target. In particular, Coordinated Turn (CT) models have been found especially useful in this work in order to provide a satisfactory description of the motion for all targets of interest (e.g. ships, helicopters, civil and military aircrafts) with a limited number of models (just three), thus implying an acceptable on-line computational burden.

Consider the target state vector $X = [x, y, v_x, v_y]'$, where x,y are the cartesian coordinates of the target position and v_x, v_y are the cartesian components of the velocity vector in the x and y directions. Further, consider the following linear state equations:

$$\begin{bmatrix} x^+ \\ y^+ \\ v_x^+ \\ v_y^+ \end{bmatrix} = \begin{bmatrix} x + \dfrac{\sin(\omega_0 T)}{\omega_0} v_x - \dfrac{1-\cos(\omega_0 T)}{\omega_0} v_y \\ y + \dfrac{1-\cos(\omega_0 T)}{\omega_0} v_x + \dfrac{\sin(\omega_0 T)}{\omega_0} v_y \\ v_x \cos(\omega_0 T) - v_y \sin(\omega_0 T) \\ v_x \sin(\omega_0 T) + v_y \cos(\omega_0 T) \end{bmatrix} + \mathbf{w}, \tag{8.2}$$

where: T is the scan period; x, y, v_x, v_y denote the state variables at scan k; x^+, y^+, v_x^+, v_y^+ are the state variables at scan $k+1$; $\mathbf{w} = \mathbf{w}(k)$ is the process disturbance, assumed zero mean and with covariance $\mathbf{Q} = \text{diag}\{0, 0, T\sigma_{v_x}^2, T\sigma_{v_y}^2\}$. Notice that in Equation 8.2 the angular speed ω has been considered as a fixed parameter ω_0. In fact, the inclusion of the angular speed ω in the state vector and its estimation are not convenient for short-duration maneuvers (e.g. military aircraft or helicopter), as there is little time for a reliable estimation of ω. For $\omega_0 = 0$, model 8.2 describes a motion with constant velocity and constant heading (straight motion). Conversely for $\omega_0 \neq 0$ it describes a maneuver (*turn*) with constant angular speed ω_0, a *left turn* ($\omega_0 > 0$) or a *right turn* ($\omega_0 < 0$) depending on the sign of ω_0. It has been found that the following three models provide an adequate and self-contained model set for our tracking purposes:

1. *Constant Velocity* (CV) model (8.2) obtained for $\omega_0 = 0$;
2. *Left Coordinated Turn* (CT_+) model (8.2) obtained for $\omega_0 > 0$;
3. *Right Coordinated Turn* (CT_-) model (8.2) obtained for $\omega_0 < 0$.

Hence, the model set for IMM has been selected as follows:

$$\mu = 3, \quad \mathcal{M}_1 = CV, \quad \mathcal{M}_2 = CT_+, \quad \mathcal{M}_3 = CT_-.$$

Notice that all three models have the same state variables; this greatly simplifies the re-initialization and fusion operations of the IMM filter. The state Equation 8.2 must be paired with the output (measurement) equation, which, obviously, is common to all the models and turns out to be

$$\mathbf{z} = \begin{bmatrix} \sqrt{x^2 + y^2} \\ \angle(x + jy) \end{bmatrix} + \mathbf{v}$$

where $\mathbf{v} = \mathbf{v}(k)$ is the measurement noise, assumed zero mean and with covariance $\mathbf{R} = \text{diag}\{\sigma_r^2, \sigma_\theta^2\}$. As the models are nonlinear in the output equations, the IMM filter requires a bank of three EKFs for the models CV, CT_+, and CT_-.

8.2.2 Data Association

Data association aims at assigning a given set of measurements (plots) to a given set of tracks. There exists a great number of data association techniques in the literature [12]. A first dichotomy is between *single-target* and *multi-target* techniques. The former proceed track-by-track in the association so that when a given track is considered, the presence of the other tracks is ignored. Conversely, a multi-target technique carries out the association procedure jointly for all tracks so that multiple (nearby) targets actually compete for the same measurements. A second important distinction is between *soft-decision* and *hard-decision* techniques. Soft-decision techniques do not select a specific measurement for a given track, but update the track's state with a combination of all measurements suitably weighted by their association probabilities. Conversely, hard-decision techniques associate to each track at most a single measurement, selected as the one maximizing the association probability. Another relevant distinction is between *single-scan* and *multi-scan* techniques. The former only consider the measurements at the present scan, but the latter also consider measurements collected over a given number of past scans in order to remedy possible previous association errors [13–15].

The choice of an appropriate data association algorithm for radar tracking in a complex scenario must take into account the following two features.

1. There can be multiple, possibly nearby, targets within the radar surveillance area.
2. The number of targets is not fixed a priori and a new target can enter the surveillance area at any time.

To cope with the above two difficulties and also to limit the computational complexity, a *multitarget hard-decision single-scan* data association algorithm is preferable. Let $Z = \{\mathbf{z}_1, \mathbf{z}_2, \ldots, \mathbf{z}_m\}$ and $\mathcal{T} = \{\tau_1, \tau_2, \ldots, \tau_n\}$ denote the set of measurements and, respectively, the set of tracks to be associated at a given scan. For data association purposes, each track is represented by a pair $\tau_j = (\hat{\mathbf{z}}_j, \mathbf{S}_j)$ where $\hat{\mathbf{z}}_j$ is the measurement prediction and $\mathbf{S}_j$ the prediction error covariance, which are both updated via a tracking filter as described in Section 8.2.1. Hence, a multitarget hard-decision single-scan association is a relation $\mathcal{A} \subseteq \{1, 2, \ldots, m\} \times \{1, 2, \ldots, n\}$ such that

- each measurement $\mathbf{z}_i$ is associated to at most one track;
- at most one measurement is associated to each track τ_j.

The specific data association algorithm adopted in this work is the so-called NNCJPDA (Nearest Neighbor Cheap Joint Probabilistic Data Association) (2, ch. 1) as it provides a good trade-off between on-line computational burden and performance in terms of low probability of wrong associations. Let $\mathcal{N}(\mathbf{z}; \hat{\mathbf{z}}, \mathbf{S}) \triangleq \det(2\pi\mathbf{S})^{-\frac{1}{2}} e^{-\frac{1}{2}(\mathbf{z}-\hat{\mathbf{z}})'\mathbf{S}^{-1}(\mathbf{z}-\hat{\mathbf{z}})}$ denote the Gaussian pdf with mean $\hat{\mathbf{z}}$ and covariance $\mathbf{S}$ and

$$\mathcal{E}(\hat{\mathbf{z}}, \mathbf{S}, \gamma) \triangleq \{\mathbf{z} : (\mathbf{z} - \hat{\mathbf{z}})'\mathbf{S}^{-1}(\mathbf{z} - \hat{\mathbf{z}}) \leq \gamma\} \tag{8.3}$$

denote the corresponding *confidence ellipsoid* with *gating probability* $P_g = \Pr(\mathbf{z} \in \mathcal{E}(\hat{\mathbf{z}}, \mathbf{S}, \gamma))$, related to $\gamma > 0$ via the χ^2 distribution. Then the NNCJPDA algorithm operates as follows.

1. For all pairs $(i,j) \in \{1, 2, \ldots, m\} \times \{1, 2, \ldots, n\}$ compute the probability β_{ij} of associating measurement $\mathbf{z}_i$ to track τ_j by

$$\beta_{ij} = \frac{e_{ij}}{\sum_{k=1}^{n} e_{ik} + \sum_{k=1}^{m} e_{kj} - e_{ij} + b} \tag{8.4}$$

   where

$$e_{ij} = \begin{cases} 0, & \mathbf{z}_i \notin \mathcal{E}(\hat{\mathbf{z}}_j, \mathbf{S}_j, \gamma) \\ \mathcal{N}(\mathbf{z}_i; \hat{\mathbf{z}}_j, \mathbf{S}_j), & \mathbf{z}_i \in \mathcal{E}(\hat{\mathbf{z}}_j, \mathbf{S}_j, \gamma) \end{cases} \tag{8.5}$$

   and $\gamma > 0$, $b > 0$ are suitably chosen parameters.
2. Set $\mathcal{L} = \{(i,j) : \beta_{ij} > 0\}$ and $\mathcal{A} = \emptyset$.
3. Repeat the following steps until $\mathcal{L}$ is empty:

$$(i^*, j^*) = \arg \max_{(i,j)\in\mathcal{L}} \beta_{ij} \tag{8.6}$$

$$\mathcal{A} = \mathcal{A} \cup (i^*, j^*) \tag{8.7}$$

$$\mathcal{L} = \mathcal{L} \backslash \{(i,j) : i = i^* \text{ or } j = j^*\} \tag{8.8}$$

Notice that at the end of the above procedure there can be tracks τ_j with no associated measurements (*missed detections*). Moreover, there can also be measurements $\mathbf{z}_i$ not associated to any track; such measurements can be passed to another set of lower priority tracks for a further data association round or can initiate new tracks (see the next section, on track initiation). A few remarks on the NNCJPDA algorithm are in order.

- Equations 8.4 and 8.5 provide a computationally cheap and approximate evaluation of the true association probabilities (CJPDA). CJPDA has been employed, in place of the exact JPDA (12, ch. 6, 16) as it implies a significantly lower computational load and also provides a satisfactory association performance.
- A null association probability is assigned by Equation 8.5 whenever the measurement $\mathbf{z}_i$ falls outside the validation region $\mathcal{E}(\hat{\mathbf{z}}_j, \mathbf{S}_j, \gamma)$ of the track τ_j.

In this way, only measurements that lie inside the ellipsoid $\mathcal{E}(\hat{\mathbf{z}}_j, \mathbf{S}_j, \gamma)$ can be assigned to the track τ_j. Clearly $\gamma > 0$, or equivalently the related *gating probability* P_g, tunes the size of the validation region and is, therefore, a parameter of paramount importance in the association algorithm. Another key parameter is $b > 0$ which, according to Equation 8.4, accounts for both the probability of missed detection and the probability of wrong association. The parameter b is related to the *false alarm probability* P_{fa} by

$$b = \frac{P_{\rm fa}}{A_{\rm rc}} \frac{1 - P_{\rm d} P_{\rm g}}{P_{\rm g}} \tag{8.9}$$

where A_{rc} is the area of the radar cell.

- If the above data association algorithm is used in conjunction with an IMM tracking filter that actually provides, for a given track, μ prediction–covariance pairs corresponding to the different models $\mathcal{M}_\ell$, a partial association probability $\beta_{ij}^{(\ell)}$ is evaluated for each model $\mathcal{M}_\ell$ and then an overall association probability β_{ij} is obtained by combining the partial association probabilities with the model probabilities p_ℓ provided by the IMM, that is

$$\beta_{ij} = \sum_{\ell=1}^{\mu} p_\ell \, \beta_{ij}^{(\ell)}$$

8.2.3 Track Initiation

The logic of initiation/promotion/termination of tracks is described by the state diagram of Fig. 8.1. Notice that *new*, *tentative*, and *confirmed* tracks have different priorities in the data association (see Fig. 8.3). In fact, confirmed tracks get, at

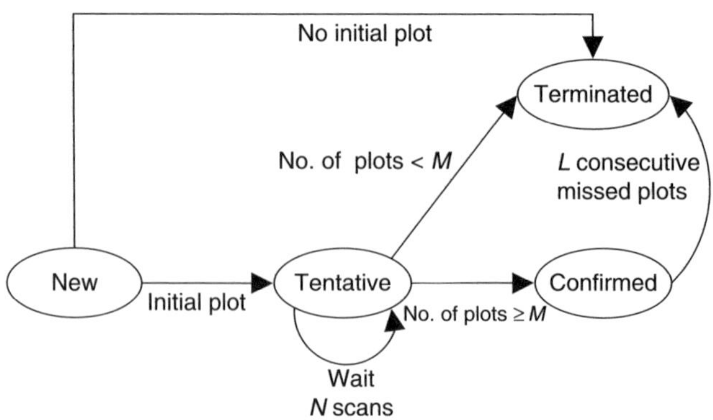

Figure 8.1 M/N logic.

each scan k, the whole set of measurements $Z(k)$, while tentative tracks get only the subset of measurements $Z_{nc}(k)$ that have been discarded by confirmed tracks and new tracks get the subset of measurements $Z_{nt}(k) \subset Z_{nc}(k)$ that have been discarded by tentative tracks. Further, let $Z_{nn}(k) \subset Z_{nt}(k)$ denote the subset of measurements that are not associated to any existing track, then any measurement in $Z_{nn}(k)$ initiates a new track at scan $k+1$. The promotion of a track from new to tentative, at scan k, requires a pair of measurements $\mathbf{z}(k-1) = [r(k-1), \theta(k-1)]' \in Z_{nn}(k-1)$ and $\mathbf{z}(k) = [r(k), \theta(k)]' \in Z_{nc}(k)$ that satisfy the following gating condition:

$$\left| r(k)\, e^{j\theta(k)} - r(k-1)\, e^{j\theta(k-1)} \right| \leq v_{\max} T, \tag{8.10}$$

where $v_{\max}$ is the maximum speed of a target and T is the scan period. Clearly, Equation 8.10 amounts to requiring that the distance of the two positions is not greater than $v_{\max}T$. A new track that fails to meet the condition of Equation 8.10 for some $\mathbf{z}(k) \in Z_{nc}(k)$ is immediately terminated. Conversely, whenever Equation 8.10 is satisfied, a tracking filter for the tentative track is initialized in order to provide, at subsequent scans $i > k$, the track prediction $\hat{z}(i) \triangleq \hat{z}(i|i-1)$ required for data association. A simpler EKF based on a (single) CV model is used for tentative tracks in place of the computationally more demanding IMM-filter adopted for confirmed tracks; notice that this implies a considerable saving of computational power as the tentative tracks are by far more numerous than the confirmed tracks. The initial state of this filter is, therefore, set equal to

$$\begin{cases} x(k) & = & r(k) \sin \theta(k) \\ y(k) & = & r(k) \cos \theta(k) \\ v_x(k) & = & \frac{x(k)-x(k-1)}{T} \\ v_y(k) & = & \frac{y(k)-y(k-1)}{T} \end{cases}. \tag{8.11}$$

A tentative track undergoes, at each scan, the data association procedure described in the previous section and the update of the associated tracking filter. The data association can have two possible outcomes for a given track: either a successful plot association or a missed plot. The promotion from tentative to confirmed track is ruled by the well known M/N logic [3, 6]. Two integer parameters M and N such that $1 \leq M < N$ are chosen; then the tentative track is confirmed if the number of plots over N consecutive scans is at least M, otherwise the track is terminated. The choice of M and N must be related to the target detection probability P_d as well as to the false alarm probability P_{fa}; for instance, reasonable values should satisfy the constraints

$$A_{vr} P_{fa} \leq \frac{M}{N} \leq P_d, \tag{8.12}$$

where A_{vr} is the average area of the validation region. A confirmed track is terminated whenever it goes through L consecutive missed detections (see Fig. 8.1).

8.3 TRACKING WITH GEOGRAPHICAL INFORMATION

A major problem in target tracking is the ambiguity of the sources of the radar echoes; in fact a radar echo can be produced either from a target (ship, aircraft, helicopter) or from clutter. Clutter echoes are mainly due to mountains, shores, buildings, but also vehicles on roads, highways, and railways. The sources of the clutter are located in specific points of the surveillance region so that the clutter density is not uniform, as shown in Fig. 8.2. Clutter sources are troublesome as they generate false measurements and this can

- cause errors in track initiation, with the consequent formation of false tracks;
- cause errors in data association, with the consequent loss of target tracks;
- prevent the termination of false tracks in the high clutter zones, wherein such tracks are very likely to be updated with clutter measurements.

To overcome these problems, the idea pursued in this work is to use the prior knowledge of the clutter distribution in order to improve the performance of the tracking system described in the previous section. In fact, since the clutter density depends on the characteristics of a given zone (e.g. sea, ground, urban zones), it is possible to use the geographical maps of the surveillance region to quantify the clutter density and hence the probability of finding a clutter echo in a given location. These maps, along with the information about the targets to be discussed later, form the so-called knowledge base *(KB)*. The target tracking algorithm that exploits the KB will be referred to as KB Tracker (KBT). Figure 8.3 shows the architecture of the

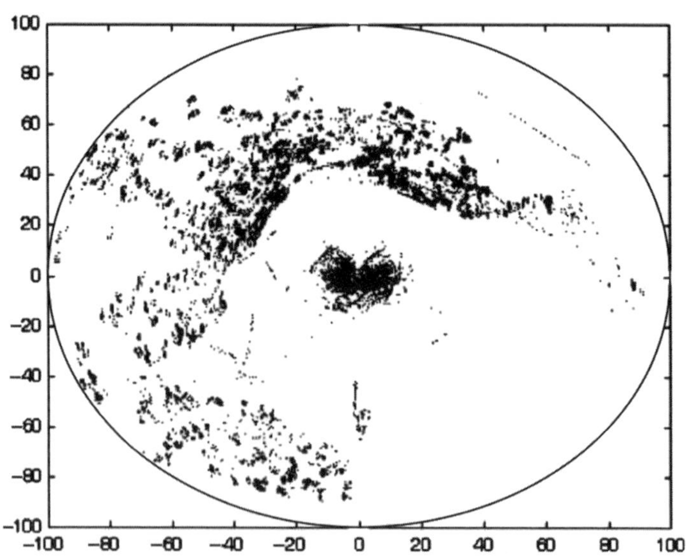

Figure 8.2 Clutter map.

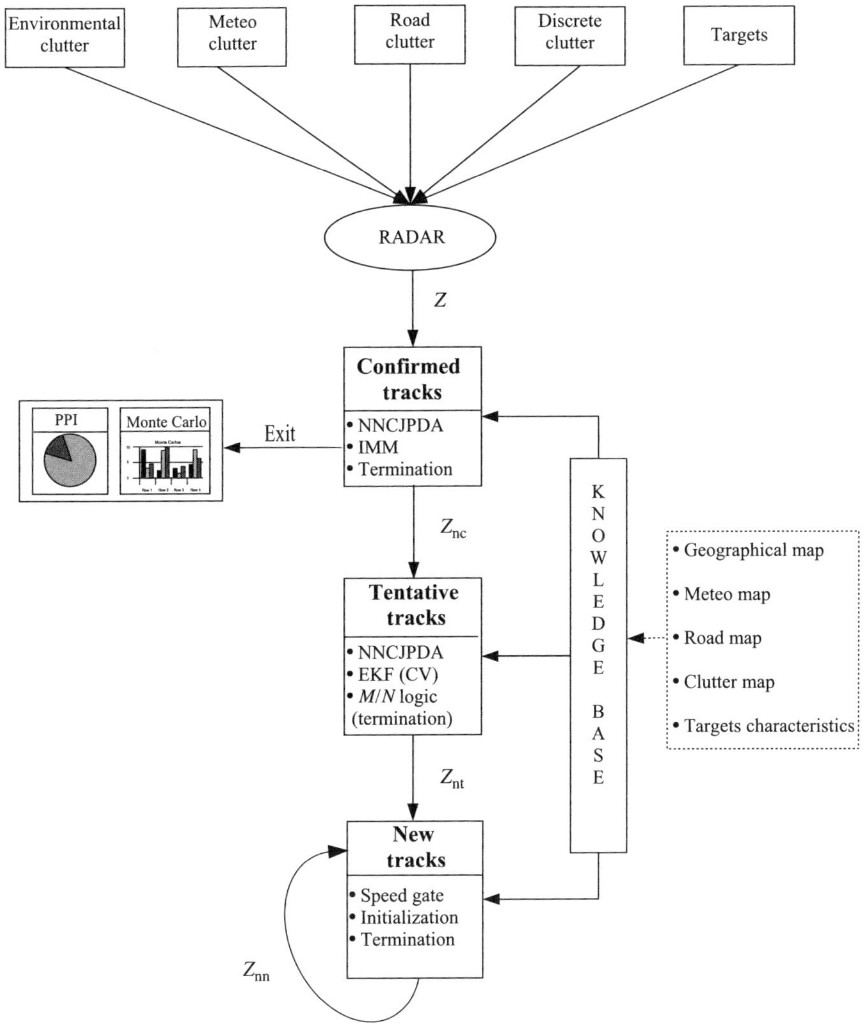

Figure 8.3 Architecture of the KB tracking system.

tracking system presented in Section 8.2 and illustrates how the KB can be included in this architecture. In particular it can be observed that

- new, tentative, and confirmed tracks have different priorities in the data association;
- tentative and confirmed tracks use two different filtering algorithms, respectively EKF and IMM;
- the KB utilizes the environmental maps and the information about the targets.

8.3.1 Processing of Geographical Maps

In this work, two types of geographical maps, that is DTED level 1 and GIS maps, have been used. In a digital format a DTED (Digital Terrain Elevation Data) level 1 map is a uniform matrix of terrain altitude values with post spacing every 3 arc seconds (approximately 100 m). A GIS (Geographical Information System) is a software technology that represents a geographic scene through attributes describing its content. The GIS (raster) data type consists of rows and columns of cells, where each cell stores a single value. The value recorded for each cell may be a discrete value, such as terrain cover, a continuous value, such as rainfall, or a null value if no data are available.

There are three main issues for implementing a KB tracking system:

1. How to process the maps in order to extract the useful information;
2. How to fuse the information provided by the single maps;
3. How to exploit the fused information in the tracking system.

The first problem has been tackled in the following way. The altitude values, provided by the DTED map, have been used to define a slope index for the (i, j) element of the DTED matrix:

$$N(i, j) = \frac{\sum_{k=-1}^{1} \sum_{h=-1}^{1} |a(i, j) - a(i-k, j-h)|}{n}, \tag{8.13}$$

where $a(i, j)$ is the altitude of the (i, j) element. Higher values of the slope index imply higher clutter density. Actually, the probability that a zone of the surveillance region generates a clutter echo depends not only on the slope index, but also on the terrain cover. For example, sea generates less echoes than land, and rural zones produce less echoes then urban zones (towns, roads, etc.). The terrain cover of a zone in the surveillance region can be deduced from the GIS map, which, therefore, helps to identify critical zones with higher clutter concentration.

The second problem, information fusion, has been addressed in two steps. First, the DTED map has been used to distinguish the shadow zones (i.e. areas behind an obstruction like a mountain that cannot be detected from the radar; see Fig. 8.4) from the illuminated zones. Each radar cell is classified either as illuminated, if it

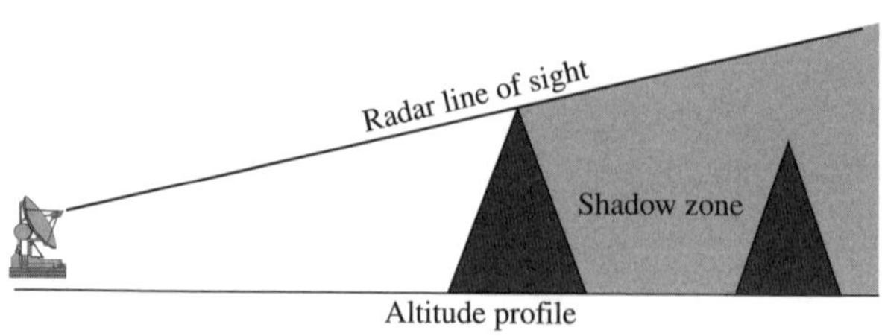

Figure 8.4 Illuminated/shadowed zones.

is along the radar line of sight, or as a shadow cell. Notice that, due to the radar wavelength, the radar line of sight is not a true line of sight, such as defined for optical instruments. To account for this, the altitude of the areas in the optical shadow zones has been increased by a given amount, for example, 20 m, before deciding if such areas are also in the radar shadow zones. Secondly, the maps (slope index, critical zones, and illuminated/shadowed zones) have been fused together and converted to a common resolution (the radar cell) to gather all available information and to classify the zones of the surveillance environment according to their capability of generating clutter echoes. Two types of classification are presented in the chapter:

- hard classification;
- fuzzy classification.

8.3.2 Hard Classification

In this technique the information is fused in a hard way, with the objective being to classify the cells of the radar surveillance region into a given number of types according to the clutter density. In particular, three types, that is,

- low clutter (LC) cells,
- medium clutter (MC) cells, and
- high clutter (HC) cells

have been considered. The classification takes into account the information gathered from the maps processing stage (i.e. slope index, critical/noncritical attribute, illuminated/shadow attribute) in the following way:

1. If the cell is shadowed, it is classified as LC;
2. Otherwise, if the cell is critical (based on the GIS map), it is classified as HC;
3. Otherwise, according to the slope index $N(i,j)$ of the cell,

    - the cell is LC if $N(i,j) < N_1$;
    - the cell is MC if $N_1 \leq N(i,j) < N_2$;
    - the cell is HC if $N(i,j) \geq N_2$;

where N_1 and N_2 are suitably chosen thresholds.

8.3.3 Fuzzy Classification

This is a soft classification method that exploits fuzzy logic in order to calculate a reliability index of each radar cell in the surveillance environment. The reliability index α is a real number ranging between zero and one. $\alpha = 0$ indicates that each echo that the radar cell produces is a clutter echo with probability 1. Conversely,

$\alpha = 1$ means that the echo is due to a target with probability 1. Intermediate values in the range [0, 1] correspond to different degrees of membership between clutter and target. In Fig. 8.5 a scheme of the fuzzy classification has been reported. From the scheme it can be seen that the fuzzy classification uses four inputs:

- slope index;
- distance from critical zones;
- distance from illuminated zones;
- distance from targets (confirmed tracks).

Notice that the above distances are Euclidean distances between the radar cell under consideration and the nearest critical cell, nearest illuminated cell, or, respectively, nearest target. Once the inputs are fuzzified, several appropriate fuzzy rules are exploited to calculate partial reliabilities of the radar cells (outputs). These outputs are then defuzzified so as to obtain the reliability index. Some possible fuzzy rules are reported below for exemplification.

- *Rule 1* If the slope is low and the distance from critical zones is high and the distance from illuminated zones is low, then the reliability is high.
- *Rule 2* If the slope is medium and the distance from critical zones is medium and the distance from illuminated zones is medium, then the reliability is medium.
- *Rule 3* If the distance from targets is low, then the reliability is high.
- *Rule 4* If the distance from illuminated zones is high then the reliability is high...
- and so on ...

8.3.4 Application of the KB to the Tracking System

Once the zones in the surveillance environment have been classified (hard or fuzzy method), one can exploit this information to improve the performance of the tracking system. The first application of the KB to the tracking algorithm consists of tuning

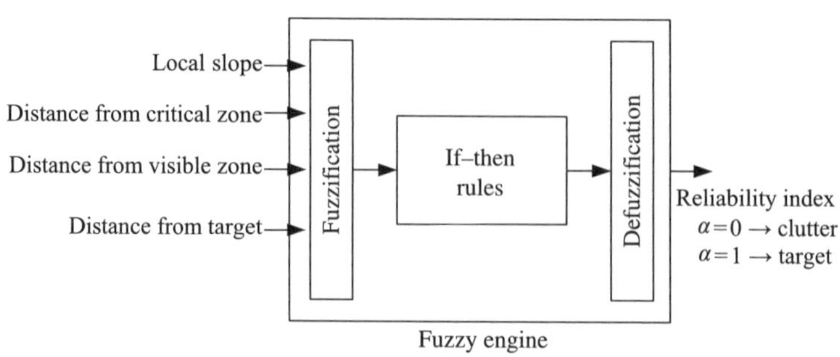

Figure 8.5 Fuzzy classification.

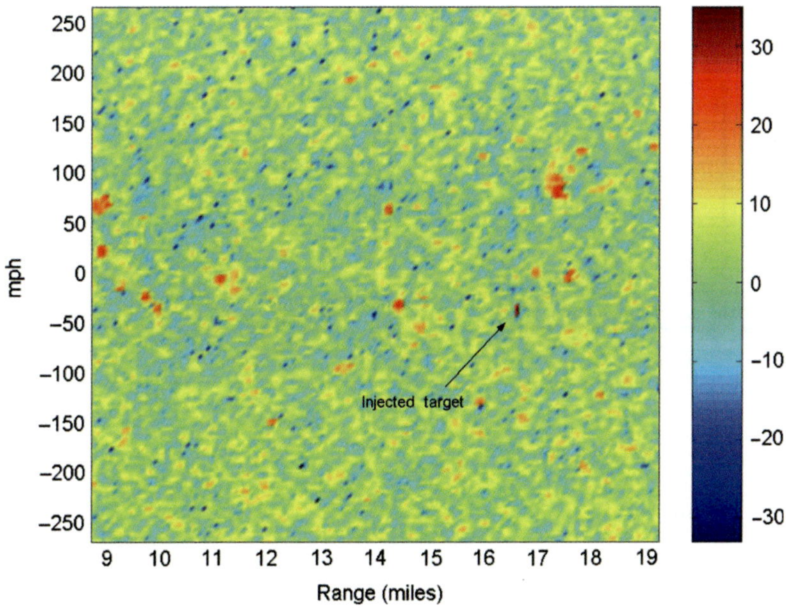

Figure 5.6 JDL processing ignoring array effects and nonhomogeneities.

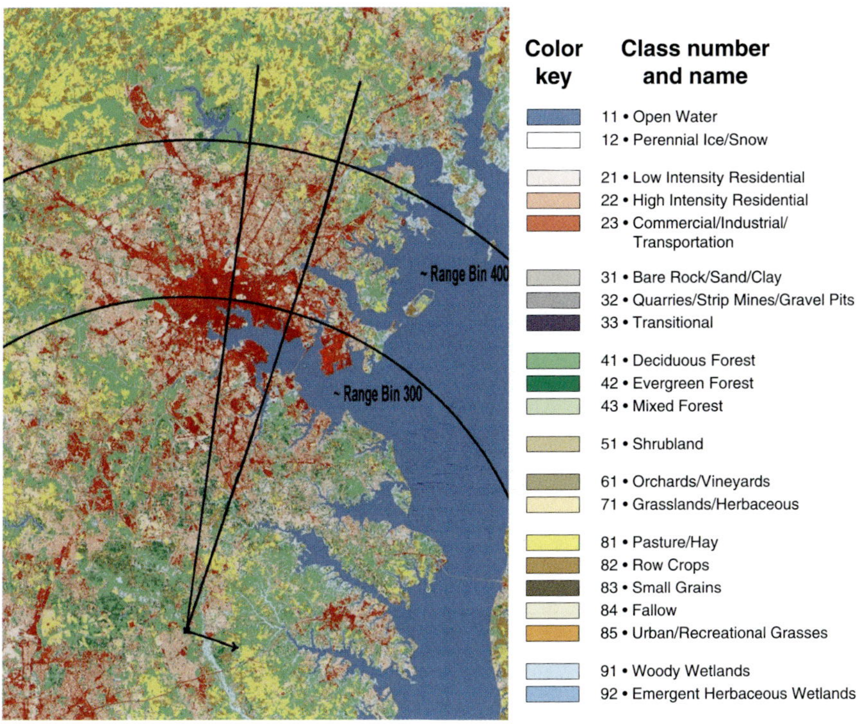

Figure 6.7 Terrain map and its legend for MCARM flight 5, acquisition 151.

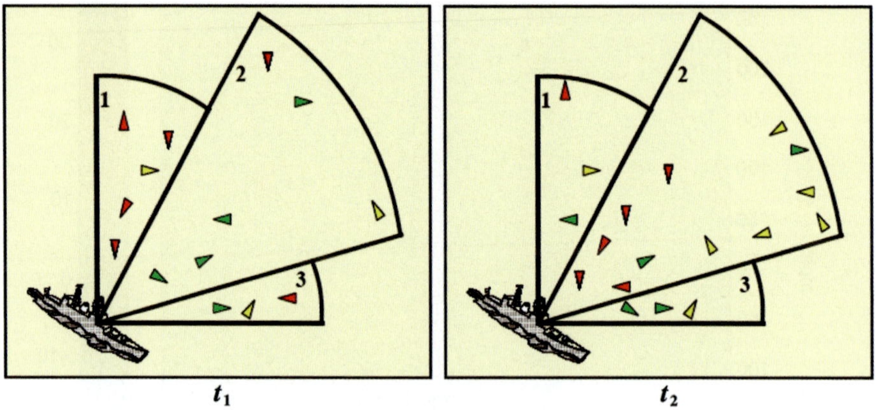

Figure 10.1 Example of a changing tactical scenario.

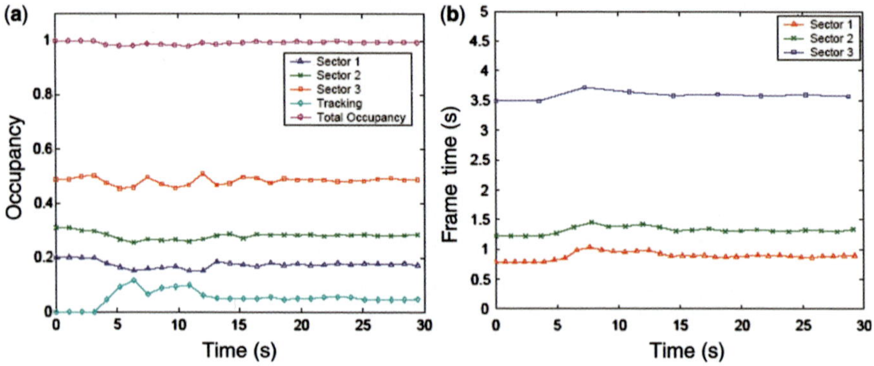

Figure 10.11 (**a**) Achieved surveillance occupancy and (**b**) frame times in an overload situation using the Orman et al. type scheduler.

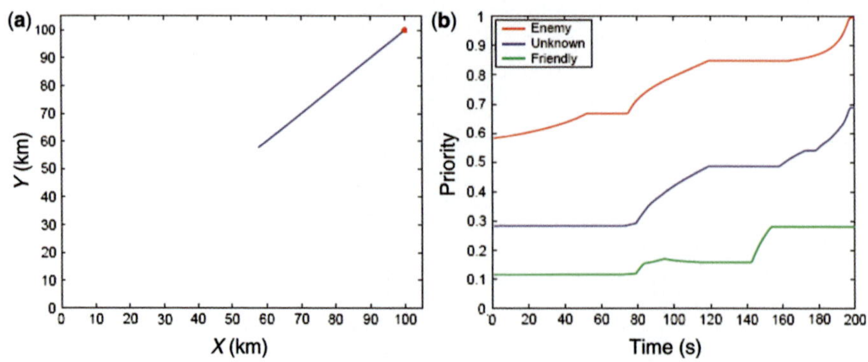

Figure 10.21 Resulting priorities for three targets with different threat levels of being hostile, moving on the same trajectory.

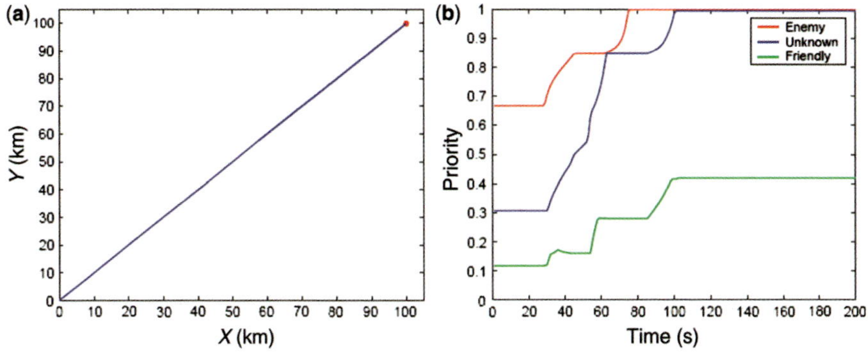

Figure 10.22 Resulting priorities for three targets with different threat levels of being hostile, moving on the same trajectory. Target velocity: 800 m/s.

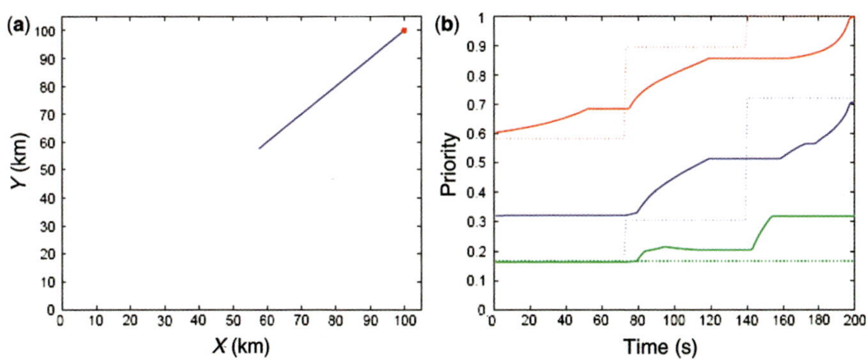

Figure 10.25 Comparison of fuzzy (solid curves) and hard logic (dotted lines) when evaluating target priorities.

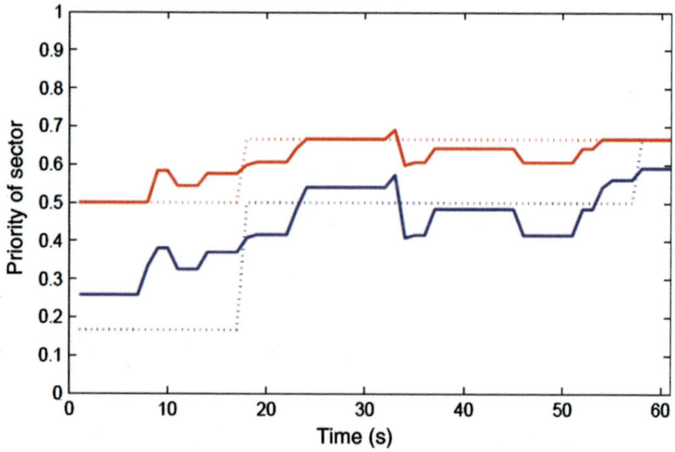

Figure 10.26 Results of the prioritization of sectors of surveillance. (Solidlines, fuzzy logic; dottedlines, hard logic.)

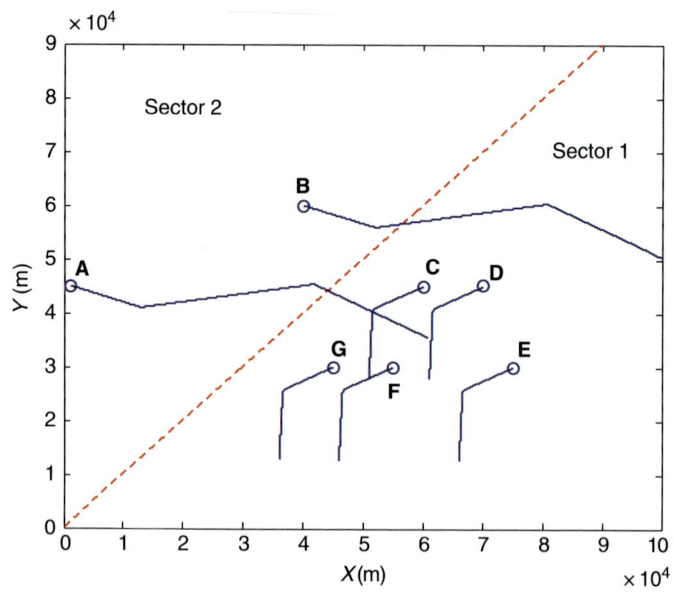

Figure 10.27 Trajectories of seven hostile targets.

some parameters of the data association and track initiation algorithms according to the classification results. In fact, as these parameters depend on the clutter density, it is natural to adapt them according to the track position in the surveillance region.

In particular, the parameters managed according to the KB are the following:

- M, N for track initiation (M/N logic),
- b for data association,
- γ for the validation gate.

These parameters — as can be seen in Equations 8.3, 8.9, and 8.12 — depend on the false alarm probability P_{fa} and on the detection probability P_{d}. The KB parameter tuning must take into account two conflicting aims:

1. to maximize the probability of true target tracking P_{TTT}, and
2. to maximize the probability of false track rejection P_{FTR}.

For example, as far as the choice of M and N is concerned, the first aim requires the selection of low values for M and N, while the second calls for high values. The designer must, therefore, select the parameters according to the specifications (P_{TTT} and P_{FTR}) on the performance of the tracking system, and also according to the location of the track in the surveillance region, which, clearly, affects the value of P_{fa} and, in turn, must affect the parameter choices. Tables 8.2 and 8.3 report the specific choices of the parameters M, N, b, γ adopted for hard classification (LC, MC, and HC cells) and, respectively, fuzzy classification; other parameters L and L_{s}, to be defined later, are also reported. Notice that for the parameters M, N, L, L_{s}, a

TABLE 8.2 Parameter values for tracking without KB and for KB hard tracking.

	NKB						DMHC						DTPHC					
	M	N	b	γ	L	L_{s}	M	N	b	γ	L	L_{s}	M	N	b	γ	L	L_{s}
LC	5	7	10^{-4}	4.6	3	—	3	4	0	9	3	—	3	4	0	9	3	—
MC	5	7	10^{-4}	4.6	3	—	4	7	10^{-5}	6	3	—	4	7	10^{-5}	6	3	—
HC	5	7	10^{-4}	4.6	3	—	—	—	—	—	8	—	6	8	10^{-3}	4.6	3	8

TABLE 8.3 Parameter values for KB fuzzy tracking.

	FKB					
	M	N	b	γ	L	L_{s}
$\alpha \geq 0.6$	3	4	0	9	3	—
$0.3 \leq \alpha < 0.6$	4	7	10^{-5}	6	3	—
$\alpha < 0.3$	6	8	10^{-3}	4.6	3	—

worst-case approach has been pursued in the tuning; that is, the parameters are tuned according to the highest clutter density (or lowest reliability index) that the track has encountered over the last scans. The location dependence of the above parameters, unfortunately, is not enough to provide good tracking performance in HC zones. For the hard approach the following further strategies have been devised:

1. Delete measurements in HC zones (DMHC),
2. Delete tracks persisting in HC zones (DTPHC).

Conversely, for the fuzzy approach, the following are adopted:

1. Delete measurements with low reliability (DMLR),
2. Use a modified data association algorithm (α-NNCJPDA).

8.3.5 Hard Classification: DMHC and DTPHC

The DMHC strategy consists of deleting all measurements located in HC zones of small size, before they are processed by the tracking algorithm. To accomplish this task, the KB is used. Unfortunately the deleted areas create artificial shadow zones and, therefore, enforce missed detections for the tracks crossing HC zones. Hence, in order to prevent unmotivated terminations of such tracks, the idea is to increase the parameter L (number of consecutive missed plots for the termination of a confirmed track) in the HC zones. Figure 8.6 depicts the two tasks of the DMHC strategy, that is,

- the elimination, from the radar measurement set Z, of the measurements located in the HC zones (via the *Measurements filter* block);
- the increase, for the tracks in the HC zones, of the parameter L.

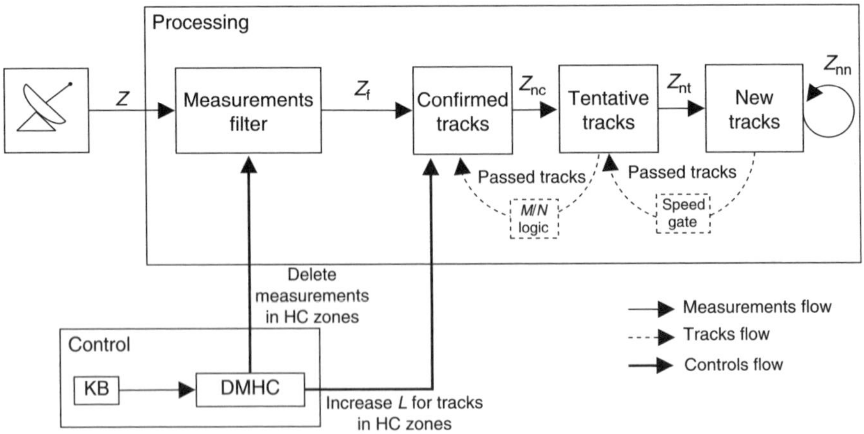

Figure 8.6 DMHC strategy.

The DTPHC strategy affects the tracking algorithm in the following three tasks:

- it prevents the initiation of new tracks in the HC zones;
- it does not confirm tentative tracks that persist for several scans in the HC zones;
- it deletes confirmed tracks that have persisted in HC zones for a sufficiently high number of consecutive scans.

A track is classified as *persistent* in an HC zone (small and large size) whenever it stays in such a zone for L_s consecutive scans. The integer L_s is, therefore, a key parameter of the DTPHC strategy. The idea underlying the DTPHC approach is that a track persisting for a long time in an HC zone is very likely to be false. For example, along a road or highway there can be many false tracks due to motor vehicles. These tracks, clearly, move along the road/highway (HC zone) and, thus, are easily recognized as false and terminated. The DTPHC strategy is illustrated in Fig. 8.7. In particular, it can be observed how this strategy manages the promotion/maintenance/termination of the tracks in the HC zones.

8.3.6 Fuzzy Classification: DMLR and α-NNCJPDA

The DMLR strategy deletes the measurements characterized by a low reliability, before they are processed by the tracking algorithm. A threshold $\alpha_{\min}$ is fixed, and all measurements located in cells with reliability index $\alpha \leq \alpha_{\min}$ are deleted. Unlike the DMHC strategy, in this case the parameter L is not increased, because the fuzzy classification also considers the tracks positions to calculate the reliability index and this prevents the strategy from creating artificial missed detections. Further,

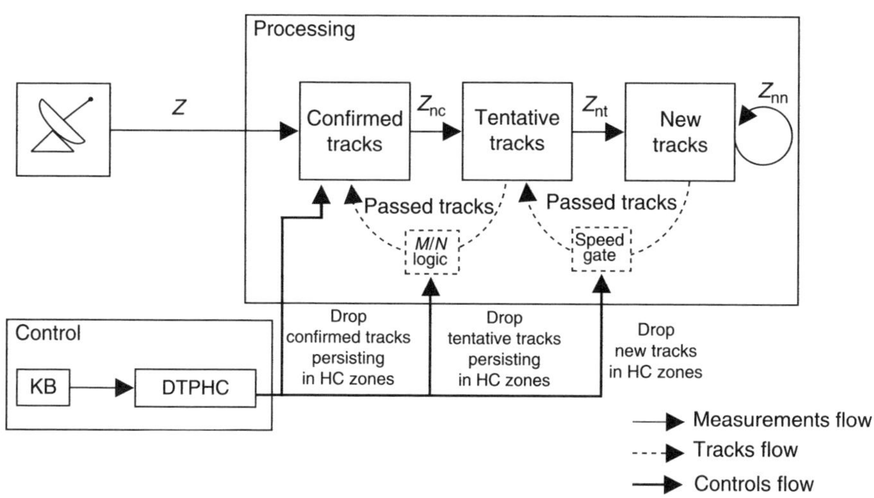

Figure 8.7 DTPHC strategy.

a new data association algorithm has been devised to take into account the reliability index. The α-NNCJPDA algorithm is a modified version of the NNCJPDA, in which the data association weight e_{ij} in Equation 8.5 is scaled by the reliability index α_i relative to the corresponding measurement. The lower is α_i, the smaller weight is given to the ith measurement. Therefore, in this case the data association weight depends on both the likelihood function and the reliability index. The latter strategy is particularly effective when the targets move near the border between zones with different reliability indices.

8.4 KNOWLEDGE-BASED TARGET ID

The identification (ID) of the target type (e.g. ship, civil aircraft, military aircraft, helicopter) is another important task of target tracking. Performing target classification can provide an improvement in the system performance. In fact, because the targets of interest have different characteristics of speed and maneuverability, it is possible to exploit such differences to fit the tracking algorithms to the target type. To identify the target, one can use the a priori information on the target characteristics as well as the environmental maps (which are part of the KB) using a set of rules. Typical rules based on the (estimated) target speed and the (estimated) target location on the maps include the following:

- *if* the target speed is $<40\,\text{km/h}$ *and* the target is located on the sea, *then* the target is a ship;
- *if* the target speed is $<40\,\text{km/h}$ *and* the target is located on the land, *then* the target is a helicopter;
- *if* the target speed is $>1200\,\text{km/h}$, *then* the target is a military aircraft;
- and so on...

The above rules are all *memoryless* and are, in most cases, insufficient to classify a target unambiguously. Hence, rules with memory can also be included in the KB such as, for instance,

- *if* the target was a helicopter *and* the target speed is now $<40\,\text{km/h}$ *and* the target is in the sea, *then* the target is a helicopter;
- *if* the target was an aircraft *and* the target speed is now $<v_e\,\text{km/h}$, *then* the target is an aircraft;
- and so on...

where v_e is the maximum speed for a helicopter. The rules with memory must have priority over memoryless rules. Clearly, the inclusion of rules with memory cannot completely avoid possible ambiguities in target classification. For instance, it does not prevent a helicopter, never classified as such, flying on the sea being still classified as a ship. This ambiguity could, for instance, be eliminated using information on the amplitude of the radar echo, as will be discussed in the next section. Also notice that the ship–helicopter ambiguity is obviously circumvented if the radar also provides elevation measurements, which is not the case considered in this work.

Once the target has been classified, the information on the target's type can be properly exploited in the various steps of tracking.

1. In the *data association* it can, for instance, prevent a track of a ship sailing along the coast from being updated with measurements on the land.
2. In the *filtering*, the knowledge of the target's type can suggest modifications to the IMM models such as

   - a change of the parameter ω_0, in the models CT_+ and CT_-, according to the type of target;
   - a change of the model set depending on the type of target [8] (e.g. a slowly maneuvering ship can be effectively tracked using a single CV model instead of the three models CV, CT_+, CT_-).

In summary, target classification is improved by the KB (maps and characteristics of the targets) and this, in turn, along with the KB can be used in data association and filtering to improve tracking performance.

8.5 TRACKING WITH AMPLITUDE INFORMATION

The radar echo wave is characterized by a power given by the *radar equation*

$$P_{\mathrm{r}} = \sigma \frac{G_{\mathrm{t}}^2 \, G_{\mathrm{r}}^2 \, P_{\mathrm{t}} \, \lambda^2 \, F_{\mathrm{t}}^2 \, F_{\mathrm{r}}^2}{64 \, \pi^3 \, R^4 \, L_{\mathrm{tot}}}, \tag{8.14}$$

where G_{t} and G_{r} are the gains of the transmitting and receiving antenna, respectively; λ is the wavelength; σ is the target's radar cross-section (RCS); R is the target range; L_{tot} represents the total losses of the radar system; F_{t} and F_{r} are the transmitter and receiver propagation factors, respectively; P_{t} and P_{r} are the transmitted and received power, respectively. The amplitude of the received signal is $A = \sqrt{2P_r}$. In this way, each measurement provided by the radar is characterized by $[r, \theta, A]$. Given two targets with the same position $[r, \theta]$, the echo's amplitude might differ because of the RCS, which depends on the shape, area, orientation, and material of the target. Echoes arise not only from the targets of interest but also from clutter; the echo amplitude can be exploited to distinguish the source of the radar echo. A simple way to use such information is to eliminate, in the data association, all measurements that are not characterized by an RCS compatible with the track under examination.

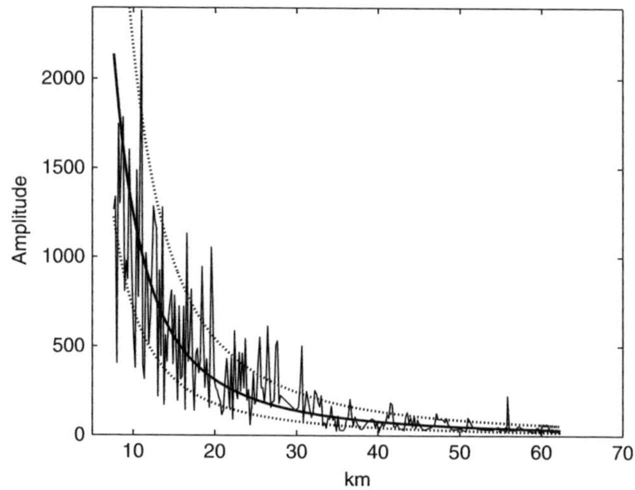

Figure 8.8 Typical plot of target echo amplitude versus target range.

In an urban environment, for instance, there exist clutter measurements having high amplitudes compared to the targets of interest (e.g. aircraft and helicopters). The elimination of such measurements, which are incompatible with the RCS of the targets, clearly reduces the probability of false alarms and, thus, favours the association. The difficulty in applying this strategy is due to the fluctuations of the RCS. In Fig. 8.8, a typical plot of the target echo amplitude versus target range is reported. This amplitude decays to zero with the range R as $1/R^2$, but exhibits large fluctuations due to target glints, surface multipath effects, and so on. Figure 8.8 also displays the ideal (average) amplitude behavior (solid line) and the 80% confidence band (dotted lines). By means of simulation experiments, it has been checked that a precise prior knowledge of the range of the RCS (e.g. 80% confidence band) of a given target allows an effective elimination of the incompatible measurements and, thus, a significant performance improvement. The main problem is, clearly, how to infer information about the range of variation of the RCS. There exist two possible alternatives for the determination of such a range:

- on-line estimation;
- off-line estimation and consequent inclusion in the KB.

The on-line estimation of the RCS of a track is a difficult task, because the RCS is fluctuating randomly and also because clutter measurements can erroneously be used in the estimation. As a consequence, a large number of measurements is required for a reliable on-line estimate of the RCS. Another approach is to include in the KB the range of the RCS for each type of target, provided that this has been suitably estimated off-line. In this case, however, the key point for a correct choice of the RCS is a reliable target classification. A possible way to overcome both difficulties is to adopt the following mixed strategy:

1. Perform classification in the first few scans in order to establish the target's class;
2. Get the RCS's *range* (i.e. the interval of RCS variations) relative to the estimated target's class from the KB;
3. Iteratively refine the RCS's range via on-line estimation.

It turns out that even a rough target classification can help in eliminating a large number of measurements with out-of-range amplitudes.

8.6 PERFORMANCE EVALUATION

In this section, simulation experiments, carried out in a realistic scenario, are discussed in order to demonstrate the superior performance of KBT systems with respect to a standard tracking system that does not exploit the KB. To evaluate tracking performance, the following metrics will be adopted:

- the percentage of successful target tracks;
- the number of false confirmed tracks;
- the number of tentative tracks.

These are natural indices for the performance of a radar tracker, as the main objectives are

- to maximize the probability of recognizing true tracks;
- to maximize the probability of rejecting false tracks;
- to minimize the number of scans for track initiation.

These objectives are related to the first two metrics, while the third metric is essentially related to the computational load of the tracking algorithm. To make the concepts of true and false tracks more precise, a target of interest is considered *under track* at a given scan if the distance between the true target position and the estimated position does not exceed vT, which represents the maximum distance traveled by that target during a scan period. Hence, at a given scan, all confirmed tracks that either are not associated to true targets or do not satisfy the above distance condition, will be considered as *false tracks*. Further, whenever a confirmed track loses the target, new tracks assigned to the target are not counted as successful target tracks until the original confirmed track is recognized as a false track and terminated. This makes the "percentage of successful target tracks" a good performance metric also for track swaps and track continuity. To better understand this, let us consider a tracking experiment with a single target present for all the duration of the experiment and let us associate with such an experiment a binary sequence defined in this way: at a given scan, 1 indicates that there is a successful target track and 0 indicates that there is no successful target track according to the above described logic. Suppose that a confirmed track, say track A, has been tracking the true target up to a given scan k, then loses the target and is terminated after S scans, and suppose that another confirmed track, say track B, is tracking the target from scan $k+1$. We then obtain the following sequence:

$$\cdots 11\underbrace{00\cdots 0}_{S}11\cdots$$

The percentage of successful target tracks at each scan is obtained by averaging the above sequences over several Monte Carlo runs. Clearly, a track swap induces a sequence of zeros and hence penalizes such a metric. Further, the time-averaged percentage of successful target tracks gives a measure of *track continuity*, as it provides the average fraction of time in which the target has been under track. Finally, the number of tentative tracks will count only *false tentative* tracks.

The following five trackers will be compared in this section:

- NKB: tracker that does not exploit contextual information at all;
- DMHC: KBT using the DMHC strategy;
- DTPHC: KBT using the DTPHC strategy;
- DTPHC-AI: KBT using the DTPHC strategy and AI (amplitude information);
- FKB: KBT using the fuzzy strategies.

All the above listed trackers use the M/N logic for track initiation, the NNCJPDA algorithm (α-NNCJPDA for FKB) for data association, and the IMM algorithm with three models (CV, CT_{+}, CT_{-}) for filtering. Notice that NKB represents a standard tracker that does not use the KB. Conversely, the other four trackers (DMHC, DTPHC, DTPHC-AI, FKB) use the KB in different ways; more precisely, they all adopt zone-dependent parameters (M, N, γ, b) and use respectively the DMHC strategy, the DTPHC strategy, the DTPHC strategy with AI and the FKB strategy (DMLR + α-NNCJPDA).

The trackers have been evaluated on a simulated scenario characterized by live clutter (Fig. 8.2) and by artificial target tracks that have been located on zones of the surveillance environment with different clutter density. Moreover, in the experiments pertaining to the use of AI (see Section 8.6.3), real aircraft tracks have also been considered. The live data pertain to an X-band SELEX-SI naval radar capable of two-dimensional air and surface surveillance, over the horizon detection, detection of low flying targets, autonomous capability of target velocity estimation via MTD, and track while scan. The radar gives accurate measures of plot range and azimuth plus a rough estimate of plot amplitude, which have been used for the application described in this work. The utilized data refer to clutter recorded during the internal test of the system. The following case studies will be examined in detail:

- aircraft ($P_{\mathrm{d}} = 0.8$) crossing an HC zone without maneuvers;
- aircraft ($P_{\mathrm{d}} = 0.8$) crossing an HC zone with a maneuver.

In all simulations the target's detection was generated as a random draw with probability P_d. For each of the above listed case studies, the previously defined trackers were compared by means of Monte Carlo simulations (60 runs). Monte Carlo simulations were carried out by varying, trial-by-trial, the clutter plot positions (randomly selected among the recorded radar scans pertaining to live data), the measurement noise, and the missed plots for the test target. Tables 8.2 and 8.3 report the choice of the various parameters M, N, b, γ, L, L_s used in the simulations for the various strategies.

The threshold values N_1 and N_2 (see Section 8.3.2) that define the LC, MC, and HC zones have been set via trial and error so as to get less than 1–2 false confirmed tracks per hour. In the two subsequent subsections, for each case study, the performance of the trackers NKB, DMHC, DTPHC, and FKB will be compared, while the performance obtained with the use of amplitude information also (DTPHC-AI tracker) will be examined in the last subsection.

8.6.1 Aircraft Simulation Results

Two cases concerning respectively a maneuvering and a nonmaneuvering aircraft crossing the coastline over an HC zone have been considered. For the aircraft a speed of 0.3 km/s, an angular speed during the maneuver of 0.2 rad/s, and a detection probability $P_d = 0.8$ have been assumed. Figure 8.9 displays the trajectories of the aircraft used in the simulations, and Figs. 8.10 and 8.11 show the results of the simulations for the crossing of the HC zone.

During the crossing, the aircraft remains in the HC zone for three scans in the maneuverless case and for six scans in the maneuver case. Moreover, in the case of maneuverless crossing, the aircraft comes from an LC zone before reaching the HC zone, and then proceeds into an MC zone. Conversely, in the maneuvering case the aircraft comes from an LC zone before the crossing and then again enters an LC zone.

Looking at the performance of the NKB tracking system (see Figs. 8.10 and 8.11) it can be observed that the track is first detected at scan 6, as expected, as a tentative

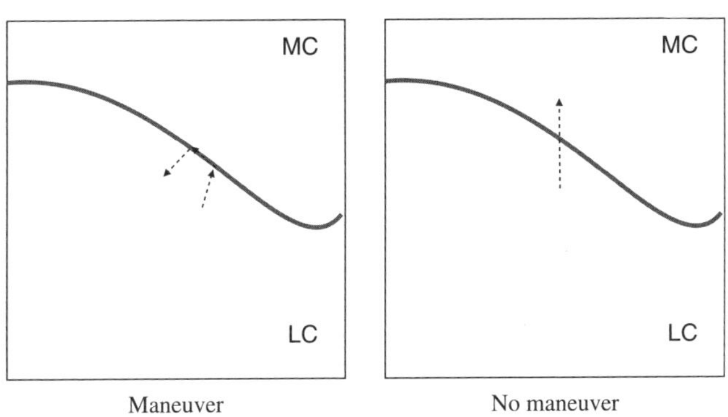

Figure 8.9 Aircraft paths.

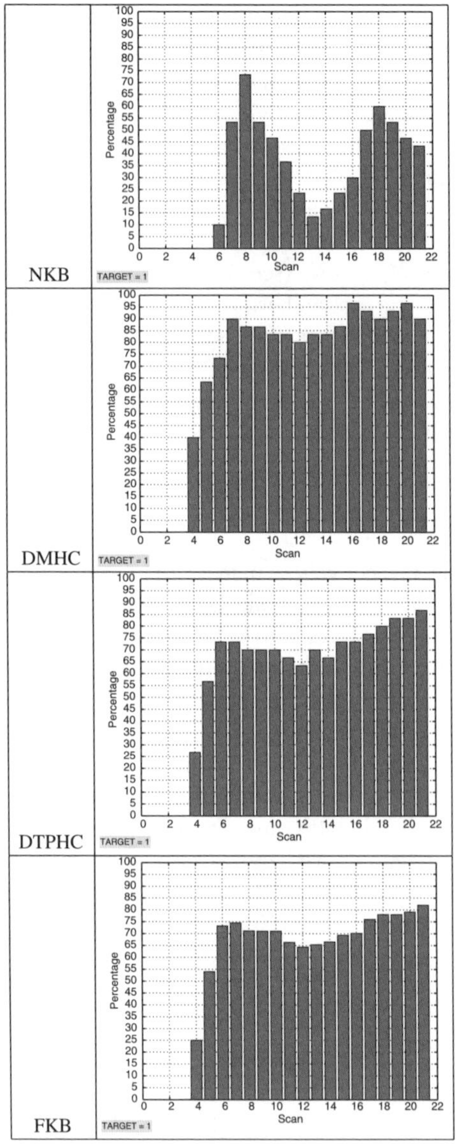

Figure 8.10 Aircraft crossing the shore in an HC zone without maneuvers: percentage of successful target tracks.

track is promoted after reaching M (in this case $M = 5$) plots. Conversely, for the KB trackers DMHC, DTPHC, and FKB $M = 3$ has been chosen and, in fact, the track is promoted after 4 scans. A closer examination of Fig. 8.10 reveals the difficulty of the standard tracking system NKB even in presence of a transverse crossing of the HC zone, because the high concentration of clutter makes the target's loss an event

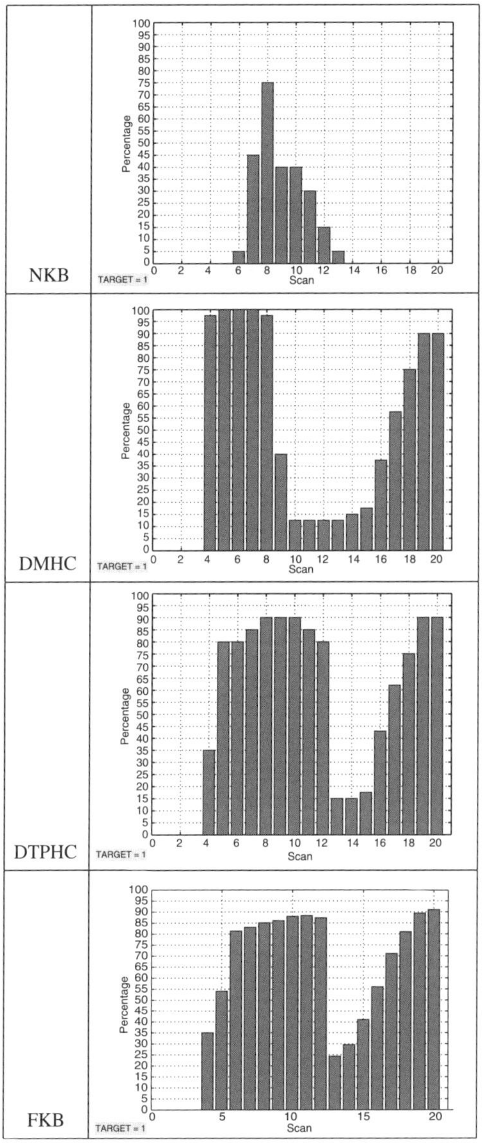

Figure 8.11 Aircraft crossing the shore in an HC zone with maneuver: percentage of successful target tracks.

with very high probability. On the other hand, the three KBTs have much higher percentages of successful tracking. In particular, DMHC yields a better performance, as the cancellation of all clutter measurements falling in the coastal area totally avoids association errors. DTPHC and FKB give similar results.

The maneuver considered in Fig. 8.9 is a critical one since the target, coming from the sea (LC zone), turns in the proximity of the coastline, then proceeds for a while along the coastline and then turns again, so that it finally proceeds in the opposite direction with respect to the original one. Clearly the standard tracker NKB yields the worst performance, for the same reasons as in the maneuverless case (see Fig. 8.11). The DMHC KBT loses the track as soon as the target enters the HC coastal zone (scan 9), because the shadowing of the coastal zone enforces missed detections that, in the maneuvering case, lead to the target's loss. For the DTPHC tracker, the critical phase is the exit from the coastal HC zone (scan 13) rather than the entry into such a zone. FKB gives a better performance than DTPHC when the target moves out of the coastal HC zone (after scan 13); this is due to the α-NNCJPDA algorithm.

The comparison between DMHC, DTPHC and FKB in Figs. 8.10 and 8.11 reveals that DMHC is preferable in the case of a nonmaneuvering target, and DTPHC and FKB are preferable in the maneuvering case. Between DTPHC and FKB, the latter yields better performance.

8.6.2 Number of False Tracks and Tentative Tracks

The proposed KB strategies have also been evaluated in terms of the numbers of false confirmed tracks and of false tentative tracks due to clutter. Table 8.4 reports the average values of such indices. It can be observed that the use of contextual information in the KBTs allows the significant reduction of both the number of false tracks and the number of tentative tracks. In particular, DMHC yields the best results, via total elimination of measurements in the HC zones. However, the DTPHC and FKB also give a good performance in successfully recognizing false tracks. From the examination of the simulation results reported in Table 8.4 and in Figs. 8.10 and 8.11, it can be observed that

- the use of KBTs yields better results, compared to NKB, for all the considered metrics;
- in the maneuverless case, DMHC provides better performance than DTPHC and FKB and the converse holds in the maneuvering case;
- DMHC provides a reduced number of false confirmed tracks and of tentative tracks.

TABLE 8.4 Number of false confirmed and tentative tracks.

	Confirmed False Tracks (no./hour)	Tentative Tracks (no./scan)
NKB	2.5	350
DMHC	0.3	200
DTPHC	0.4	250
FKB	0.3	215

Hence, the following conclusions can be drawn:

- In the small-extension HC zones it can be convenient to adopt the DMHC strategy so as to eliminate plots and thus reduce the computational load, because targets cannot perform critical maneuvers inside such zones due to the small extension.
- In the large-extension HC zones, the DMHC strategy does not work properly, and the DTPHC or FKB strategies are more convenient.

From Fig. 8.11 it can be seen that, in the case in which the target crosses an HC zone with a maneuver, the performance of KBTs, although improved with respect to NKB, is nevertheless unsatisfactory. The use of AI can overcome this problem.

8.6.3 The Use of Amplitude Information

Figure 8.8 shows a typical trend of the target echo amplitude versus target range, for a sample of 240 real data plots. This amplitude decays to zero with the range R as $1/R^2$, but exhibits large fluctuations. Figure 8.8 also displays the ideal (average) amplitude behavior (solid line) and the 80% confidence band (dotted lines).

In order to evaluate the benefits arising from the use of amplitude information (AI), real data concerning an aircraft that performs a critical maneuver in an HC zone (see Fig. 8.9) have been considered. The real values of the amplitude are reported in Fig. 8.8. The AI, as specified in Section 8.5, is used in the data association in order to eliminate all the (clutter) plots whose amplitude values are incompatible with the target amplitude confidence band of Fig. 8.8. In order to carry out this elimination, it is assumed that the range of the amplitude is a priori given for each type of target. Clearly, the ambiguity in target classification implies an uncertainty on the amplitude interval.

Table 8.5 reports the percentage of successful target tracking (averaged over the time period of HC crossing) obtained for the tracking of the above-mentioned target with the following KBTs:

- DTPHC without AI,
- DTPHC with exact knowledge of the confidence band,
- DTPHC with a 30% confidence bound,
- DTPHC with a 50% confidence bound.

TABLE 8.5 Percentage of successful target tracks.

DTPHC	44%
DTPHC + AI exact	80%
DTPHC + AI (30% error bound)	60%
DTPHC + AI (50% error bound)	48%

In particular, the last two cases allow the evaluation of the effect on the performance of the amplitude range uncertainty. From Table 8.5 it can be observed how the performance is significantly improved by a precise knowledge of the amplitude range and how a rough estimate of such a range can still yield benefits. This motivates the approach proposed in Section 8.5, which consists of using all measurements associated to the track, scan by scan, in order to iteratively refine the estimate of the amplitude and, thus, increase the effectiveness of the amplitude-based elimination as far as the track's life proceeds.

8.7 CONCLUSIONS

A KB system for multitarget tracking has been studied. The main ingredients of the tracker are

- Extended Kalman Filtering to take into account nonlinearities;
- Interacting Multiple Model for managing the target maneuvers;
- Nearest Neighbor Cheap Joint Probabilistic Data Association for robust plot–track association;
- *M* out of *N* logic for track initiation;
- Use of the Knowledge Base (geographical maps and targets characteristics) and of Amplitude Information;
- Use of fuzzy logic for classification of the surveillance region.

The technical solution has been tested against simulated and live data pertaining to an SELEX-SI naval surveillance radar, and it has been demonstrated that the KB approach provides meaningful advantages, allowing the reduction of false and tentative tracks while permitting the continuous track of useful targets. Topics for future investigation about the KB tracking system will include

- the inclusion of statistical models for the echo amplitude of the various sources (e.g. targets and clutter) using the maximum likelihood estimation technique;
- use of the echo amplitude to improve the track initiation logic;
- an end-to-end KB system, which combines the KB techniques in data processing, considered in this chapter, with KB techniques in the signal-processing stage studied in references 17–19.

ACKNOWLEDGMENTS

The authors wish to thank Sergio Romagnoli for his valuable help on algorithmic implementation and data processing during his Master Thesis work at the DSI, Università di Firenze.

REFERENCES

[1] H.A.P. Blom, Y. Bar-Shalom. The interacting Multiple Model Algorithm for systems with Markovian switching coefficients. *IEEE Trans. Automatic Control* 1970;6:473–483.

[2] Y. Bar-Shalom. *Multitarget-Multisensor Tracking: Advanced Applications.* Vol. I. Norwood, MA: Artech-House; 1990.

[3] F.R. Castella. Sliding window detection probabilities. *IEEE Trans. Aerospace and Electronic Systems* 1976;12:815–819.

[4] A. Farina. Application of knowledge-based techniques to tracking function. *NATO RTO Lecture Series 233*, Neully Sur Seine Cedex, France, 2003.

[5] C. Morgan, L. Moyer. Knowledge-based applications to adaptive space–time processing. Vol. V: Knowledge-based tracking rule book. AFRL-SN-TR-2001-146 Vol. V (of Vol. VI), Final Technical Report, 2001.

[6] A. Farina, F.A. Studer. *Radar Data Processing, vol. I: Introduction and Tracking.* New York: John Wiley & Sons; 1985.

[7] Y. Bar-Shalom, X.R. Li, T. Kirubarajan. *Estimation with Applications to Tracking and Navigation.* New York: John Wiley & Sons; 2001.

[8] Y.X.R. Li. Engineer's guide to variable-structure multiple-model estimation for tracking. In Y. Bar-Shalom, editor. *Multitarget-Multisensor Tracking: Applications and Advances.* Vol. III, Chapter 10. Norwood, MA: Artech-House; 1996.

[9] G.A. Ackerson, K.S. Fu. On state estimation in switching environments. *IEEE Trans. Automatic Control* 1970;15:10–17.

[10] C.B. Chang, M. Athans. State estimation for discrete systems with switching parameters. *IEEE Trans. Aerospace and Electronic Systems* 1978;14:418–425.

[11] R.A. Singer. Estimating optimal tracking filter performance for manned maneuvering targets. *IEEE Trans. Aerospace and Electronic Systems* 1970;15:10–17.

[12] Y. Bar-Shalom, X.R. Li. *Multitarget-Multisensor Tracking: Principles and Techniques.* Storrs, CT: YBS Publishing; 1995.

[13] A. Farina, M. De Feo, A. Graziano, R. Miglioli. IMMJPDA versus MHT and Kalman filter with NN correlation: performance comparison *IEEE Proc. Radar, Sonar Navigation* 1997; 144:49–56.

[14] R. Torelli, A. Graziano, A. Farina. IM3HT algorithm: a joint formulation of IMM and MHT for multitarget tracking. *European Journal of Control* 1999;5:46–53.

[15] D. Reid. An algorithm for tracking multiple targets. *IEEE Trans. Automatic Control* 1979; 24:843–854.

[16] T.E. Fortmann, Y. Bar-Shalom, M. Scheffe. Sonar tracking of multiple targets using joint probabilistic data association. *IEEE Journal of Oceanic Engineering* 1983;8:173–184.

[17] E. Conte, A. De Maio, A. Farina, G. Foglia. Design and analysis of a knowledge-aided radar detector for Doppler processing. *IEEE Trans. Aerospace and Electronic Systems* 2006;42:1058–1079.

[18] A. De Maio, A. Farina, G. Foglia Knowledge-Based Recursive Least Squares techniques for heterogeneous clutter suppression *IEE Proc. Radar and Sonar Nav.* 2007;1:106–115.

[19] A. De Maio, A. Farina, M. Wicks. KB-GLRT: Exploiting knowledge of the clutter ridge in airborne radars. *IEE Proc. Radar and Sonar Nav.* 2005;152:421–428.

[20] B. Ristic, S. Arulampalam, N. Gordon. *Beyond the Kalman Filter: Particle Filters for Tracking Applications.* Norwood, MA: Artech House; 2004.

[21] A. Benavoli, L. Chisci, A. Farina, S. Immediata, L. Timmoneri. Knowledge-based system for multi-target tracking in a littoral environment. *IEEE Trans. Aerospace and Electronic Systems* 2006;42:1100–1119.

9

KNOWLEDGE-BASED RADAR TARGET CLASSIFICATION[1]

Igal Bilik and Joseph Tabrikian

In this chapter, knowledge-based algorithms are developed for the problem of target classification for ground surveillance Doppler radars. Two sources of knowledge are presented and incorporated within the classification algorithms: (1) statistical knowledge on radar target echo features and (2) physical knowledge, represented via the locomotion models for different targets. An algorithm that combines both sources of knowledge is proposed. Various methods to incorporate these sources of knowledge are presented. Maximum-likelihood (ML) and majority-voting decision schemes were applied for target classification. The proposed classification approaches were tested using real data of radar echo recorded by ground surveillance radar, which include various target classes, such as walking person(s), tracked or wheeled vehicles, animals, and clutter. The combined approach, which implements both statistical and physical prior knowledge, provides the best classification performance, and it achieves a classification rate of 89%. The majority-voting decision scheme, which is suboptimal, is found to perform better than the ML criterion due to modeling mismatch.

9.1 INTRODUCTION

The main tasks of radar systems are target detection, localization/tracking, and classification [1]. In typical radar systems, these tasks are fully automated. Target

1. Reprinted, with permission, from *IEEE Trans. Aerospace and Electronic Systems*, references 6 and 29.

Knowledge-Based Radar Detection, Tracking, and Classification. Edited by Fulvio Gini and Muralidhar Rangaswamy

classification is the least studied among other radar functions. Therefore, many radar systems essentially require human involvement for this purpose. For example, in most applications of ground surveillance radars, target classification is performed using sound representation of the target Doppler characteristics [1–4]. This concept requires huge human resources and high expenses. Moreover, the operational performance of border control and monitoring systems are limited by human senses and performance. For example, systems that require an operator's involvement in the target classification process are slow and restrict the real-time field missions. In addition, overall system performance is not fully reliable due to instable operator performance that varies between operators [5, 6].

Limitations of the human operator-based classification concept motivate the extensive effort invested in development of automatic target recognition (ATR) systems for radar application [5, 6]. The task of an ATR system is to decide on the target type, based on the sensor data. Demands for efficient ATR systems are readily apparent [7]. The battlefield capabilities would be greatly enhanced if the surveillance systems could provide automated and reliable classification of objects in the surveyed areas. With the use of radar across the wide range of land, sea, and air applications, the need for automated radar target classification is becoming increasingly critical [8]. Classifying objects with high resolution radars is a well known and commonly used technique. For example, Zyweck et al. [9] proposed to classify the detected aircraft using high resolution signatures provided by the radar system. Another approach is to use a two-dimensional inverse synthetic aperture radar (ISAR) image from objects. Ships from different classes were classified in reference 10 based on this approach.

Classification of moving vehicles on the road is addressed in reference 11, using both high range resolution profiles (HRRP) and ISAR. In these applications the sensor provides two-dimensional data. Therefore, the majority of current ATR algorithms deal with image data. Regardless of the sensor type, ATR for such systems is implemented via image processing algorithms. The image-based ATR algorithms are based on the information in the received signal echoes, representing different geometrical structures of targets. This means that fine details of the target geometrical structure are crucial for automatic recognition. Therefore, implementation of image-based recognition is suitable for high resolution sensors, and cannot be used in low resolution applications such as ground surveillance radars. ATR based on synthetic aperture radar (SAR) images, is the most studied application [12]. Generally, objects of the same shape reflect radio waves in a similar manner (phase and amplitude); that is, the object has a unique signature according to its shapes and orientations toward the radar. This signature can be used to identify objects within the SAR scene in the same way as object identification in optical photographs. In references 13 and 14 the target signature was identified using radar cross-section (RCS) measurement by step-frequency radar. In this technique, the scattering parameters, which are directly related to target geometry, are assigned to each target for classification. In reference 15, human walking motions were estimated from frequency modulated continuous wave (FMCW) radar echoes using the time-varying Doppler signature. Human walking motions were simulated using the human locomotion model, presented in reference 16. Any classification process requires some kind of prior information. This information can be incorporated within the classification process using statistical or parametric models. To exploit this information in a target recognition context, two approaches are widely used: *statistical* and *parametric* models.

The statistical knowledge-based (KB) approach provides a systematic framework for incorporating prior knowledge about the different target classes within the observational models. Given these statistical models, it is possible to obtain an optimal solution to the classification problem by application of basic principles of statistical inference [17, 18]. Statistical prior knowledge is usually obtained by learning the statistical model of different classes [8] from a set of training databases. Using these statistical models, the decision regions can be determined. The main disadvantage of this approach is that it requires a training database. The corresponding classification performance strongly depends on the size and degree of fidelity of the training database. In many applications, such a training database is not available, such as the design of new radar systems or in cases where the database construction is expensive and complicated. Additional limitations of the statistical KB classification approach include (a) inability to extend the class spaces beyond those spanned by the training database and (b) inability to explore the relations between the target properties and the modeled stochastic process.

Parametric modeling, derived from unique physical properties of targets within each class, is another approach for prior knowledge utilization. This approach is useful in cases where a training database is not available [19]. A similar approach was also used in reference [20], where classification of HRR range profiles was performed using a training set of simulated profiles. The physical prior knowledge is usually expressed by different parametric models for target classes. Physical KB classifiers can be implemented in direct or indirect concepts. In the direct concept, the model parameters are estimated and a score function under each hypothesis is calculated. In the indirect concept, the physical properties of the target classes are used in order to generate radar echoes from which a synthetic database can be constructed. This synthetic database is used for training a statistical-based classifier. In contrast to the statistical KB classifiers, this classifier does not involve intensive database collection, which in many cases can be expensive or impossible. Moreover, the size of the synthetic database can be enhanced to an unlimited degree in order to generate a database that reliably reflects the prior physical knowledge. The disadvantage of utilizing physical prior knowledge is the presence of modeling mismatch. In many radar applications, accurate models for the different target classes cannot be provided. The disadvantages of statistical and physical KB classifiers can be overcome by a combined approach. In this solution, the classifier can be trained by both the synthetic and collected databases. Therefore, both the statistical and physical knowledge sources are incorporated within the classifier.

In this work, the proposed KB classification approaches are analyzed for the application of ATR in ground surveillance Doppler radars. The problem of classification of a walking person, a pair of walking persons, and a slowly moving vehicle is studied. In the considered problem the targets move with similar radial velocity toward the radar, and therefore they cannot be distinguished using instantaneous Doppler information. However, time-varying velocities of target segments generate time-varying Doppler signatures [15, 19], which are used for target classification.

The walking person class is modeled using biomechanical locomotion models [16, 21–24], from which the radar echo signals are generated. The human locomotion model, presented in reference 16, is adopted in this work. This model is built from experimental data, and describes the dynamics of the human body parts over time. The model of a pair of persons is approximated by superposition of two asynchronous person locomotion models. The slowly moving vehicle was modeled as a point target with piecewise constant acceleration motion. Classification performances of the presented approaches are evaluated using real data of ground surveillance radar echo signals. The main contribution of this work is in application of the physical locomotion models in the context of ground surveillance radar target classification. The physical KB classification approach, derived in this work, can be very useful when real data are unavailable. The performance of the physical KB classifier is found to be close to the performance of the statistical classifier, without requirements for the real training database. Additionally, the combined approach derived in this work outperforms the classification performance of the statistical KB approach.

The rest of this chapter is organized as follows. A collected database is presented in Section 9.2. Human operator performance, evaluated in this work, is presented in Section 9.3. Two studied classification schemes are presented in Section 9.4. The KB classification approach is presented in Section 9.4.1. The statistical KB approach is studied in Section 9.4.2. The physical KB approach for classification is presented in Section 9.5. In Section 9.6, the combined classification approach is derived. The classification performances according to various approaches are evaluated and compared in Section 9.7. Finally, our conclusions are drawn in Section 9.8.

9.2 DATABASE

This section summarizes the database that was collected during this work and is available at www.ee.bgu.ac.il/~ spl. The training and testing databases were obtained using records of a low resolution ground surveillance radar. For the recording procedure, the target was detected and tracked automatically by the radar, allowing continuous target echo records. Targets from the following classes were recorded.

1. *Relevant targets*
   a. Person and group of persons. Combinations of the following cases were represented in the collected database:
      - Number of persons: 1, 2, and 3
      - Speed of motion: slow (2–3 km/h), normal (3–5 km/h), fast (5–8 km/h), walking and running (8–9 km/h)
      - Continuous and piecewise locomotion

- Locomotion without hand motions
- Person motion with long and short communication antenna
- Straight/zigzag locomotion
- Angles of motion toward the radar: 0°, 15°, 30°, 45°, 60°
- Synchronous/asynchronous motion of persons in a group

b. Vehicle
- Wheeled/tracked vehicle
- Speed of motion: slow (10–20 km/h), normal (20–30 km/h), and fast (30–90 km/h)
- Angles of motion toward the radar: 0°, 15°, 30° 45°, 60°

2. *Irrelevant targets*
   a. Animal (dog, cow, horse, sheep, pig)
   b. Vegetation clutter (trees, bush, wheat)
   c. Rain
   d. Fixed location rotating bodies (motors, water sprinklers)

For the recording procedure, the targets were detected and tracked automatically by the radar, allowing continuous target echo records.

Each scenario was recorded at least three times (30 frames of 4 s each). In total, 31 tests were carried out in 21 different sites. Fourteen different persons were recorded in these tests. The training and testing databases contained various targets and scenarios, such as different velocity, different locomotion types and different directions [6]. For each target class, the data in many scenarios were recorded. At least 120 s of each scenario was recorded. In total, the database of the collected real data for each target class contained 525 records of 4 s duration.

The sensor used in the database collection process, was a 9 GHz ground surveillance pulse-Doppler radar. The radar parameters were receiver bandwidth (3 MHz), transmit peak power (5 W), planar array antenna gain (31 dB) pulse width (12 μs,) duty cycle (10%), range resolution (125 m), and azimuth resolution (4°). In order to obtain an accurate target model, it was important to conduct the data collection experiments at high signal-to-noise ratio (SNR) conditions. Therefore, the range between the radar and the target was set to be short (200–600m). The recorded targets were within the line-of-sight, in the presence of ground clutter with low vegetation and without any interference. The target motions were fully controlled. One target at a time was recorded in each scenario. Sampled radar target echoes of one, two, and three persons, vehicle, clutter (consisting of 80% vegetation records, 10% rain records, and 10% fixed-location rotating bodies), and animal are presented in Fig. 9.1. Spectrograms of these targets, presented in Fig. 9.2, show that each target class has unique time–frequency characteristics, which can be used for classification.

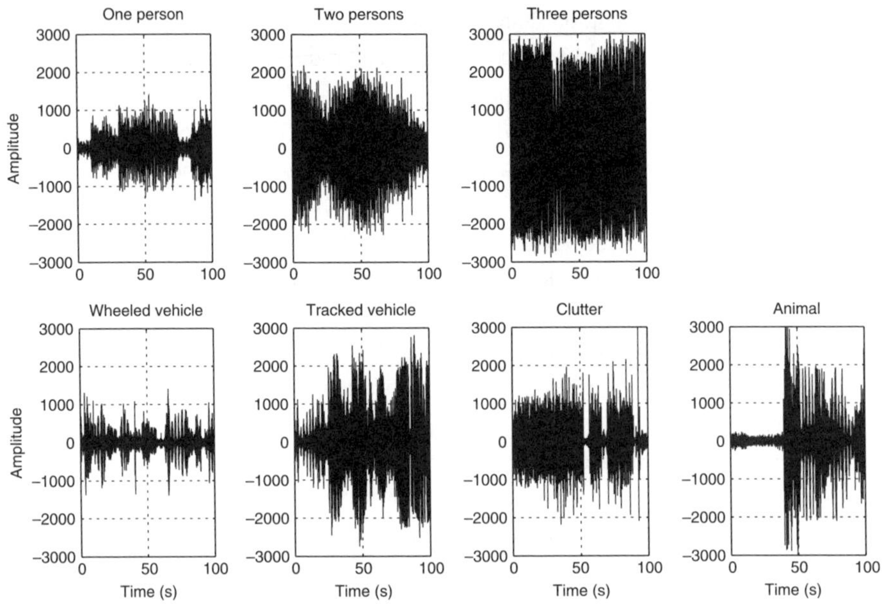

Figure 9.1 Radar echo samples of moving targets.

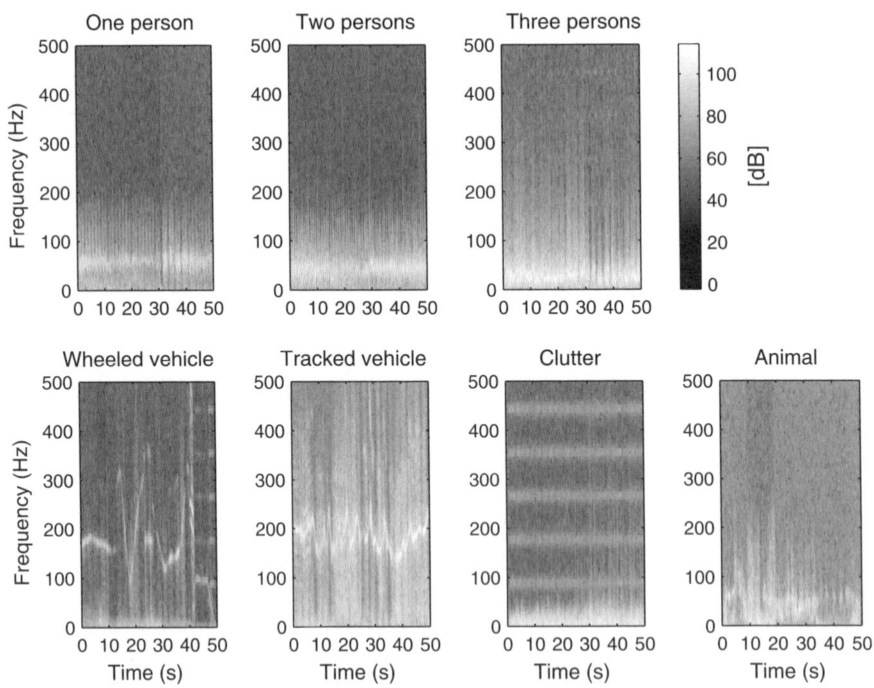

Figure 9.2 Spectrograms of radar echo target samples.

9.3 TARGET RECOGNITION BY HUMAN OPERATOR

In designing ATR systems, it is helpful to understand first the mechanism used by the human radar operator in target recognition. An operator identifies the targets using two types of information. The first one is the so-called high level information and a priori knowledge. This information is not related to the detected target. Instead, it is concerned with local topology, terrain profile, a priori known regions with potential hazards, and so on. This information is also concerned with operator experience and knowledge about previous target detections and their a posteriori identifications. High level information is very helpful for the human operator and improves its recognition performance. Unfortunately, it is complicated to implement an automatic system that uses this information.

The second type of useful information for recognition is so-called low level information. This information, which is related to the physical structure of the detected target and its motion. The major challenge of the ATR systems stems from the requirements to classify targets using low level information alone.

Human operator performance has been evaluated as a reference for the ATR performance evaluation alone. Human operator were "trained" using samples from the training database (used in the Gaussian mixture model (GMM) training). More than 100 sound records of different moving targets from the testing database (used in the ML and majority-voting based classifiers performance evaluation) were played to 20 experienced operators. The length of each record was 12 s. Table 9.1 presents the classification performance (in %) obtained in this test.

The average human operator classification rate was found to be 37%. The average rejection rate (unknown) was 11%. No significant discrimination between wheeled and tracked vehicle classes was obtained by human operators. Therefore, these two classes were combined into "Vehicle" class.

9.4 CLASSIFICATION SCHEME

In the classification scheme described in this section, the detected target was classified into one of M possible classes using estimated target models. The maximum

TABLE 9.1 Human operators classification performance.

Class/ Decision	One Person	Two Persons	Three Persons	Vehicle	Animal	Clutter	Unknown
One Person	**38**	3	5	9	23	15	9
Two Persons	34	**11**	11	3	20	10	11
Three Persons	18	5	**18**	14	3	27	15
Vehicle	0	1	0	**82**	1	10	6
Animal	20	2	6	11	**12**	35	15
Clutter	2	0	0	6	6	**76**	9

likelihood (ML) and majority-voting classification schemes are investigated in this section. In the ML classification approach, the decision is made by finding the target class $\hat{m}$, which maximizes the likelihood function

$$\hat{m} = \arg \max_{m=1,\ldots,M} f_{\mathbf{y}}(\mathbf{y}; H_m), \tag{9.1}$$

where $f_{\mathbf{y}}(\mathbf{y}; H_m)$ is the probability density function (pdf) of the classification features $\mathbf{y}$, extracted from the data $\mathbf{z}$ under hypothesis H_m. In the case that the target classes have equal a priori probabilities, the ML decision rule is optimal with respect to the Bayes criterion.

A possible alternative to the ML classifier in the multiple-hypotheses problem is the combination of binary classifiers [25–27]. Implementation of this concept in the target classification application, referred to as majority-voting was discussed in reference 6. This concept proposes to decompose the complicated M-ary hypotheses problem into a set of $\binom{M}{2} = \frac{M!}{2!(M-2)!}$ simple binary classification problems [28]. According to this concept, binary likelihood ratio tests (LRT) for discrimination between every pair of classes m and n are employed:

$$\frac{f_{\mathbf{y}}(\mathbf{y}; H_m)}{f_{\mathbf{y}}(\mathbf{y}; H_n)} \underset{H_n}{\overset{H_m}{\gtrless}} \gamma_{mn} \quad \forall (m \neq n) = 1, 2, \ldots, M. \tag{9.2}$$

If the prior probabilities of the hypotheses are equal, the majority voting and ML concepts coincide, because both of them minimize the same Bayesian cost function. In practice, the pdfs for the different classes are unknown and their parameters need to be estimated. The estimation process introduces modeling errors, and therefore the theoretical thresholds for a decision between each pair of hypotheses are no longer optimal. The majority voting decision rule enables one to adjust the thresholds for each pair in order to minimize the Bayesian cost function. Thus, the LRT is performed between each pair of hypotheses, and the corresponding threshold can be optimized separately for each test. The threshold, γ_{mn}, is determined to minimize the classification error for the pair of target classes (m, n). The pairwise decisions are combined by voting; that is, the class with the most pairwise wins is selected. This decision rule is schematically shown in Fig. 9.3. The majority voting decision rule is expected to outperform the ML decision rule in the presence of modeling errors, which can be caused due to small training database or modeling mismatch, such as erroneous model order. Note that the majority voting decision rule involves determination of multiple thresholds. In cases where there is a small database, the performance of this solution may degrade due to overfitting, especially in problems with a large number of classes, in which the number of threshold parameters is large.

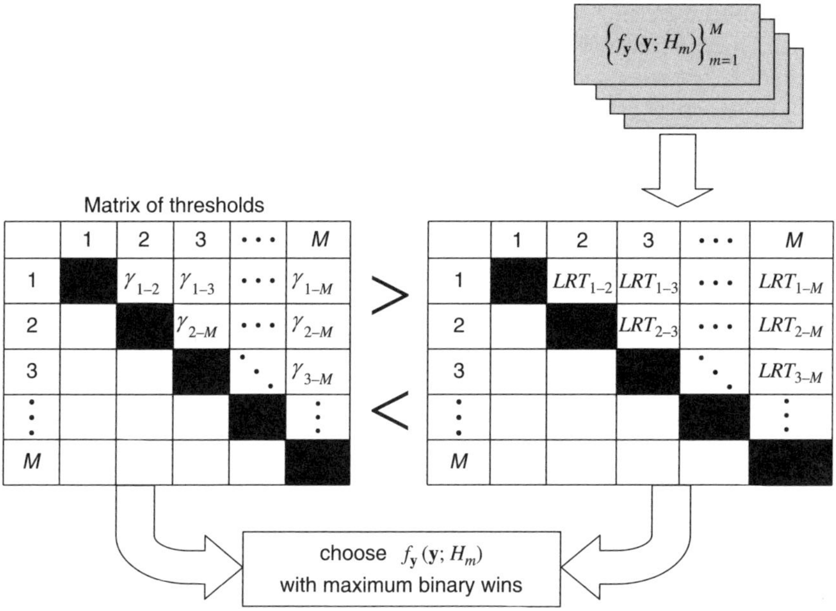

Figure 9.3 Majority-voting decision scheme.

9.4.1 Knowledge-Based Models

According to the ML and majority voting decision rules, the conditional pdfs $f_{\mathbf{y}}(\mathbf{y}; H_m)$ for each target class should be obtained. The main question addressed in this section is how the prior knowledge may be efficiently incorporated in the classification scheme. Two different types of prior knowledge are available [5]. The first is statistical knowledge about the target echo signal, extracted from the training data via a training procedure.

The second is the knowledge about the physical target properties and target locomotion. According to the physical KB classification approach, dynamics target models are derived using available physical knowledge about the target classes. The physical models are used for synthetic data generation. These data contain a variety of target motions, represented by the vector of parameters $\boldsymbol{\theta}_m$ under hypothesis H_m. The a priori knowledge about the target, represented by physical models, is translated into a synthetic database. This physical model is incorporated into the classification scheme via two concepts: *indirect* and *direct*. The main difference between them is in the way that the synthetic database is used in the classification. According to the indirect concept, a generated synthetic database replaces the training database in the statistical KB approach. The direct concept allows incorporation of the physical knowledge directly into the distribution of the target echo signal. The model describing the target measurements under hypothesis H_m is given by

$$H_m : \mathbf{x} = \mathbf{h}_m(\boldsymbol{\theta}_m) + \mathbf{n}, \tag{9.3}$$

where the vector $\mathbf{x}$ denotes the radar echo, and the vector $\mathbf{n}$ is assumed to be white complex circular Gaussian noise. The likelihood under the mth hypothesis can be written as

$$f_{\mathbf{x}}(\mathbf{x}; H_m) = \int_{\boldsymbol{\theta}_m} f_{\mathbf{x}}(\mathbf{x}|\boldsymbol{\theta}_m; H_m) f_{\boldsymbol{\theta}}(\boldsymbol{\theta}_m; H_m) \mathrm{d}\boldsymbol{\theta}_m, \tag{9.4}$$

where $f_{\boldsymbol{\theta}}\ (\cdot; H_m)$ is the pdf of the model parameters under hypothesis m, and $f_{\mathbf{x}}(\mathbf{x}|\boldsymbol{\theta}_m; H_m)$ is the conditional pdf of the radar echoes, modeled with set of parameters $\boldsymbol{\theta}_m$ under hypothesis H_m. In the direct concept, target synthetic data are generated according to each target class model, $\mathbf{h}_m(\boldsymbol{\theta}_m)$ using the vector parameter $\boldsymbol{\theta}_m$. These synthetic data are used for the estimation of the conditional pdf $f_{\mathbf{x}}(\mathbf{x}|\boldsymbol{\theta}_m; H_m)$ calculation.

Both direct and indirect concepts require the a priori pdfs of the model parameters $f_{\boldsymbol{\theta}}(\boldsymbol{\theta}_m; H_m)$ within the subspace of the parameters that represents valid target locomotion. In the case that the subspace of the parameters and the a priori pdf of the model parameters are unknown, they can be estimated from a training database. The required database in this case is small compared to the statistical KB approach. This limited database is used only once for each target class, and it is object-dependent and not radar-dependent. The a priori pdf of the model parameters estimated from the limited training database is modeled by GMM.

Another approach that combines the statistical knowledge with the physical target locomotion knowledge is described here. The combined database contains "real" data from the training database along with the synthetic data (from the physical KB approach). Therefore, the statistical knowledge is combined with the indirect physical KB approach. The synthetic data can contain an enhanced variety of target motions over the "real" data, and therefore this approach may improve the classification performance of the statistical-based approach. The use of the "real" data may compensate for the physical model mismatch, and therefore may improve the classification performance of the physical approach. The following sections describe in detail the proposed KB classification approaches.

9.4.2 Statistical Knowledge-Based Approach

In this approach the statistical prior knowledge, extracted from the training database, is incorporated into the statistical target models [5, 29]. The statistical models are implemented using features that allow discrimination between different target classes. The classification features used in this work are cepstrum coefficients [30]. Human operators classify the targets using an audio representation of the Doppler effect caused by the target motion as it is described earlier. Therefore, the cepstrum coefficients, which are widely used in speech and speaker recognition applications [30, 31], are used as a feature in this approach. The cepstrum coefficients $y[\cdot]$ of a sequence $x[\cdot]$ is defined in reference 32 as

$$y[\cdot] = F^{-1}\{\log|F\{x[\cdot]\}|\}, \tag{9.5}$$

where $x[\cdot]$ is the sampled target echo, and $F[\cdot]$ and $F^{-1}[\cdot]$ are the discrete Fourier and the inverse discrete Fourier transforms, respectively. It was shown in references 5 and 6 that other features such as linear predictive coding (LPC) coefficients result in lower classification performance.

GMMs are commonly used for non-Gaussian pdf representation [31]. In reference 33 it was proved that a large family of densities can be estimated to any finite degree of approximation, using a finite Gaussian mixture model. This justifies the use of Gaussian mixtures in classification problems [33].

In this work, the statistical knowledge for each target class is represented by a GMM for the features vector $\mathbf{y}$. A Gaussian mixture density, defined as a weighted sum of Gaussian components, is a useful tool for non-Gaussian pdf modeling. The pdf of $\mathbf{y}$ under hypothesis H_m is modeled by a GMM of order K_m:

$$f_{\mathbf{y}}(\mathbf{y}; H_m) = \sum_{k=1}^{K_m} \pi_m^{(k)} \phi_{\boldsymbol{\theta}_m^{(k)}}(\mathbf{y}) \quad \forall m = 1, \ldots, M, \tag{9.6}$$

in which $\phi_{\boldsymbol{\theta}_m^{(k)}}(\mathbf{y})$ represents the kth Gaussian mixture component, where the set of unknown parameters $\boldsymbol{\theta}_m^{(k)}$ includes its mean vector and covariance matrix, and $\pi_m^{(k)}$ is its mixing weight, satisfying $\sum_{k=1}^{K_m} \pi_m^{(k)} = 1$.

The expectation-maximization (EM) algorithm [34] is an efficient tool for estimating the GMM parameters. The algorithm allows iterative optimization of the mixture parameters, under nondecreasing likelihood requirement. The performance of this algorithm depends on its initialization, due to its tendency to converge to local maxima. In addition, the EM algorithm is not useful for estimating the number of Gaussian components, which has a dramatic effect on the classification performance. An efficient solution for the initialization problem is provided by the "greedy" learning of GMM [35]. The GMM order M can be obtained using a model order selection algorithm such as the minimum description length (MDL) [36]. The main idea behind the greedy learning algorithm is to start with one Gaussian component and build the mixture component-wise, rather than applying the EM on all the mixture components simultaneously. The first mixture component is obtained by assuming a single Gaussian. The ML estimation of its parameters (mean vector and covariance matrix) is straightforward. The algorithm is performed in two steps: insert a component into the mixture, and run an EM algorithm to estimate the updated mixture parameters. This iterative procedure is stopped if a given convergence condition is satisfied.

The classification algorithm using the statistical KB approach is schematically presented in Fig. 9.4.

9.5 PHYSICAL KNOWLEDGE-BASED APPROACH

The main idea of this approach is to use a priori knowledge about the target locomotion model [5, 28]. Target locomotion causes time-varying Doppler shifts in the

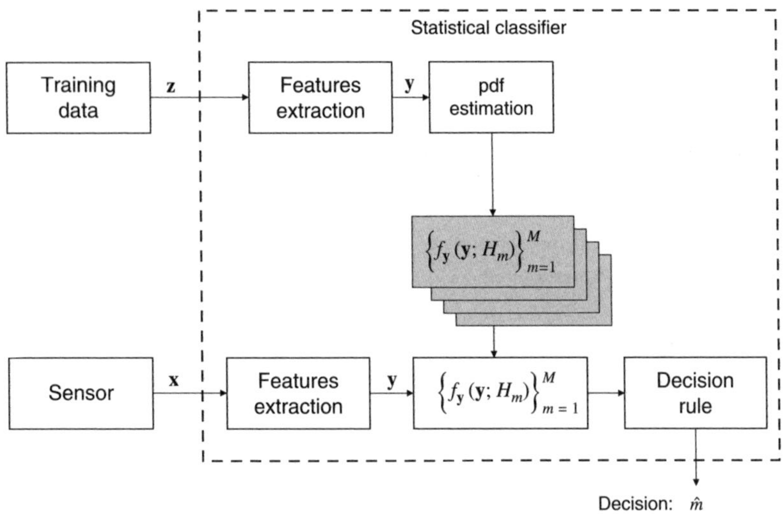

Figure 9.4 Statistical classification approach.

frequency of the transmitted radar signal. Therefore, the capability to model target locomotion allows simulating the corresponding Doppler shifts. Physical KB classifiers are based on predetermined locomotion models and the classification can be performed using the difference between motion characteristics of various targets.

9.5.1 Physical Model Construction

In this section, the target modeling for ground surveillance radar is studied. For this purpose, the locomotion models of person and vehicle targets are developed.

Human locomotion models have been developed based on biomechanics [16, 21–24]. Most of these models have been developed for computer animation, robotics, and biomechanics research. In order to create valid human motions, it is essential to take into account the geometric, physical, and behavioral aspects. In reference 16, a human walking model is obtained using experimental data including a wide range of normalized velocities. The human body was modeled by 12 body segments, connected by time-dependent translations and rotations.

In the following, the simulated human locomotion will be translated to radar echo signals. Human walking is a form of locomotion in which the body's center of gravity moves alternately on the right and the left sides. At all times at least one foot is in contact with the ground and, during a brief phase, both feet are in contact with the ground. For modeling human locomotion, a "stick" model of the human body used in animation, was developed [16, 21–24]. This model consists of nine elements: torso, arms, hands, thighs, and legs (see Fig. 9.5).

Human locomotion consists of rotatory motion of its segments. Thus, each reflection point along the segment has a different radial velocity measured by the radar. The weighted summation of the point reflections with random fluctuations create

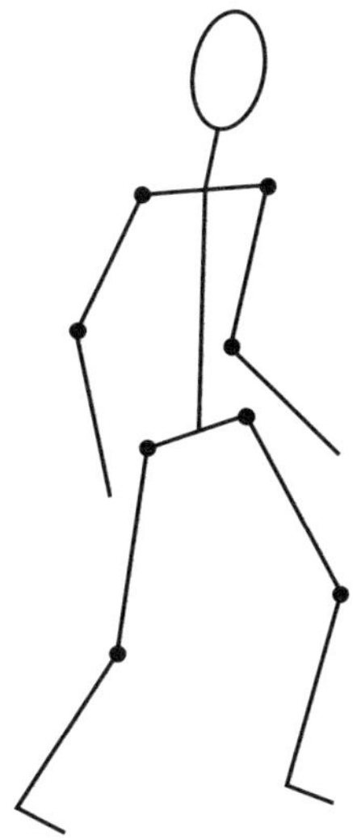

Figure 9.5 "Stick" model of human locomotion.

target echo phase shift changes. These phase shift changes in the received radar echo are determined by considering the interelement target dynamics. A target echo Doppler signature can be determined by phase changes in the received radar echo.

Consider a Doppler radar transmitting a narrowband signal. The transmitted signal is reflected from the moving target and received by the radar. For a point target, the baseband signal at the output of a noiseless synchronous receiver is modeled by

$$r_0(t) = A_R(t)e^{i\phi_d(t)}, \tag{9.7}$$

where $A_R(t)$ is the time-varying amplitude of the received signal and $\phi_d(t)$ is the Doppler phase given by $\phi_d(t) = 2\pi\int_0^t f_d(\tau)d\tau$, where $f_d(\cdot)$ is the instantaneous Doppler frequency. The Doppler frequency linearly depends on the target radial velocity $v_r(t) : f_d(t) = \frac{2v_r(t)}{\lambda}$, where λ is the transmitted signal wavelength. Thus, the instantaneous Doppler phase can be expressed as

$$\phi_d(t) = \frac{4\pi}{\lambda}\int_t^0 v_r(\tau)\,d\tau. \tag{9.8}$$

Using this expression, Equation 9.7 can be rewritten as

$$r_0(t) = A_R(t)e^{i\frac{4\pi}{\lambda}\int_0^t v_r(\tau)d\tau}. \tag{9.9}$$

Figure 9.6 shows the dynamics of a human body segment (such as a hand). The resulting segment echo is the superposition of point target echoes. A composed

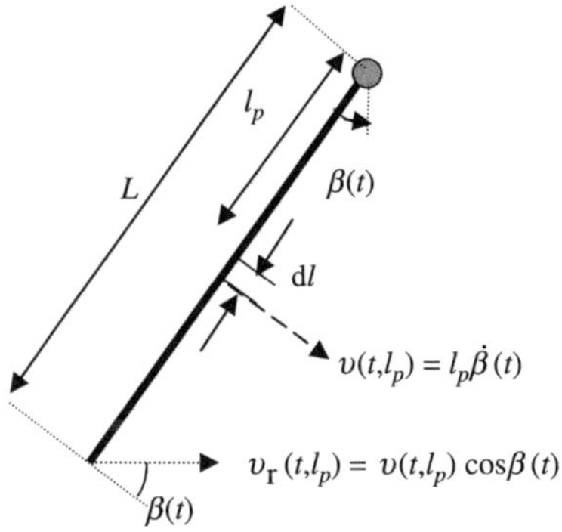

Figure 9.6 Simulation of human body segment motion.

target, such as the human body, is divided into J segments. Each segment is represented by $P = L/\mathrm{d}l$ point targets, of the infinitesimally small length dl. The displacement of the pth point along the segment is $l_p = p\mathrm{d}l$, $\forall p = 1, \ldots, P$.

Using Equation 9.9, the radar echo of the pth point target with displacement l_p is

$$r_0(t, l_p) = A_\mathrm{R}(t, l_p)e^{i\frac{4\pi}{\lambda}\int_0^t \upsilon_\mathrm{r}(\tau,p)\,\mathrm{d}\tau} = A_\mathrm{R}(t, l_p)e^{i\frac{4l_p\pi}{\lambda}\int_0^t \dot{\beta}(\tau)\cos(\beta(\tau))\,\mathrm{d}\tau} \tag{9.10}$$

where $\upsilon_\mathrm{r}(\tau, p)$ is the radial velocity of the pth point along the segment, $\beta(t)$ is the instantaneous angle of the segment from the zenith and $\dot{\beta}(\tau) = v_\mathrm{r}(\tau, p)/l_p$ stands for the instantaneous angular velocity. $A_\mathrm{R}(t, l_p)$ is a stationary two-dimensional zero-mean, white complex circular Gaussian process, representing amplitude fluctuations of the segment.

The radar echo of the jth segment can be stated as

$$r_j(t) = \sum_{p=1}^{P_j} r_{0j}(t, l_p) = \sum_{p=1}^{P_j} A_{\mathrm{R}j}(t, l_p)e^{i\frac{4l_p\pi}{\lambda}\int_0^t \dot{\beta}_j(\tau)\cos(\beta_j(\tau))\,d\tau}. \tag{9.11}$$

A composed target, such as a human, consists of several segments with various velocities and motions; the total target echo is the superposition of segment echoes. Finally, the baseband signal at the output of the receiver is modeled by

$$x(t) = \sum_{j=1}^{J}\sum_{p=1}^{P_j} A_{\mathrm{R}j}(t, l_p)e^{i\frac{4l_p\pi}{\lambda}\int_0^t \dot{\beta}_j(\tau)\cos(\beta_j(\tau))\,\mathrm{d}\tau} + \text{noise}. \tag{9.12}$$

The locomotion model of a walking human target is graphically presented in Fig. 9.7. The radar echo signal from all human body segments is modeled according

Figure 9.7 Graphical illustration of the human locomotion simulation.

to Equation 9.12. Each step cycle was divided into five substages according to different segments of human body motion. This model is characterized by the following 28 parameters:

- lengths of segments (5 parameters): $L_j = \mathrm{d}l \cdot Pj$;
- extreme angles between segments in different stages of the step (13 parameters) $\beta_j(t)$;
- duration of each stage of the step (5 parameters);
- variance of amplitude for each segment (5 parameters): $\mathrm{var}(A_{\mathrm{R}j}(t, l))$.

These parameters uniquely characterize human locomotion. Various locomotion types can be modeled by these parameters. The human locomotion model used in this work is partially based on the model proposed in reference 16. Some relations between the locomotion model parameters presented in reference 16, such as temporal and spatial characteristics, are used here. The relative cycle length and the maximal angle between feet are obtained from the relative velocity, defined as the average walking velocity normalized by the thigh length. The temporal characteristics are obtained from the relative length of cycle and the relative velocity [16].

In reference 15 is the mean sizes of the human body parts are presented. However, the subspace of the valid range of parameters required to obtain a valid human locomotion model is not defined in the biomechanical literature. Additionally, the valid subspace of circular motion of the human body segments, required for Doppler signature estimation, has not been explicitly defined. The knowledge of this subspace is not required for the human walking estimation, but dramatically influences the classification performance.

Therefore, a limited database of recorded radar echoes can be used to determine the valid subspace of the model parameters. This subspace can be obtained by estimating the locomotion parameters from the limited database using a genetic algorithm (GA) [38]. The fitness function of the GA was the mean-square-error (MSE) between the recorded radar echo and the signal model. The obtained subspace of locomotion parameters is used to avoid generation of invalid motions when a synthetic database of walking human echoes is generated.

For synthetic data generation, the physical model parameters are randomly generated using a priori knowledge of their pdfs. In the case of unknown pdf parameters, a limited database can be used, as schematically shown in Fig. 9.8. This stage requires only a limited database for determining the distribution of the locomotion model parameters. The estimated pdf of the locomotion model parameters $\tilde{f}_{\boldsymbol{\theta}}(\boldsymbol{\theta}_m; H_m)$, modeled by GMM, reflects only those motions included in the database. In order to extend the model beyond the motions included in the database, the covariance matrices of the GMM components are artificially expanded by a factor of 3. The vector parameters $\boldsymbol{\theta}_m$ are generated from the new GMM, $f_{\boldsymbol{\theta}}(\boldsymbol{\theta}_m; H_m)$, with expanded covariance matrices. Next, N sets of parameters $\{\boldsymbol{\theta}_m^{(n)})\}_{n=1}^{N}$ are randomly generated according to the new pdf, $f_{\boldsymbol{\theta}}(\boldsymbol{\theta}_m; H_m)$ under each hypothesis. Then, their vector parameters are used to obtain the physical locomotion models $\{\mathbf{h}_m(\boldsymbol{\theta}_m^{(n)})\}_{n=1}^{N}$ generating synthetic radar echoes. The limited database used in this synthetic database generation process, is used only once for each target class, and is not radar-dependent, but object-dependent.

The precise Doppler signature of a group of asynchronous walking persons contains multiple interactions. In this work, we consider radars with a wavelength of a few centimeters. Complex targets with physical dimensions larger than the radar wavelength consist of many scatterers, and therefore effects of multiple interactions can be approximately modeled as independent scatterers. In this work, a simplified

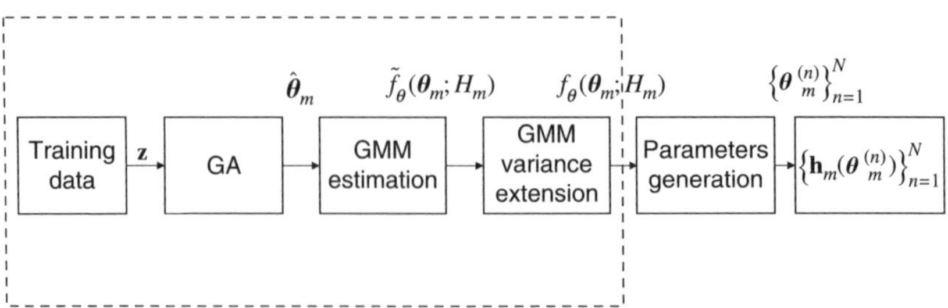

Figure 9.8 Synthetic data generation.

model neglecting the effect of multiple interactions is adopted. The group of asynchronous walking persons is modeled as a superposition of human locomotion models. In this work, the pair of walking persons is modeled as a superposition of two human locomotion models. This model is characterized by $57 = 28 \cdot 2 + 1$ parameters, obtained by the GA from the radar echoes database. The additional parameter characterizes the relative time delay between two human locomotion models in asynchronous locomotion.

The vehicle is modeled as a point target with constant acceleration:

$$v_{\mathrm{r}}^{(v)}(t) = v_{\mathrm{r}0} + a_{\mathrm{r}}t, \tag{9.13}$$

where $v_{\mathrm{r}0}$ is the initial velocity and a_{r} is the radial acceleration. The above model is justified, because during the observation period (about 90 ms) the acceleration of the vehicle is approximately constant. Using Equation 9.9, this model can be written as

$$r^{(v)}(t) = A_{\mathrm{R}}^{(v)}(t)e^{i\frac{4\pi}{\lambda}\int_0^t (v_{\mathrm{r}0}+a_{\mathrm{r}}\tau)d\tau} + \epsilon(t), \tag{9.14}$$

where $\epsilon(t)$ is the measurement and modeling errors, which is assumed to be zero-mean, white complex circular Gaussian. The valid subspace of the vehicle locomotion model parameters is assumed to be known:

- initial velocity $v_{\mathrm{r}0}$ — uniformly distributed in the range 0–100 km/h;
- radial acceleration a_{r} — uniformly distributed in the range 0–10 m/s^2;
- variance of the model and measurement noise: $\mathrm{var}(\epsilon(t)) = 0.1$–$10$;
- variance of the Gaussian-distributed echo signal amplitude $\mathrm{var}(A_{\mathrm{R}}^{(v)}(t)) =$ 1–100.

The results of the modeling process are locomotion models for each target class. These models are used in the classification process. Knowledge of the locomotion models can be employed by the classification algorithm in two different concepts, described in the following subsections.

9.5.2 Indirect Concept

In this concept, knowledge of the target models is used to construct a synthetic database, which can contain variety of target motions by simulating their locomotion with different model parameters. This synthetic database can be used for the training stage of the statistical KB classification algorithm. The indirect concept is schematically presented in Fig. 9.9. The implementation of the indirect physical KB approach is identical to the statistical KB approach, except for the synthetic data-generation stage. The synthetic database generation capability is an important benefit of the proposed locomotion modeling. The capability to generate the radar echo can improve the classifier robustness, because a wider database can be employed for model development in the training procedure. A variety of target motions can be

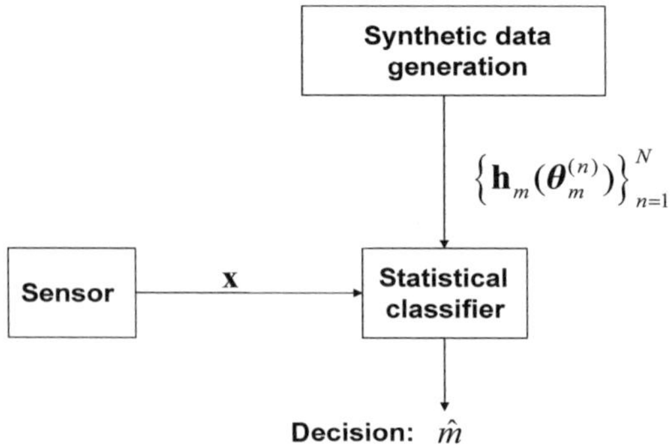

Figure 9.9 Indirect concept for physical KB classification.

generated using target models with suitable parameters. This capability can be especially useful when the data collection is expensive or practically complicated, or when a new radar or new classification mode is designed.

9.5.3 Direct Concept

In this concept, the physical knowledge about target locomotion, modeled by Equation 9.3, is directly used for estimation of the radar echo pdf under different hypotheses required for the classification. This concept is schematically presented in Fig. 9.10. As mentioned above, the model parameters under each hypothesis $\boldsymbol{\theta}_m$ are limited to the subspace of valid locomotion models. Therefore, the Bayesian approach is implemented to obtain the radar echo pdf under each hypothesis:

$$f_{\mathbf{x}}(\mathbf{x}; H_m) = \int_{\boldsymbol{\theta}_m} f_{\mathbf{x}}(\mathbf{x}|\boldsymbol{\theta}_m; H_m) f_{\boldsymbol{\theta}}(\boldsymbol{\theta}_m; H_m)\, \mathrm{d}\boldsymbol{\theta}_m, \qquad \forall m = 1, \ldots, M. \tag{9.15}$$

Assuming zero-mean, white complex circular Gaussian measurement noise in Equation 9.3, the conditional distribution of $\mathbf{x}$ under hypothesis H_m given

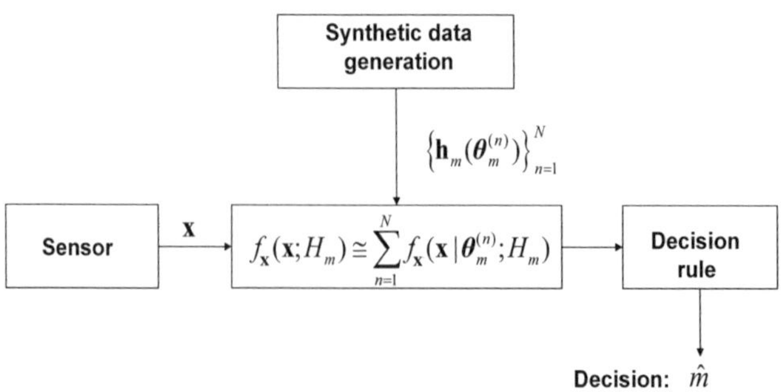

Figure 9.10 Direct concept for physical KB classification.

model parameter $\boldsymbol{\theta}_m$ is

$$\mathbf{x}|\boldsymbol{\theta}_m; H_m \sim N^c(\mathbf{h}_m(\boldsymbol{\theta}_m), \sigma^2\mathbf{I}), \tag{9.16}$$

where $N^c(\cdot, \cdot)$ stands for complex circular Gaussian distribution. The a priori pdf of the model parameters, $f_{\boldsymbol{\theta}}(\boldsymbol{\theta}_m; H_m)$, can be estimated using a training database via GMM, and the integration in Equation 9.15 can be approximated using the Monte Carlo method. For this purpose, the synthetic data generation stage presented in Fig. 9.8 is used. At the output of this stage, $\{\mathbf{h}_m(\boldsymbol{\theta}_m^{(n)})\}_{n=1}^N$ is obtained, providing the pdf $f_{\mathbf{x}}(\mathbf{x}|\boldsymbol{\theta}_m^{(n)}; H_m)$. The Monte Carlo estimation of $\log f_{\mathbf{x}}(\mathbf{x}; H_m)$ is given by

$$L_m = \sum_{n=1}^{N} \log [f_{\mathbf{x}}(\mathbf{x}; \boldsymbol{\theta}_m^{(n)}; H_m)]. \tag{9.17}$$

Finally, target classification can be performed via either the ML or the majority voting concepts.

9.6 COMBINED APPROACH

In the combined KB classifier, both statistical and physical knowledge sources are utilized. The only difference between the statistical KB classification approach and the indirect physical KB approach is in the source of the training database. Therefore, the statistical and physical knowledge can be easily utilized by combining the two training databases.

The statistical knowledge is available from the training database, while the physical knowledge is provided by the target locomotion model. By combining the "real" and synthetic databases, it is expected to obtain improved classification performance because of two reasons. First, the physical KB approach allows the limited training database to be extended to different targets not represented in the real database. Therefore, the combined approach performance is expected to be higher than the statistical KB approach. The second reason for the expected classification improvement is that the real database compensates for the modeling errors in the physical KB approach. This approach is schematically presented in Fig. 9.11.

9.7 EXPERIMENTAL RESULTS

In this section, classification performances according to various approaches were evaluated and compared. First, the classification performance of the statistical KB classifier with various classification features and decision rules was evaluated in a seven-class problem. Next, the classification performance of the statistical KB classifier was compared to the results published in the literature in a three-class problem. Finally, the physical KB classifiers were compared to the statistical classifier in the

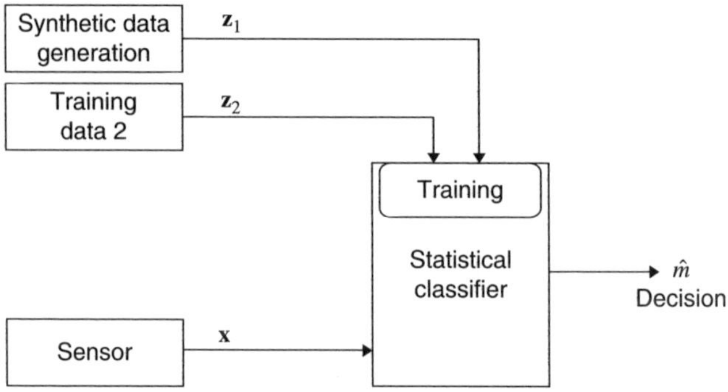

Figure 9.11 Combined classification approach.

three-class problem. In the performed classification tests, the system input was a series of signal frames, which were classified as one of the possible targets. Each classification frame of 4 s length consists of 44 nonoverlapping segments of 90 ms. For the statistical KB approach, the database of each target class was arbitrarily divided into three equal-size parts (of 175 classification records), and the classification performance was evaluated using three-fold cross-validation [39]. For the physical KB approach, a limited database of 50 records was used for determination of the valid subspace of the physical model parameters.

9.7.1 Statistical Knowledge-Based Classifier for the Seven-Class Problem

In this section, the classification performances of the statistical KB classifier with ML and majority voting decision rules are evaluated.

The confusion matrix of the statistical KB classifier with ML decision rule for the seven-class problem is presented in Table 9.2. An average classification rate of 88% was obtained in this test. The statistical KB classifier with majority voting decision rule and cepstrum coefficients was tested under the same conditions, and its confusion matrix appears in Table 9.3.

TABLE 9.2 Confusion matrix of the greedy GMM-based classifier with ML decision rule for the seven-class problem.

Class/Decision	One Person	Two Persons	Three Persons	Wheeled Vehicle	Tracked Vehicle	Clutter	Animal
One Person	**86**	14	0	0	0	0	0
Two Persons	0	**100**	0	0	0	0	0
Three Persons	0	0	**100**	0	0	0	0
Wheeled Person	0	0	0	**92**	8	0	0
Tracked Vehicle	0	0	0	23	**77**	0	0
Clutter	0	0	0	10	0	**96**	4
Animal	4	0	7	6	0	17	**66**

TABLE 9.3 Confusion matrix of the greedy GMM-based classifier with majority voting decision rule for the seven-class problem.

Class/Decision	One Person	Two Persons	Three Persons	Wheeled Vehicle	Tracked Vehicle	Clutter	Animal
One Person	**97**	3	0	0	0	0	0
Two Persons	0	**100**	0	0	0	0	0
Three Persons	0	0	**100**	0	0	0	0
Wheeled Vehicle	0	0	0	**100**	0	0	0
Tracked Vehicle	0	0	0	11	**89**	0	0
Clutter	0	0	0	0	0	**100**	0
Animal	0	0	0	0	0	11	**89**

In the confusion matrix of the majority voting based classifier in the seven-class test presented in Table 9.3, an average classification rate of more than 96% was obtained. The target models for each decision rule are optimized with respect to the classification rate criterion. Therefore, each target class is modeled differently for the ML and the majority voting decision rules.

In order to compare the performance of the proposed statistical KB classifier with references 2 and 4, the three-class problem consisting of one person, a wheeled vehicle, and a tracked vehicle was considered. The statistical KB classifier with ML and majority voting decision rules achieved average classification rates of 88% and 98%. The corresponding confusion matrices appear in Tables 9.4 and 9.5. For comparison, references 2 and 4 reported average classification rates of 87% and 90%, respectively, for the same test.[2] Note that classification between walking

TABLE 9.4 Confusion matrix of the statistical KB classifier with ML decision rule and cepstrum coefficients in the three-class problem.

Class/Decision	One Person	Wheeled Vehicle	Tracked Vehicle
One Person	**100**	0	0
Wheeled Vehicle	4	**64**	32
Tracked Vehicle	0	0	**100**

TABLE 9.5 Confusion matrix of the statistical KB classifier with majority voting decision rule and cepstrum coefficients in the three-class problem.

Class/Decision	One Person	Wheeled Vehicle	Tracked Vehicle
One Person	**100**	0	0
Wheeled Vehicle	0	**100**	0
Tracked Vehicle	0	7	**93**

2. This comparison is limited, because the classifiers were tested on different databases.

person and slowly moving vehicle is impossible using Doppler information only, because the vehicle radial velocity can be comparable with walking person velocity.

The performance of the statistical KB classifier with both decision rules was evaluated under additional scenarios with different class combinations, and the results are presented in Table 9.6. Although a good classification performance between wheeled and tracked vehicles was obtained in a binary test, as presented in Table 9.6, discrimination between these classes is the most difficult problem in the seven-class test case (see Table 9.2). Discrimination between classes of animal, clutter, and group of three persons is also a difficult problem. A good performance was obtained in classification between classes of a single person and groups of two and three persons.

9.7.2 Physical Knowledge-Based Classifier for the Three-Class Problem

A three-class classification problem of walking person, pair of walking persons, and slowly moving vehicle was studied. The radial velocity of the moving vehicle was considered to be small, such that the walking person and the moving vehicle cannot be distinguished using instantaneous Doppler information.

The training and testing databases were obtained using records of a low resolution ground surveillance radar. For the statistical, the indirect physical, and the combined approaches, a cepstrum feature vector was extracted from each segment of 90 ms length, and each classification frame of 4 s length was classified using 44 feature vectors. For the statistical KB approach, the database of each target class was arbitrarily divided into three equal-sized a parts (of 175 classification records), and the classification performance was evaluated using threefold cross-validation [39]. For the physical KB approach, a limited database of 50 records was used for determination of the valid subspace of the physical model parameters. A synthetic database was generated according to the physical KB approach scheme with random physical model parameter vectors. The synthetic database contained 600 records for each target class. The classification performance of the physical KB approach

TABLE 9.6 Summary of the classification performance of the statistical KB classifier with both ML and majority voting decision rules with cepstrum coefficients.

Classes	ML Classification Rate	Majority Voting Classification Rate
Tracked, wheeled vehicle	94%	95%
1, 2, 3 persons	98%	99%
1, 2, 3 persons, wheeled vehicle	94%	99%
1, 2, 3 persons, wheeled vehicle, clutter	92%	99%
1, 2, 3 persons, wheeled vehicle, clutter, animal	84%	97%
1, 2, 3 persons, wheeled and tracked vehicles, clutter, animal	88%	96%

was evaluated on 175 classification frames of the real data, according to the partition used for the statistical KB approach.

The sensitivity of the proposed approaches to SNR level was tested. For the SNR tests, the recorded radar echoes were contaminated with additive white complex circular Gaussian noise. The SNR was calculated at the output of the matched filter in the radar receiver by assuming that the recorded radar echo signals contain no noise. This assumption is justified, because the radar echoes were recorded at high SNRs. Figures 9.12 and 9.13 show the classification performance of various classification approaches versus SNR. Figure 9.12 shows the performance of classification between a walking person and a vehicle. The upper and lower subplots show the classification rate for various classifiers using the ML and majority voting decision rules, respectively. In a binary classification problem, the only difference between the ML and the majority voting is the threshold. In the presence of modeling errors under each hypothesis, optimization of the threshold in the majority voting decision rule is expected to improve the classification performance. Optimal threshold determination is important especially when the pdfs of the tested hypotheses overlap, as occurs at low SNRs. Therefore, the majority-voting-based classifier outperforms the ML-based classifier at low SNRs, and their performances coincide at high SNRs.

Figure 9.13 shows the average classification rate for a three-class problem, including one person, two persons, and a vehicle target. It can be observed that by using

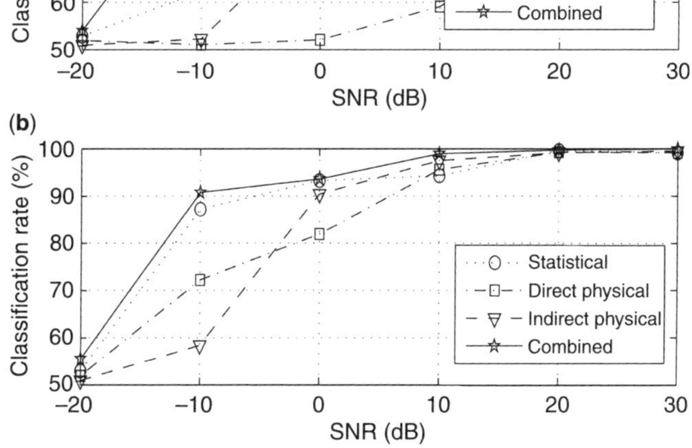

Figure 9.12 Classification rate versus SNR of the different classifiers in the binary problem: (**a**) ML classifier; (**b**) majority voting classifier.

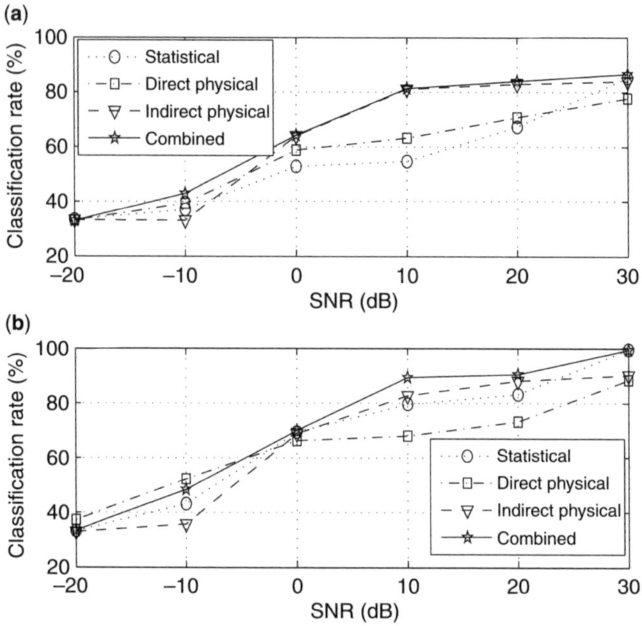

Figure 9.13 Classification rate versus SNR of the different classifiers in the three-class problem: **(a)** ML classifier; **(b)** majority voting classifier.

both decision rules, the combined KB classifier outperforms other classifiers at most SNR levels. Multiple thresholds in the majority voting decision scheme allow additional degrees of freedom for optimization of classification. Accordingly, the threshold SNR for the ML decision rule is higher than for the majority voting decision rule. The direct physical KB approach has the worst classification performance compared to other approaches, due to the modeling mismatch of the physical models.

For further analysis of the binary classification problem with a walking person and vehicle classes, the decision error trade-off (DET) is used. Figure 9.14 shows the DET curves for different classification approaches. These figures support the abovementioned conclusion that the combined approach provides the best classification performance. Additionally, at high SNR levels all of the methods achieve good classification performance.

The confusion matrices for the three-class problem for different classifiers are presented in Tables 9.7–9.10. The majority voting decision rule was used for the different classification approaches. The tests were performed with noise at an SNR of 10 dB and without noise. These tables show that in the noiseless case, the classification performance of the statistical classifier coincides with the combined KB classifier. However, at 10 dB SNR level, the combined KB classifier outperforms the other classifiers. In addition, the statistical and combined KB classifiers outperform direct and indirect physical KB classifiers with and without noise. As expected,

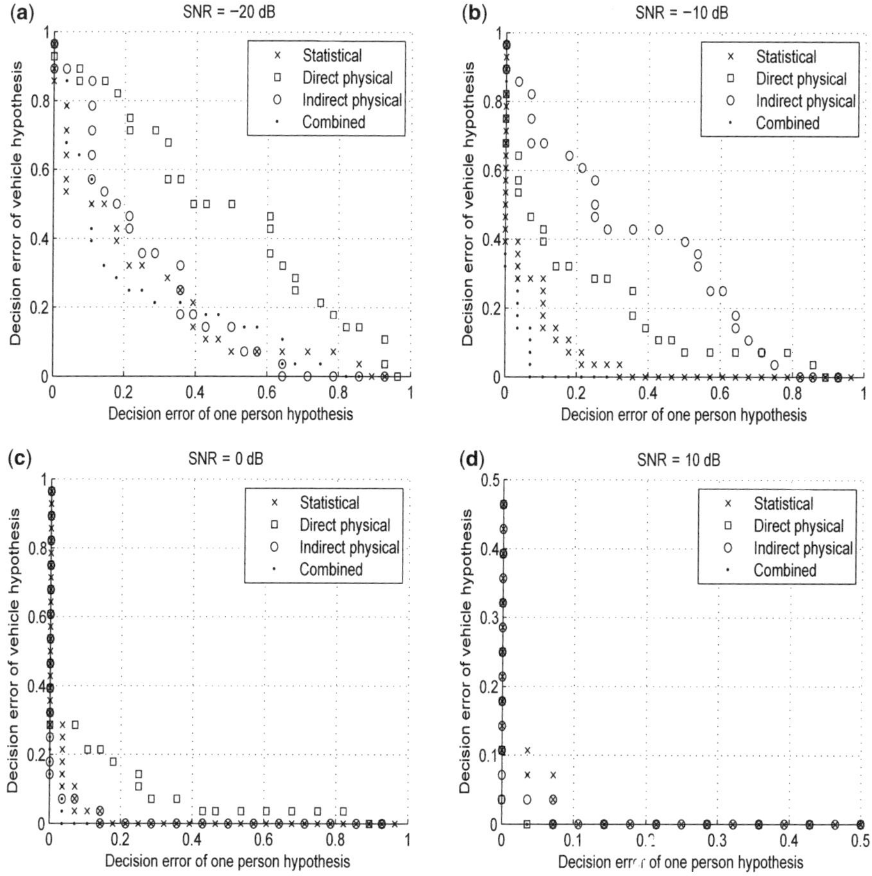

Figure 9.14 DET curves of the different classifiers at different SNRs: (**a**) SNR = −20 dB; (**b**) SNR = −10 dB; (**c**) SNR = 0 dB; (**d**) SNR = 10 dB.

TABLE 9.7 Confusion matrix of the statistical classifier using majority voting decision rule at SNR of 10 dB and without noise (in parenthesis).

Class/Decision	One Person	Two Persons	Vehicle
One Person	**83** (**100**)	14 (0)	3 (0)
Two Persons	16 (0)	**79** (**100**)	5 (0)
Vehicle	14 (3)	9 (0)	**77** (**97**)

TABLE 9.8 Confusion matrix of the direct physical KB classifier using majority voting decision rule at SNR of 10 dB and without noise (in parenthesis).

Class/Decision	One Person	Two Persons	Vehicle
One Person	**80** (**100**)	16 (0)	4 (0)
Two Persons	37 (21)	**61** (**79**)	2 (0)
Vehicle	29 (15)	7 (0)	**64** (**85**)

TABLE 9.9 Confusion matrix of the indirect physical KB classifier using majority voting decision rule at SNR of 10 dB and without noise (in parenthesis).

Class/Decision	One Person	Two Persons	Vehicle
One Person	**85** (**92**)	8 (4)	7 (4)
Two Persons	12 (7)	**80** (**89**)	8 (4)
Vehicle	10 (8)	6 (4)	**84** (**88**)

TABLE 9.10 Confusion matrix of the combined KB classifier using majority voting decision rule at SNR of 10 dB and without noise (in parenthesis).

Class/Decision	One Person	Two Persons	Vehicle
One Person	**91** (**100**)	8 (0)	1 (0)
Two Persons	10 (3)	**88** (**97**)	2 (0)
Vehicle	6 (0)	5 (0)	**89** (**100**)

classification between classes of a walking person and a pair of walking persons is a more difficult problem.

9.8 CONCLUSIONS

Knowledge-based target classification for ground surveillance Doppler radar was investigated in this work. The ML and the majority voting decision rules were applied to the proposed classification problem. The majority voting decision rule allows additional degrees of freedom, and therefore it was shown that the investigated classifiers with the majority voting decision rule outperform the corresponding classifiers with the ML decision rule. However, the drawback of the majority voting decision rule is that in the case of a small training database the overfitting problem due to multiple threshold determinations may cause performance degradation, especially in problems with a large number of classes.

Two sources of knowledge were considered for target classification: statistical and physical. The statistical knowledge was obtained from a training database of recorded target echo and the physical knowledge was obtained by developing locomotion models for the different targets. The statistical classifier was applied to a seven-class problem of radar targets such as walking person, group of walking persons, tracked vehicle, wheeled vehicle, animals, and clutter. The human operator's performance has also been evaluated and it was found that the derived classifiers outperform the human operator's classification.

In many cases, a training database may not be available, and in some cases, it may be insufficient to represent the different classes. On the other hand, the inaccuracy in the locomotion models results in limited classification performance. The best performance is achieved using a combined approach, which incorporates both the statistical and physical knowledge sources. The performances of the physical, statistical, and combined KB algorithms were tested using real data records from the three classes of one person, two persons, and vehicle. The results show that the combined approach outperforms both the statistical and physical model-based target classification algorithms. Further work can focus on the development of locomotion models for additional ground moving targets using a similar technique as presented in this chapter.

REFERENCES

[1] M. Skolnik. *Introduction to Radar Systems.* McGraw-Hill, New York, NY, 2000.

[2] M. Jahangir, K. Ponting, J.W. O'Loghlen. A robust Doppler classification technique based on Hidden Markov Models. *IEE Proc. Radar Sonar and Navigation*, 2002, p. 162–166.

[3] M. Jahangir, K. Ponting, J.W. O'Loghlen. Application of HMM to the classification of ground moving targets. *Proc. IGARSS 2000*, p. 33–36.

[4] A.G. Stove, S.R. Sykes. A Doppler-based automatic target classifier for a battlefield surveillance radar. *Proc. RADAR 2002*, p. 419–423.

[5] I. Bilik. Automatic target recognition using Doppler information. PhD dissertation, Ben-Gurion, University of the Negev, Israel, 2007.

[6] I. Bilik, J. Tabrikian, A. Cohen. GMM-based target classification for ground surveillance Doppler radar. *IEEE Trans. Aerospace and Electronic Systems* 2006;42:267–278.

[7] B. Bhau. Automatic target recognition: State of the art survey. *IEEE Trans. Aerospace and Electronic Systems* 1986;22:364–377.

[8] K. Copsey, A. Webb. Bayesian gamma mixture model approach to radar target recognition. *IEEE Trans. Aerospace and Electronic Systems* 2003;39:1201–1217.

[9] A. Zyweck, R.E. Bogner. Radar target classification of commercial aircraft. *IEEE Trans. Aerospace and Electronic Systems* 1996;32:598–606.

[10] C.M. Bachman, M. Charles, et al. Classification of simulated radar imagery using lateral inhibition neural network. *Proc. IEEE International Joint Conference on Neural Networks*, Baltimore, 1992, p. 279–288.

[11] N. Leslie. A comparison of 1-D and 2-D algorithms for radar target classification. *Proc. IEEE International Conference on Systems Engineering*, August 1991.

[12] M. Cooper, A. Srivastava. Asymptotic analysis of pattern-theoretic object recognition. *Proc. SPIE* 2000;4050:371–376.

[13] I. Jouny, C. Wu. Classification of radar clutter using features extracted from the time-frequency domain. *SPIE* 1997;3069:49–56.

[14] I. Jouny. Radar target identification using genetic algorithms. *SPIE* 1998;3371:152–158.

[15] P. van Dorp, F. Groen. Human walking estimation with radar. *IEE Proc. Radar Sonar Navigation* 2003;150:356–365.

[16] R. Boulic, M. Thalmann, D. Thalmann. A global human walking model with real-time kinematic personification. *Vis. Comput.* 1990;6:344–358.

[17] H.V. Poor. *An Introduction to Signal Detection and Estimation.* New York: Springer-Verlag; 1994.

[18] K. Fukunaga. *Statistical Pattern Recognition.* Boston: Academic Press; 1990.

[19] I. Bilik, J. Tabrikian, A. Cohen. Doppler-based ATR for ground surveillance radar. *Proc. RADAR 2004*, p. 221–226.

[20] J. Zwart, R. van der Heiden, S. Gelsema, F. Groen. Fast translation invariant classification of HRR range profiles in a zero phase representation. *IEE Proc. Radar, Sonar and Navigation*, 2003;150(6):411–418.

[21] S. Chen. Animating human locomotion. *Computer Animation*, 2003.

[22] S. Bandi, D. Thalmann. A configuration space approach for efficient animation of human figures. *Proc. Nongrid and Articulated Motion Workshop*, 1997, p. 38–45.

[23] M. Garcia. The simplest walking model: stability, complexity and scaling. *ASME: Journal of Biomechanical Engineering*, 1998;120(2):281–288.

[24] L. Liu, A.B. Wright, G.T. Anderson. Trajectory planning and control for a human-like robot leg with coupled neural-oscillators. *Proc. of Mechatronics*, 2000.

[25] T. Hastie, R. Tibshirani. Classification by pairwise coupling. *The Annals of Statistics* 1998;26(1):451–471.

[26] T. Hamamura, H. Mizutani, B. Irie. A multi-class classification method based on multiple pairwise classifiers. *Proc. ICDAR* 2003;123–128.

[27] J. Lee, I. Oh. Binary classification trees for multi-class classification problem. *Proc. ICDAR* 2003;129–133.

[28] S. Har-Peled, D. Roth, D. Zimak. Constrained classification for multiclass classification and ranking. *Proceedings on NIPS* 2002.

[29] I. Bilik, J. Tabrikian. Radar target classification using Doppler signatures of human locomotion models. *IEEE Trans. Aerospace and Electronic Systems* 2007;43(4):1510–1522.

[30] L. Rabiner, B. Juang. *Fundamentals of Speech Recognition.* Englewood Cliffs, NJ; Prentice Hall; 1993.

[31] D.A. Reynolds. A Gaussian mixture modeling approach to text-independent speaker identification. PhD dissertation, Georgia Institute of Technology, 1992.

[32] J.R. Deller, J.L. Hansen, J.G. Proakis. Discrete-time processing of speech signals. *Proc. SPIE* 1995;2484:595–602.

[33] J.Q. Li, A.R. Barron. Mixture density estimation. In: *Advances in Neural Information Processing Systems* 12, MIT Press; Boston, MIT, 2002.

[34] A. Dempster, N. Laird, D. Rubin. Maximum likelihood from incomplete data via the EM algorithm. *Journal of the Royal Statistical Society* 1977;39:1–38.

[35] N. Vlassis, A. Likas. A greedy EM algorithm for Gaussian mixture learning. *Neural Processing Letters* 2002;15(1):77–87.

[36] P. Stoica, Y. Selen. Model-order selection: a review of information criterion rules. *IEEE Signal Processing Magazine* 2004;31:36–47.

[37] J. Verbeek, N. Vlassis, B. Ǩrose. Efficient greedy learning of Gaussian mixture models. In: *Neural Computations*. MIT Press; Boston, MIT, 2002.

[38] J. Holland. Genetic algorithms. *Scientific American* 1992;267:66–72.

[39] R. Duda, P. Hart, D. Stork. *Pattern Classification.* Wiley-Intersience; New York, NY, 2001.

[40] A.W. Rihaczek, S.J. Hershkowitz. *Theory and Practice of Radar Target Identification.* Artech House; Boston, 2000.

10

MULTIFUNCTION RADAR RESOURCE MANAGEMENT

Sergio Luis de Carvalho Miranda, Chris J. Baker,
Karl Woodbridge, and Hugh D. Griffiths

Electronic scanning is set to become the norm for future radar systems. This brings about many advantages such as the ability to re-point the radar beam at will, to dwell in order to optimize detection, tracking, and classification, and to adjust spatial sensitivity to cope with sources of interference. It is now the radar that has to make many of the decisions previously carried out by a human, and this must be done on much faster and possibly pulse-by-pulse basis. This therefore gives rise to the question, "How does the radar make these decisions in order to maximize the potential offered by electronic scanning?" The answer requires the radar to move towards a degree of intelligence in the way processing is carried out; it needs to utilize maximum prior knowledge and must arrive at robust solutions to problems such as multitarget tracking, and so on. In this chapter we examine part of the answer to this question by comparing the performance of scheduling methods and of hard and soft logic approaches to the way in which volumes of surveillance space are prioritized. Examples are given for a generic naval radar system that has to cope with a number of evolving tasks to show the effects of the differing approaches.

10.1 INTRODUCTION

In recent years phased array antenna technology has been maturing rapidly, and this form of transduction is set to become the norm in complex and advanced radar systems. A phased array radar system can adapt its parameters on a near-instantaneous basis according to the way in which it perceives its operating environment. This allows the combination of functions such as tracking, surveillance, and weapon guidance, which were traditionally performed by dedicated individual radars.

Knowledge-Based Radar Detection, Tracking, and Classification. Edited by Fulvio Gini and Muralidhar Rangaswamy

Hence, phased array radar systems may be referred to as "multifunction radars." However, although being able to instantaneously and adaptively position and control the beam has clear advantages, it also brings a new set of challenges. Radar resource management is the problem of how to allocate finite available resources (e.g. power and time) in an optimal way to carry out a chosen mission, sometimes on a pulse-by-pulse basis. The intelligent management of radar resources in multifunction radar is central to unlocking the potential offered by electronic scanning. Indeed, the fact that a radar beam can be pointed in any direction or several directions at any time automatically requires the radar to make its own decisions as to how to deploy waveforms and exploit most effectively resources such as waveforms, pulse repetition frequency (PRFs), pulse lengths, and dwell times. This requires the radar to have a degree of intelligence unthought of in traditional mechanically scanned systems. Indeed, this is of fundamental importance if multifunction radar systems are to fulfill their potential. Consequently, radar resource management has been the subject of increasing research over recent years. In this chapter we concentrate attention on (1) automatic scheduling and (2) priority assignment of radar tasks.

Scheduling is an important subproblem of radar resource management due to the strong correlation between how and what tasks should be performed, as well as the time available to carry them out, and of course their timeliness. In the next sections of this chapter, two scheduling algorithms previously reported in the literature are compared in order to investigate differences in their performance. This provides new insights that may be invaluable in the development of future optimal schedulers. A detailed model of a multifunction radar has been developed to enable comparison using the same operational conditions. This can model both "underload" and "overload" situations.

Studies of scheduling have examined different approaches based on artificial neural networks, decision theoretics, information theory and mathematical programming techniques, including linear, nonlinear, and dynamic programming [1]. The analysis of radar resource management has often been divided into subproblem areas such as adaptive track updates, search scans, and scheduling [2–5]. Scheduling has been the focus of intensive research aimed at optimizing the process of forming the set of measurements to be carried out by the multifunction radar [6–12]. Although scheduling approaches tend to be dissimilar in design, little information has been reported about the differences in the resulting resource allocation performance when different scheduling algorithms are utilized in similar operational conditions. It might be expected that their performance should be similar when the radar system is operating in underload situations and there are sufficient available radar resources that can be allocated carrying out system functions. The differences between them might, therefore, be expected to become more evident when the environment changes and there are insufficient radar resources available. In these circumstances the schedulers would have to select tasks that would be undertaken based on their priorities, with lower priority tasks possibly being deferred.

In this chapter two different scheduling algorithms previously reported in the literature [10, 11] are compared in order to examine differences in the way radar resources are allocated as a function of changing environments. This also provides a degree of insight that may be exploited in the design and optimization of future schedulers. To make this comparison possible, a model of the multifunction phased array radar has been developed to represent important characteristics regarding the allocation of radar resources. This model includes two main radar functions, namely surveillance and tracking, as well as internal management functions such as priority assignment and scheduling. Other factors such as the impact of clutter and multipath are assumed to affect both schedulers in the same way and hence are not explicitly included.

The radar resource is finite and hence needs care in the way it is allocated. This subsequently leads to the necessity for appropriate prioritization in order to ensure the most important tasks are executed and those of least importance scheduled last, if at all. The job of the scheduler could be made much simpler if it were possible to pre-assign task priorities. Here we compare fixed and fuzzy logic approaches applied to sectorized zones. It is also vital that the form of prioritization and the impact on system performance is understood by both designer and operator. Thus, this chapter also addresses the development of an intelligent, adaptive prioritization assignment. The approach favored is one based on a fuzzy-reasoning algorithm used for ranking targets and sectors of surveillance in dynamically changing tactical environments. The performance of this approach is compared with other prioritization methods based upon more conventional "hard decision logic" and simple fixed prioritization.

An example of a generic tactical situation is shown in Fig. 10.1 for two different time instants of an area of under surveillance by a naval multifunction radar. It is divided into three sectors where t_1 and t_2 are different time instants. The figures identified in the sectors by the color red are hostile targets, yellow unknown targets, and green friendly targets. It is reasonable to expect that the priorities of both targets and sectors of surveillance should evolve between the two observation times and that the resource allocation will be influenced by this changing environment. For example, the number of hostile targets in sector 2 has increased over the time interval. Thus, taking

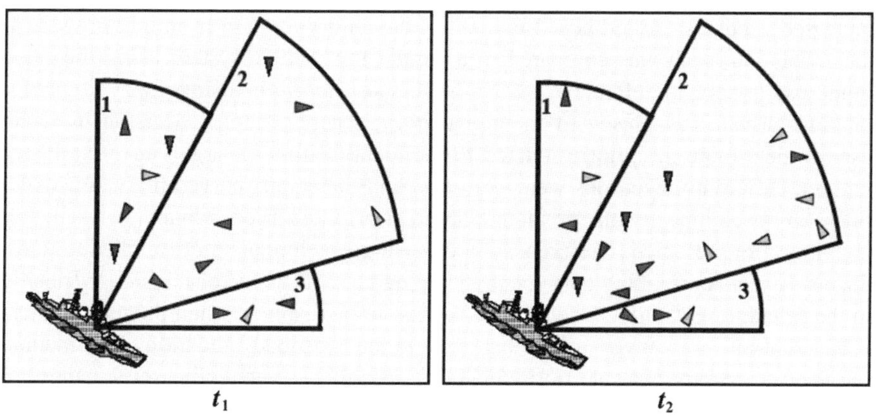

Figure 10.1 Example of a changing tactical scenario. (See color insert.)

into account the possibility of target splitting and there being more than one target, it would be desirable to increase the priority of this sector in order to detect any splitting in an overload situation. This simple situation exemplifies a concept that is accounted for in the priority assignment methods developed herein.

Radar resource managers are also likely to include a prioritization module that will have a close relationship with the scheduling function. The scheduling function has to consider a number of constraints such as time and energy in order to maximize the number of the tasks that are able to meet their desired deadline [11]. However, highly stressing scenarios may preclude some tasks from being carried out at all. Thus, the importance and priority of tasks will dictate ultimate achieved performance. For these reasons the ranking of radar tasks is an extremely important subproblem of radar resource management. It not only contributes to an efficient scheduling process by defining which conflicting tasks have to be delayed, but also influences the overall resource allocation by determining the tasks that will not be performed at all in overly stressing conditions.

There have been a number of publications in the literature addressing resource management from the perspective of the design of efficient prioritization and scheduling algorithms. For example, a system using neural networks, able to interleave radar tasks, was proposed in reference 14. Likewise, two approaches for scheduling tracking and surveillance tasks were presented in reference 9. The first is based on Operational Research (OR) theory; the other is based on temporal logic using artificial intelligence. In reference 10, heuristic algorithms were developed, applying the concept of on-line coupled-task scheduling. In this study the objective was to ensure tasks were carried out as near as possible to their required time. However, in these and other proposed resource allocation techniques, fixed prioritization ranking methodologies for radar tasks were applied [11, 15, 16].

Far fewer publications have addressed target ranking by using either a neural network or fuzzy logic based approach. These types of techniques have the advantage of making softer or slower decisions and hence not re-assigning resources instantly if they are not absolutely required. For example, a neural network approach was proposed in reference 17 for target priority assignment based on detected parameters. For increased accuracy, the proposed model has the ability to learn from training sets of data. This approach was compared to a simple fuzzy logic system using the same input parameters in reference 18. Similarly, an algorithm using fuzzy logic for ranking targets was presented in reference 19, where complex decision trees are used to determine the priority related to the identification of targets in surveillance sectors. Different trajectories were evaluated and the results proved the validity of a fuzzy approach in assessing the identification priority of targets. Finally, in a scheduling algorithm presented in reference 20, fuzzy logic techniques were applied to introduce soft decision-making concepts by labeling tasks as dangerous, agile, or friendly in the scheduling. Conflicts between tasks had to be resolved when the radar system was operating in an environment that leads to an "overload" situation. This method provided an adaptive way of prioritizing radar tasks that is potentially useful in the allocation of radar resources. Here, we develop a novel adaptive prioritization assignment, fuzzy-reasoning-based algorithm that is used for ranking targets and sectors of surveillance in dynamically changing tactical environments. The performance of this approach is compared with other prioritization methods based upon more conventional "adaptive hard decision logic" and simple fixed prioritization schemes.

This chapter is organized as follows. First, the radar model used in this study is described in Section 10.2. Section 10.3 provides descriptions of the two schedulers used in the comparison. In Section 10.4, the similarities and differences in the performance of the two schedulers are addressed. Section 10.5 examines issues regarding the choice of scheduler design for multifunction radar. Section 10.6 presents the architecture of the simulation used to carry out the analysis and describes methods of prioritizing tracking and surveillance tasks using fuzzy logic techniques. Section 10.7 examines the validity of using fuzzy logic to rank radar tasks. A number of different scenarios are examined to establish behavior when employing the fuzzy logic approach. These approaches are compared in Section 10.8 with fixed logic and fixed priority prioritization schemes. Section 10.9 assesses the relative merits of fixed and fuzzy logic approaches to prioritization and Section 10.10 provides a summary and draws overall conclusions.

10.2 SIMULATION ARCHITECTURE

To investigate and compare scheduling algorithms, a simulation model of a multifunction phased array radar is developed. This provides a better understanding of scheduler effects on the allocation of radar resources and on subsequent radar performance. Here we consider the behavior of just one single face of a multifaced phased array antenna radar system. For simplicity, this includes surveillance and tracking radar functions only. The architecture used in the radar model is presented in Fig. 10.2.

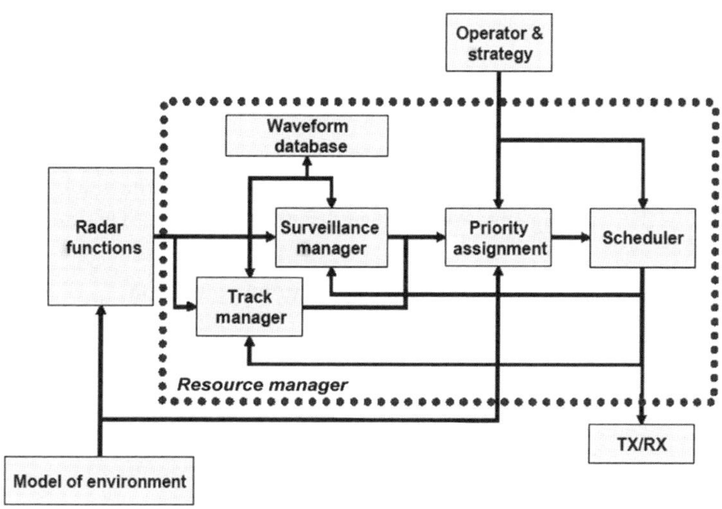

Figure 10.2 Architecture of the simulation.

This arrangement provides an environment in which different radar resource management techniques can be represented and examined against any given operational scenario. The radar model has a modular approach, adopted to allow comparison of different techniques while keeping other radar parameters fixed. In addition, different approaches can be compared assuming the same initial conditions and tactical characteristics for both targets and the environment. Apart from the scheduler, which is described in detail in Section 10.3, the main blocks of the model are briefly described in the following subsections.

10.2.1 Priority Assignment

This block assigns degrees of importance of the tasks that must be executed by the radar. Several factors must be taken into consideration when evaluating a task priority. These include evolution of the environment, the nature of the task (surveillance, tracking, and weapon guidance), degree of threat of the target under track, and so on. The fixed prioritization order presented in Table 10.1 is assumed as being representative of most scheduling approaches described in the literature. In this case, the relative importance of the tasks to be performed by the multifunction radar are determined a priori and not adapted as a result of changes in the tactical situation. Although this priority assignment may not be the most appropriate for adaptive resource allocations, it is suitable for the comparison of scheduling algorithms we examine here.

10.2.2 Surveillance Manager

This function maintains a queue of unscheduled surveillance tasks and provides the scheduler with a smaller list of requests that are close to their execution time. It also selects the parameters of the waveform to be used in the transmission of the radar pulses to meet the performance requirements in the surveillance mode.

10.2.3 Track Manager

Like the surveillance manager, the track manager keeps a list of unscheduled track requests, sending them to the scheduler when appropriate. The track requests are generated by the tracking process associated with each target. This determines the

TABLE 10.1 Priority order for scheduling used in the comparison.

Priority	Radar Job
1	Surveillance (lowest priority)
2	Track update
3	Track initiation
4	Plot confirmation
5	Track maintenance (highest priority)

next desired update time and the position of the radar beam for the measurement so that the requirements of tracking performance may be achieved.

10.2.4 Radar Functions

The surveillance function is responsible for creating a list of task requests that correspond to radar beam positions that must be adopted in order to maintain a required detection performance over a radar coverage area. The tracking function represents the filters used for predicting the positions of detected targets, and an adaptive Kalman filter was employed.

10.2.5 Operator and Strategy

The overall preferences depend on mission requirements and resource management decisions, which themselves are based upon evaluation of the tactical scenario. These are accounted for in the block "Operator & Strategy" in Fig. 10.2. This module operates as a human-machine interface allowing corrections or alterations in the system behavior.

Having introduced the main building blocks of the phased array radar system the two different types of scheduler to be examined are now described in more detail.

10.3 THE SCHEDULERS

The principle of multifunction radar scheduling is the effective planning of sequences of measurements to be executed in order to meet the performance requirements of the radar functions while observing a set of given constraints such as deadlines and available resources. Most scheduling approaches presented in the literature suggest that the decision-making regarding resource allocation must rely on the relative priorities of the radar tasks [6–11]. In situations where few resources are available, the scheduler selects the tasks that will be undertaken based on their priorities, and lower priority tasks may be deferred. To analyse how different schedulers behave in changing environments and different load conditions with respect to the allocation of radar resources, two types of scheduling algorithms described in the literature are compared here.

10.3.1 Orman et al. Type Scheduler

The first scheduling algorithm we consider is that proposed in reference 10, and is based on the concept of coupled tasks. A coupled task is a job consisting of two different operations separated in time by a specified interval. Thus, each coupled task can be represented by the processing time of the first task (transmission of radar pulses), the separation time between the tasks, and the processing time of the second task (reception of radar echoes). The scheduler can organize a queue of tasks to be executed in any way provided that two tasks do not overlap in the radar time-line. The separation time (or idle time) and the reception time can be estimated according to the radar function considered and via additional information such as range of target, boundaries of the surveillance region, and the transmission time of tasks. Once the duration of these is determined, interleaving algorithms may use the idle time for scheduling additional tasks. Hence, a better use of the radar time can be achieved. The flow diagram of this algorithm is presented in Fig. 10.3.

10.3.2 Butler-Type Scheduler

The second algorithm we consider is described in reference 11 and incorporates some improvements on a previous algorithm proposed in reference 6. A "time-balance" scheme controls the scheduling process of the requested tasks. There is a time-balance associated with every radar job, indicating to the scheduler how much time is owed by the radar to that job. A job consists of several tasks and is usually associated with surveillance of a region of coverage or with keeping a target under track.

A task is a group of activities that are noncoherently integrated to give detection and can be divided into looks, consisting of one or more activities. The algorithm used here computes the time-balance that represents the earliness or lateness of a task. When a task is late, its time-balance is positive. On the other hand, an early task has a negative time-balance. Finally, a task due to be undertaken at the exact moment has a time-balance equal to zero. If the job table receives a new task request that should not be scheduled for n seconds, the time-balance associated to that task should be set to $-n$ seconds. The time-balances of all tasks are incremented as time elapses. A new job is always inserted in the job table before its due time of execution and, hence, with negative time-balance. The flow diagram of this scheduling algorithm is presented in Fig. 10.4. In this way, the algorithm schedules the looks of a task in sequence, and the tasks are then selected according to the desired priority order.

10.4 COMPARISON OF THE SCHEDULING ALGORITHMS

Three aspects are assessed to evaluate the performance of the two schedulers These are

- the ability to schedule the tracking tasks close to their due time of execution,
- the ability to maximize the use of the radar time-line, and
- the behavior of resource allocation to very demanding tracking requirements.

It is important to stress that no task interleaving is assumed in the analyses of both schedulers. Thus, the waiting time between the transmitting and receiving periods is not considered when calculating the idle times of the radar time-line.

The surveillance performance was determined according to mission requirements and three sectors of surveillance were considered. For this comparison, a region of

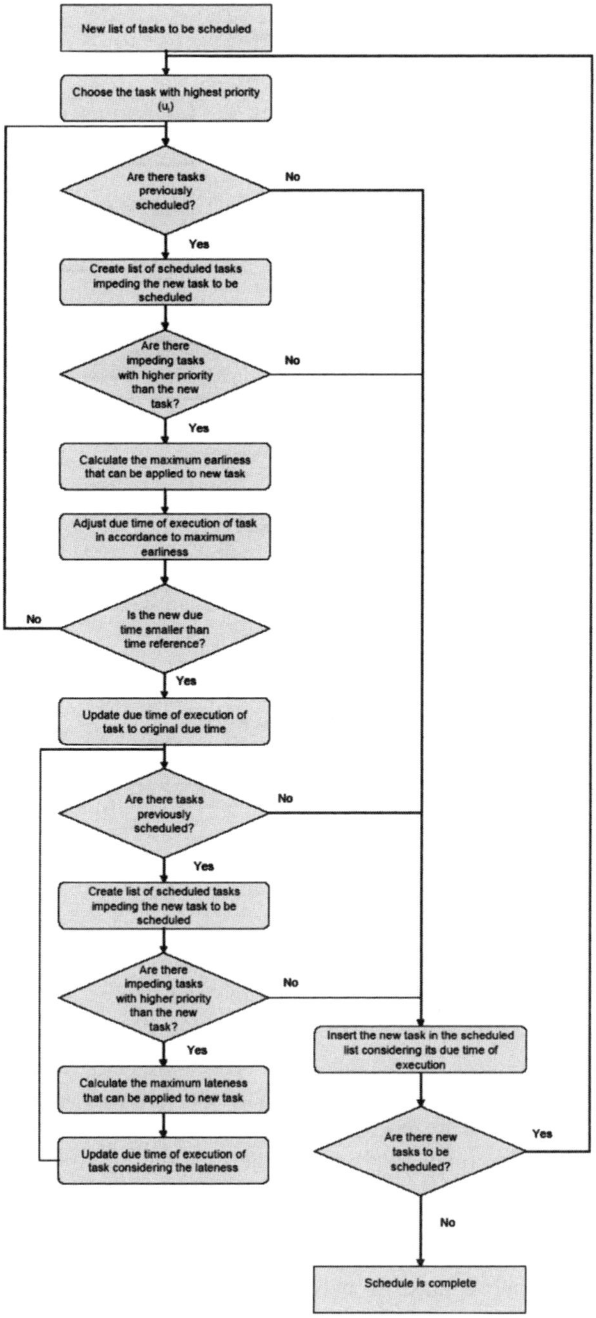

Figure 10.3 Flow diagram of the Orman-type scheduler.

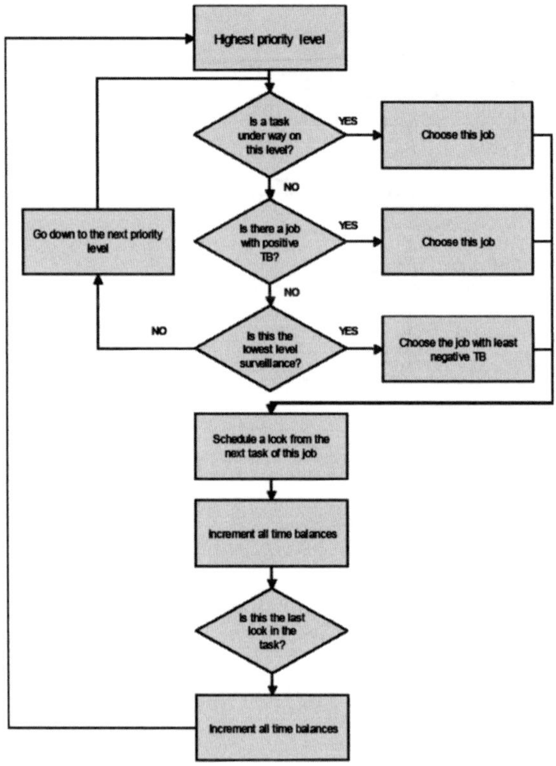

Figure 10.4 Flow diagram of the Butler scheduler algorithm.

surveillance spanning a 90° azimuth sector was assumed. The coverage volume was divided as follows:

- sector 1 spanned from 0° to 10° elevation;
- sector 2 spanned from 10° to 20° elevation; and
- sector 3 spanned from 20° to 50° elevation.

Dwell times are chosen to reflect the priority of the different sectors and to enable performance comparisons to be carried out. We now consider operation when the scenario is nonstressing (underload) and when it is stressing (overload).

10.4.1 Underload Situations

The first comparison we make is between the scheduling algorithms addressing resource allocation when the radar system is operating in surveillance-only underload configuration according to the parameters presented in Table 10.2. The parameters in Table 10.2 are indicative of the relative priorities assigned to each of the sectors. Sector 3 is assigned the highest priority and hence has the longest dwells, highest

TABLE 10.2 Scheduling parameters of the three-sector surveillance volume, using less demanding surveillance requirements.

Sector	Broadside Dwell Time (ms)	Occupancy (%)	Number of Beams	Function Time (s)	Frame Time (s)
1	1	16	108	0.108	0.675
2	2	24	135	0.270	1.125
3	3	40	360	1.080	2.700

occupancy, and so on. An underload situation is one where there is more time available than there are resources to allocate.

The first aspect we analyse is the ability of the schedulers to adapt radar function parameters in order to use all available resource. This was related to the variation of the maximum achieved surveillance occupancy according to the existing radar tracking load. Analysis of the resource allocation results presented in Fig. 10.5 suggests that the Butler scheduler used all the available radar time to execute the radar functions.

There was no period of time in which the radar was not performing any function when using this scheduler, because surveillance frame times were adapted according to the available resources. Therefore, low priority, tasks are executed earlier to avoid planning the set of measurements with idle times in the radar time-line. This figure presents results similar to those of a surveillance configuration, which demands 100% of the radar resources to meet the surveillance requirements, where frame times are smaller and surveillance occupancies are greater than those specified in the initial surveillance requirements in Table 10.2. In contrast, the Orman et al. scheduler did not offer this flexibility, as is shown in Fig. 10.6.

In underload situations, when radar functions do not demand all available resources, the scheduler maintained the surveillance allocation as initially specified, creating idle time intervals in the radar time-line. This behavior is explained by the

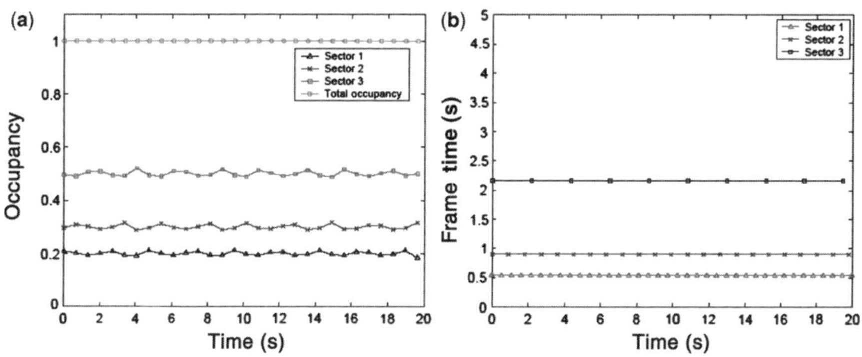

Figure 10.5 (**a**) Surveillance occupancy and (**b**) frame times after the simulation (period using the Butler scheduler, when no targets were detected) Required surveillance occupancy is 80%.

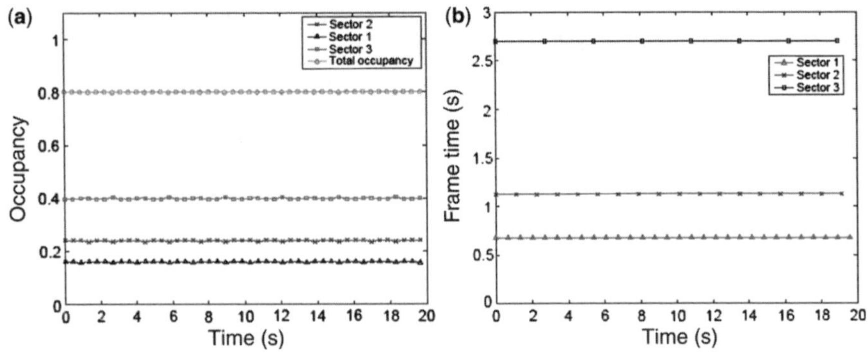

Figure 10.6 (**a**) Surveillance occupancy and (**b**) frame times after the simulation period. The required surveillance occupancy was 80%.

fact that once the desired surveillance occupancy was defined, the scheduler was only able to degrade the performance of this function in overload situations in order to transmit tracking tasks or any other high priority task that had conflicting transmitting times with the surveillance tasks [13]. The results presented in Fig. 10.6 show that, on average, approximately 20% of the radar resources were unused.

The examination of underload situations for both surveillance and tracking also gives similar results to those above. For example, Fig. 10.7a shows results of a situation using the Butler algorithm, where only 80% of radar resources were required for meeting the surveillance requirements and targets were detected.

Because the tracking load was small, the radar system does not face an overload situation. It is seen that no target is detected at first and, as a consequence, the scheduler uses all available time to perform surveillance. As targets were detected, less surveillance was carried out to allow the execution of tracking tasks. Throughout the evolution of this changing scenario, the scheduler adapted the amount of

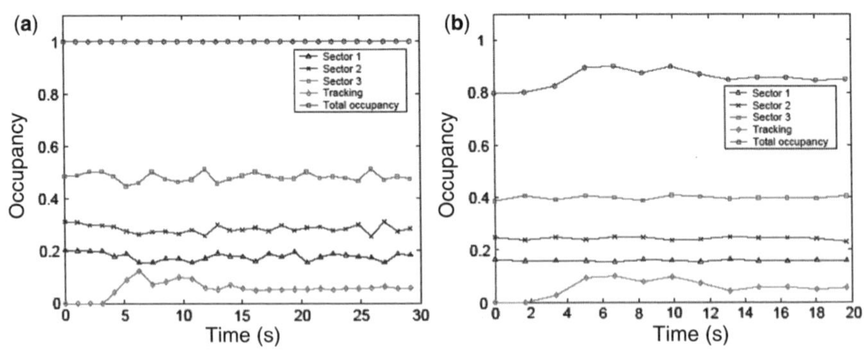

Figure 10.7 (**a, b**) Surveillance occupancy and frame times after the simulation period when targets are detected; (**a**) the Butler-type algorithm, (**b**) The Orman et al type aigorithm. Required surveillance occupancy was 80%.

surveillance performed to ensure that all available resources were applied. In contrast, when the Orman et al. scheduler was examined in the same scenario, the nonadaptive behavior shown in surveillance-only underload situations is repeated. Figure 10.7b shows an example of the resource allocation evolution when using this scheduler. The total radar occupancy remains at 80% until targets were detected. Thereafter, the total occupancy was increased; however, after the tracking load was stabilized, around 15% of the radar resources still remained unused.

This inability to use all free resources may represent a disadvantage of the Orman et al. algorithm when compared to that of the Butler algorithm. In underload situations, it might be desirable to avoid leaving resources unused when they could be used to enhance the performance of lower priority functions. However, it should be noted that both algorithms do achieve the specified performance and therefore may be considered to be satisfactory.

Another aspect of the comparison between the scheduling algorithms is the ability to schedule radar tasks close to their desired times of execution. The configuration used in this analysis was the same as that presented in Table 10.2. It is assumed that three targets are to be detected and tracked. The targets have different trajectories and are updated by an adaptive Kalman filter. For simplicity, the dwell times for the tracking updates are fixed at 5 ms.

Figure 10.8 presents histograms indicating the lateness and earliness of plot confirmation and track update requests when using the Orman et al. scheduler. This figure shows that almost all tasks were scheduled at their due times of execution. Different results are obtained when using the Butler scheduler. The results are presented in Fig. 10.9. In this case, both plot confirmation and tracking tasks are also scheduled early or late to allow a more effective use of the radar time. As track updates had lower priority than plot confirmations, more track update tasks were scheduled late. Nevertheless, the maximum earliness and lateness was about 6 ms, which is not significant in respect of track degradation, as the target would most likely remain within the radar beam. Nevertheless, in real systems this is an aspect of performance that will require care to ensure that tasks remain scheduled in a timely and well prioritized fashion.

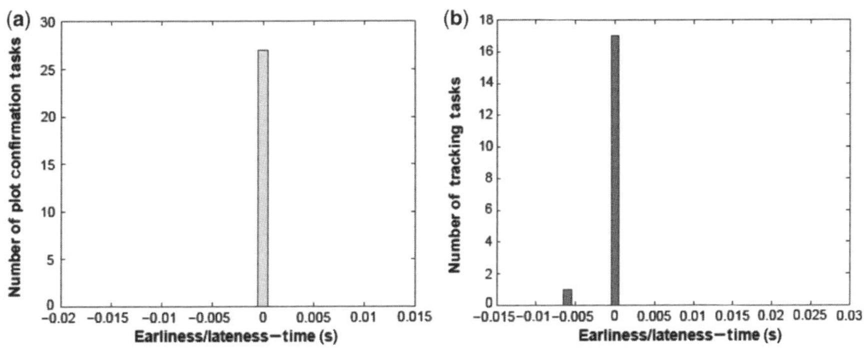

Figure 10.8 (**a, b**) Earliness/lateness of plot confirmation and tracking tasks using the Orman et al. type scheduler in an underload situation.

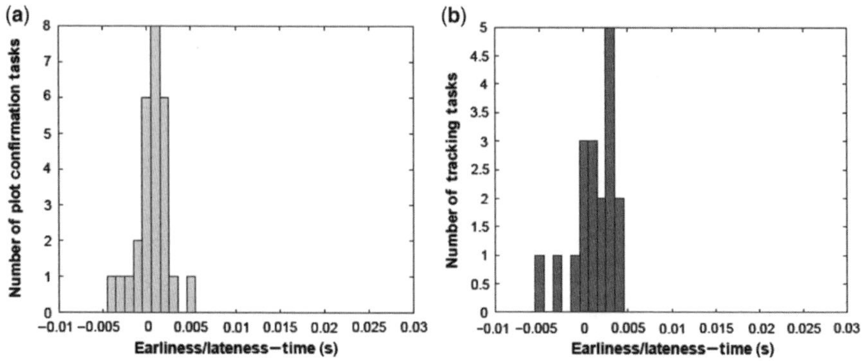

Figure 10.9 (**a, b**) Earliness/lateness of plot confirmation and tracking tasks using the Butler-type scheduler in an underload situation.

10.4.2 Overload Situations

When there are not enough resources to perform all requested tasks, the multifunction radar is operating in an overload situation. In this case, it is expected that some functions may have their performance degraded. Our initial overload configuration assumes that in order to meet the surveillance requirements, 100% of radar resources will be necessary. Therefore, when targets are detected, not all the surveillance requirements will be met. The purpose of using such high desired surveillance occupancy is to examine the Orman et al. scheduler when it has to use all available resources. The results of the resource allocation are presented in Figs. 10.10 and 10.11 for the Butler and Orman et al. algorithms, respectively. In both cases, as a consequence of target detections, the frame times in the surveillance sectors were increased and the corresponding occupancies decreased. This behavior is explained by the priority order for radar tasks assumed, forcing low priority tasks to be

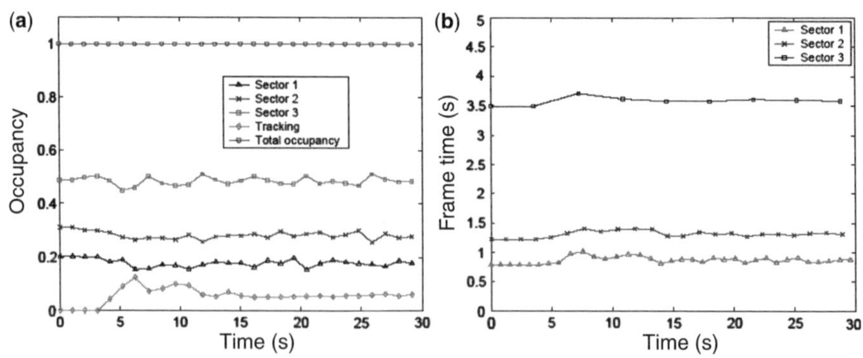

Figure 10.10 (**a**) Achieved surveillance occupancy and (**b**) frame times in an overload situation using the Butler-type scheduler.

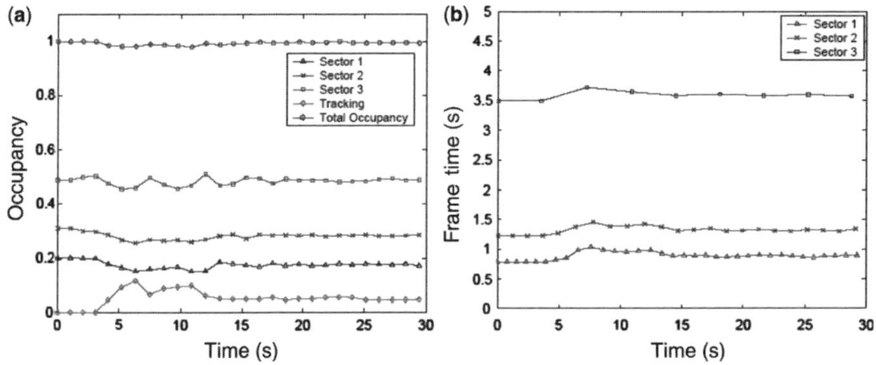

Figure 10.11 (**a**) Achieved surveillance occupancy and (**b**) frame times in an overload situation using the Orman et al. type scheduler. (See color insert.)

scheduled later than their desired execution times. In addition, the scheduler prepares the set of measurements in order to use all resources at all times.

So, low priority tasks were planned late and high priority tasks were scheduled as close as possible to their desired times of execution. The results suggest that both algorithms performed in a similar way in overload situations, allowing degradation of low priority functions in order to meet the requirements of the high priority functions. However, it is observed that, unlike the Butler algorithm, after the tracking process starts, the Orman et al. scheduler produces an average total radar occupancy around 98% after the target detections. Furthermore, the smallest achieved total radar occupancy occurs when a large number of plot confirmation and track updates have to be carried out over a small period of time (from 5 s to 11 s of simulation time in this example). This behavior may be explained by the design characteristics of the Orman et al. scheduler with respect to the planning of tracking tasks.

By using all available time resources at all times, the Butler scheduler uses radar resources more effectively, although the difference is quite small. Conversely, the Orman et al. scheduler allows idle times to be created in the queue of radar measurements even in overload situations, when there are not enough resources to perform all requested tasks. In this case, leaving the radar idle for any time interval implies additional degradation of the low priority functions, which is undesirable.

Similar conclusions to those in underload situations are made when analysing the histograms of earliness and lateness of both plot confirmation and tracking tasks in overload situations. The analysis assumes that the surveillance requirements are very demanding and, in order to meet the surveillance performance, 100% of the radar time resources should be allocated for surveillance. Several targets are successively detected and tracked. The results presented in Fig. 10.12 show the number of tasks scheduled by the Orman et al. scheduler during the simulation period and their respective earliness or lateness. It is seen that this scheduler behaves in a similar way in both underload and overload situations. Most tasks are scheduled at their due time of execution or very close to them. Once again, it is shown that the Orman et al. scheduler has a very good ability to schedule both plot confirmation and tracking

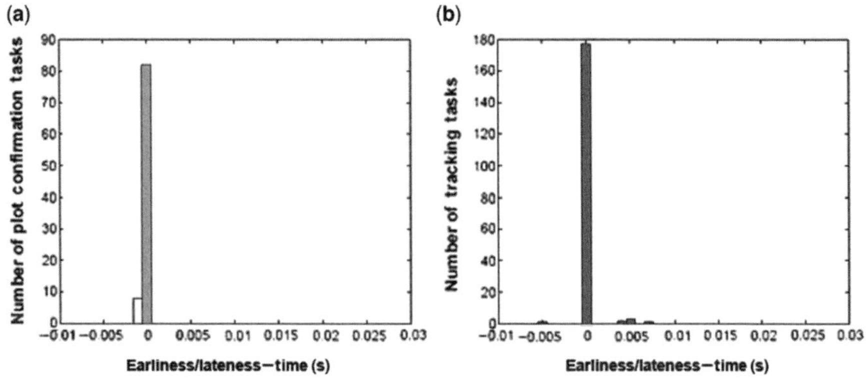

Figure 10.12 Earliness/lateness of **(a)** plot confirmation and **(b)** tracking tasks using the Orman et al. type scheduler in an overload situation.

tasks using very small earlinesses and latenesses (i.e. close to their due time). The maximum earliness and lateness observed in the queue of tasks was around 7 ms, which does not represent significant degradation in the tracking performance.

Likewise, Fig. 10.13 presents the number of tasks scheduled by the Butler algorithm during the simulation period and their respective earliness or lateness. These results show that the scheduled times for both plot confirmation and tracking tasks were distributed around their due time of execution. Once again, this scheduler allowed a greater number of tasks to be displaced from their desired execution time with the purpose of using the available time resources more effectively.

Lastly, the scheduling algorithms are compared with each other to examine how they allocate radar resources in situations where tracking requirements are so demanding that in order to achieve the desired tracking performance almost all time resources available are used for tracking. To simulate this overload condition, the surveillance configuration presented in Table 10.3 is assumed, and 30 high priority targets are

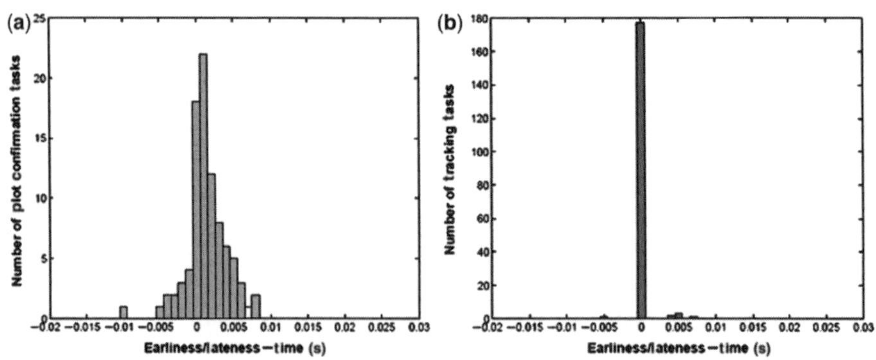

Figure 10.13 Earliness/lateness of **(a)** plot confirmation and **(b)** tracking tasks using the Butler-type scheduler in an overload situation.

TABLE 10.3 Scheduling parameters of the three-sector surveillance volume used in the analysis of a high tracking load situation.

Sector	Broadside Dwell Time (ms)	Occupancy (%)	Number of Beams	Function Time (s)	Frame Time (s)
1	1	20	108	0.143	0.715
2	2	30	135	0.359	1.125
3	3	50	360	1.739	3.478

detected and tracked. The parameters used for the tracking configuration are presented in Table 10.4. Two different cases are examined regarding the priorities of the surveillance sectors. In the first case, all sectors have the same priority, whereas in the second case the priorities differ from sector to sector. The tracking configuration represents a very demanding tactical situation that is, perhaps, rarely found in real situations; nonetheless, it provided useful insights into the behavior of the algorithms in respect of the effective allocation of radar resources.

In the first case examined, the scheduling algorithms considers only three priority levels: plot confirmation (highest level), tracking, and surveillance (lowest level). Figures 10.14a and 10.15a show the evolution of resource allocation when all surveillance tasks have the same level of priority. The results show that as new targets are successively detected, due to the very demanding tracking situation, no surveillance is performed beyond around 23 s after the simulation is started. This happens because fewer resources are available for low priority functions as the tracking load increases. When all available resources are used for tracking, all surveillance is stopped. As the surveillance tasks in all sectors had the same priority, these free resources were shared between them according to their original surveillance requirements. This analysis suggests that if the multifunction radar operates in an environment where several high priority targets have to be tracked accurately, the majority of resources might be allocated for tracking these targets and hardly any surveillance would be carried out. In real radar systems, it may be desirable to provide an additional control to monitor the achieved surveillance resource allocation in all sectors, and a minimum performance might be set. Thus, the situation where no surveillance is carried out at all would be avoided.

A different behavior was observed when the priorities assigned to the sectors of surveillance were different. The same surveillance and tracking configurations of the previous examination were considered. However, the priority order presented in Table 10.5. was assumed, where each surveillance sector has a different priority.

TABLE 10.4 Tracking configuration in an overload situation.

Number of High Priority Targets	Nominal Waveform Dwell Time (ms)	Update Interval (ms)
30	5	150

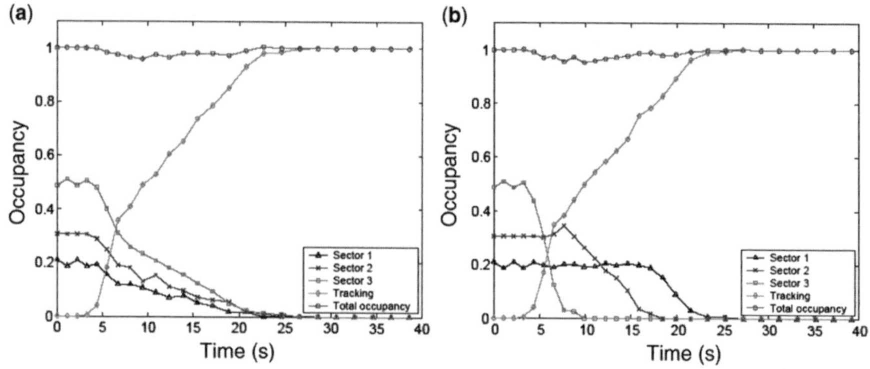

Figure 10.14 Achieved surveillance occupancy and frame times in an overload situation using the Orman et al. type scheduler: (**a**) priorities of surveillance are the same; (**b**) priorities of surveillance are different.

This could, for example, account for prior knowledge about the expected threats and the area in which the radar was operating. For this case, the final resource allocation is presented in Figs 10.14b and 10.15b.

Results of the simulations suggest that, once again, both schedulers tend to maintain the performance level of high priority functions in overload situations, only executing low priority functions if there is available time. As a consequence, rapid surveillance degradation in sector 3 is observed soon after the first few targets are detected at 4 s of simulation time, and, as more targets are detected, the surveillance performance is degraded progressively in sectors 3, 2, and 1. Previous knowledge attributed the highest surveillance priority to sector 1. This explains why resources for surveillance are maintained for this sector until around 19 s. From that time instant, this was the sector that had to free resources to allow new tracking tasks to be carried out. After some time, the tracking load was so high that no surveillance was performed at all.

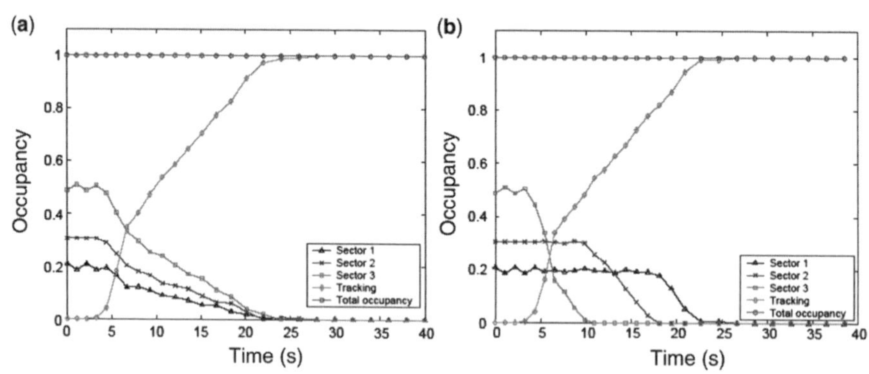

Figure 10.15 Achieved surveillance occupancy and frame times in an overload situation using the Butler type scheduler: (**a**) priorities of surveillance are the same; (**b**) priorities of surveillance are different.

TABLE 10.5 Priority order for the comparison of schedulers, when the surveillance sectors had different degrees of importance.

Priority	Radar Function
1	Surveillance in sector 3 (lowest priority)
2	Surveillance in sector 2
3	Surveillance in sector 1
4	Track update
5	Track initiation
6	Plot confirmation
7	Track maintenance (highest priority)

The analysis of resource allocation considering different surveillance sector priorities suggests that an adaptive prioritization scheme to assess the relative importance of the surveillance sectors in a changing environment may lead to a more effective use of the available resources as the surveillance performance may be maintained in important sectors. This will be the subject of future research.

As expected, the results of the Orman et al. scheduler show idle times in the radar time-line while surveillance was carried out. However, this behavior is not seen when only tracking tasks are performed. These results may be explained by the fact that all tracking tasks were assumed to have the same priority. Therefore, the available time resources were shared between all tracking tasks, allowing their update interval to be increased to enable all tracking tasks to be executed. In situations where long dwell times are used for tracking targets or narrow beamwidths are used for tracking missiles, this delay may be large and therefore undesirable.

10.5 SCHEDULING ISSUES

The scheduler is an important component of the resource manager in multifunction array radars. From the results presented, it is seen that both schedulers perform in a similar way for the conditions used in the simulations, especially when analysing the resulting radar resource allocation in overload situations and their ability to schedule the tracking tasks as close as possible to their desired times of execution. The difference in their performance is related to how the radar time-line is used by the scheduler. The Orman et al. scheduler is not able to make use of all available resources in underload situations. In addition, it left some idle times (gaps) in the radar time-line-even when operating in overload situations. As a result, the radar resources were not fully allocated to perform radar functions. This behavior is not desirable as it implies an additional performance degradation for low priority tasks, as time is not used to provide any information about the environment for the radar system. That is to say, radar time is wasted at the expense of the performance of less important radar functions. In contrast, the Butler scheduler makes use of all the available radar time resources by utilizing the entire radar time-line.

By scheduling some tasks earlier or later than their due time of execution, this algorithm provides a better utilization of radar time resources. This early or late scheduling could cause performance degradation if the tasks are performed far from their desired execution times or from their latest update times, and this may require modification of the algorithm. However, results of analyses in Section 10.4 show that the largest displacement from the desired execution time is around 7 ms, which should not cause significant performance degradation for the radar functions considered here.

Finally, it may be inferred that despite the fact that different radar schedulers may be proposed using a variety of techniques to create the set of measurements to be performed, little overall difference is observed in resource allocation when operating in overload situations. Thus, it might tentatively be concluded that the detailed implementation of a scheduler is not going to significantly impact the overall system performance. However, in real systems, whether or not tasks are performed early or late or whether all resources are used to give the most effective performance may be crucial. This can only be tested in real systems undergoing full-scale experiments, which are outside the scope of the research presented here. In general, the schedulers allocate the resources in order to allow the high priority tasks to meet their performance requirements and then share the remaining resources amongst the low priority tasks. Hence, although making resource allocation predictable, the use of fixed prioritization schemes does not take into account changes in the environment. In contrast, changing the priorities of radar functions may lead to very different final resource distribution and, as a consequence, different performance results for all radar functions. Thus, adaptive methods to assess the priorities assigned to each of the radar functions may significantly affect multifunction radar resource management. We consider this in the next sections of this chapter.

10.6 PRIORITIZATION OF RADAR TASKS

In this section, an intelligent, adaptive prioritization assignment fuzzy-reasoning-based algorithm is developed. This algorithm is used for ranking targets and sectors of surveillance in dynamically changing tactical environments. The prioritization approach is presented and resource allocation in different scenarios is assessed when subjected to three different prioritization methods:

- fuzzy logic adaptive prioritization;
- "hard logic" adaptive prioritization; and
- fixed prioritization, based on predefined criteria.

The overall resource management architecture used in the comparison is presented in Fig. 10.16. This provides an environment in which the different prioritization approaches can be represented and examined against any given scenario and scheduling system. Here again, the behavior of one face of a multiface phased array antenna radar system is considered. For simplicity, the analysis includes only surveillance and tracking radar functions. A modular approach was used in developing the simulation

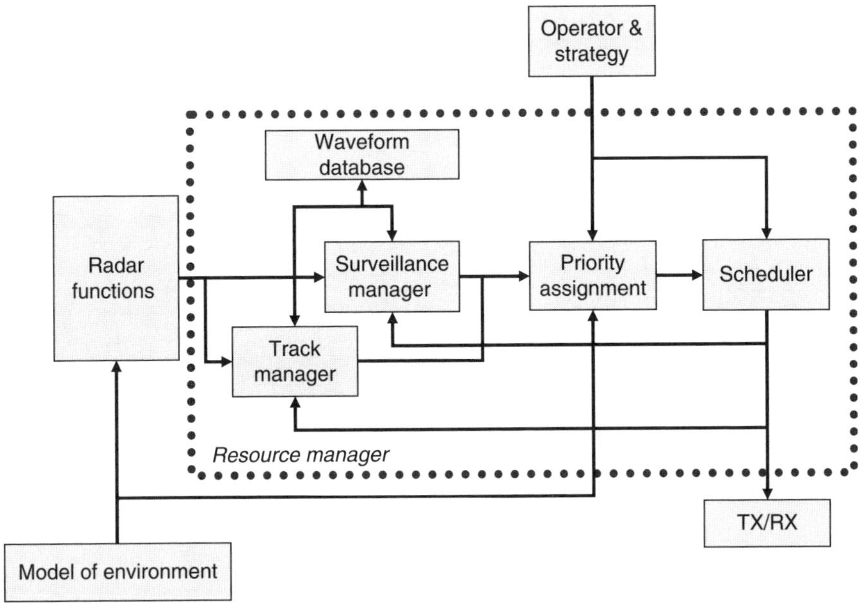

Figure 10.16 Block diagram of the simulation architecture used in the comparison.

model. The advantage of this modularity was that comparisons of different approaches can be performed keeping other radar parameters fixed. The approaches were compared under the same initial conditions and the same tactical characteristics in respect of targets and environment.

In typical naval applications, for example, different sensors coexist with a multifunction radar on the same platform. Thus a component of the prioritization approach is to use information provided by these sensors to assess the importance of a given radar function with higher fidelity. In this way, the identity of a target can be inferred by using an Identification, Friend or Foe (IFF) system. It is also reasonable to consider that, in some situations, fewer radar resources should be spent tracking friendly targets at long range than looking for new targets in a surveillance sector where an increasing number of threatening targets are being detected. Thus, as the tactical environment evolves, the radar task priorities should also be adaptively reassigned, resulting in a continuous reallocation of the radar resources.

10.6.1 Prioritization of Tracking Tasks

The priority of tracked targets was evaluated using the decision tree presented in Fig. 10.17. The information required to assign a priority is provided by a tracking algorithm, by other sensors, or by other operational modes of the multifunction radar. Five different variables provide information concerning the priority:

- quality of tracking,
- hostility,

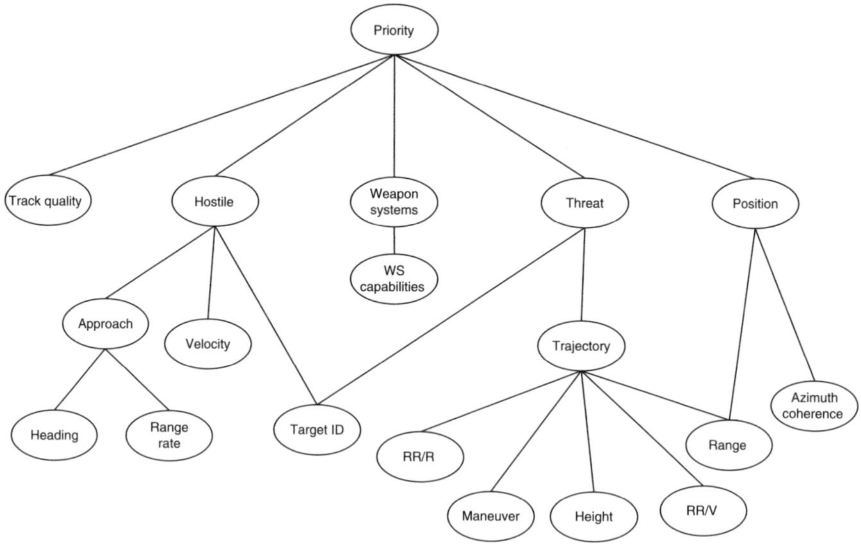

Figure 10.17 Decision tree for target priority assessment.

- degree of threat,
- weapon system capabilities of the platform, and
- relative position of the target.

"Track quality" refers to the accuracy of the predicted position of the target with respect to the desired accuracy. "Hostile" is a fuzzy variable related to four concepts: range to the targets, absolute target velocity, identity, and the way the target is approaching the radar platform. Thus, according to the way the target is approaching the radar platform, its absolute velocity, its range, and its identity, the priority for tracking may vary. The variable "Weapon systems" represents the importance of a target with respect to the weapon systems of the radar platform. In order to assess this importance, three concepts are combined: the identity of the target, the operational range of the weapon systems, and the ratio between the range rate and the absolute velocity of the target. "Threat" is the linguist variable that represents the degree of threat of a target according to its trajectory and identity. Trajectory combines four fuzzy variables: height, maneuver, absolute velocity, and range rate with respect to the trajectory on which the target is moving. Finally, "Position" is a linguistic variable whose value is given by the combination of the fuzzified values of the range and azimuth of a target.

Fuzzy values were attributed to each variable. Some examples of the fuzzy values are presented in Table 10.6. After evaluation of these variables according to a set of fuzzy rules, the importance (priority) of the target is determined.

10.6.2 Prioritization of Sectors of Surveillance

A similar methodology was applied to the surveillance function based upon the decision tree presented in Fig. 10.18. In this case, the priority of surveillance

TABLE 10.6 Examples of fuzzy variables used in the assignment of priorities for targets.

Fuzzy Variable	Fuzzy Values
Priority	Very low, low, medium low, medium, medium high, high, very high
Hostile	Nonhostile, unknown, and hostile
Weapons systems	Low, medium, and high priorities
Threat	Very low, low, medium low, medium, medium high, high, and very high
Position	Close, far, medium

Figure 10.18 Decision tree for sectors of surveillance priority assessment.

sectors was assessed through the original priorities attributed to the regions with respect to the expected tactical scenarios and the information gathered during the evolution of the actual environments. This includes aspects such as

- the original priority,
- the rate of detection of new targets,
- the number of threatening targets, and
- the rate of detection of new threatening targets.

The fuzzy variable "Original priority" is the fuzzified value of the importance of a sector of surveillance, after taking into account the a priori information about the radar environment. The linguistic variable "New target rate" is the fuzzified value of the rate of new targets that are either detected in the sector or move into it coming from another sector. Likewise, the variable "Number of threatening targets" is the fuzzified value of the number of threatening targets detected in a given surveillance sector. Sectors with a high number of threatening targets are important due to the possibility of target splitting. Finally, the linguistic variable "Threatening target rate" is the fuzzified value of the rate of detection of new threatening targets. In sectors where threatening targets already exist, this rate may represent either the results of target splitting or the detection of new threatening targets in the coverage area of the multifunction radar. However, in sectors where no such targets are detected, this rate may represent the detection of a new wave of attacking targets approaching the radar platform from a different area.

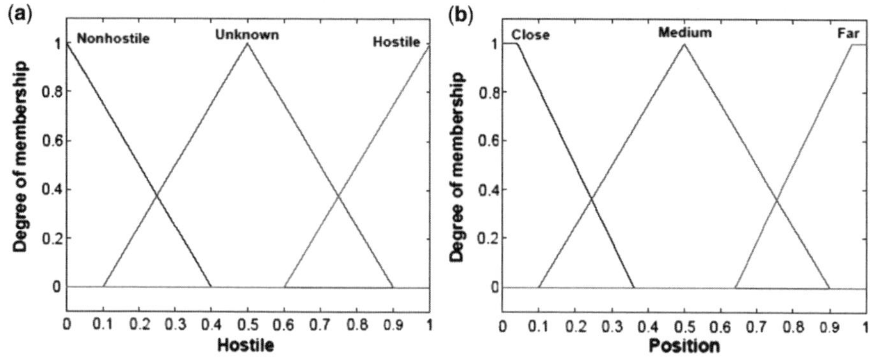

Figure 10.19 Fuzzy membership functions related to the fuzzy variables representing the degree of hostility and the position of a target.

10.7 EXAMINATION OF THE FUZZY LOGIC METHOD

Figure 10.19 is an example presenting the membership functions employed for the fuzzy values of the hostile and position variables used for evaluating the priorities of the targets. Apart from the track quality, all the variables are fuzzified in early stages of the priority evaluation. These fuzzifications explain the fact that the universes of discourse, or domains, of all variables whose membership functions are presented vary between 0 and 1. The output of the system is the evaluated priority of the target under study, which is represented by the fuzzy variable "Priority."

The fuzzy representation of the input and output parameters is considered complete when the membership functions, the domain of each variable, and the number of fuzzy sets (values) are computed [22]. Thus, the determination of the fuzzy if–then inference rules is the next step in the design of the fuzzy-logic-based prioritization system. The number of fuzzy rules required to assess the value associated with a knot in the decision tree of Fig. 10.17 is equal to the product of the number of fuzzy values that compose the fuzzy variables linked to that knot. Table 10.7 shows the number of if–then rules used to evaluate the main fuzzy variables assumed here.

TABLE 10.7 Fuzzy values related to the main variables used in the priority assignment.

Fuzzy Variable	Number of Fuzzy Rules
Priority	270
Hostile	108
Weapon systems	16
Threat	36
Position	6

Here, the inferential rules of the fuzzy system are written on the basis of intuitive and expert considerations and are then tuned by simulation tests. The actual number of rules used in the inference system may, in some cases, be smaller than the required number. This is due to the fact that particular combinations of fuzzy variables are very unlikely to be observed in real systems. The reduced number of rules does not represent a drawback, as max–min associations are used by the fuzzy inference system. These associations ensure that the truth of an assertion is not affected by the number of contributing rules but by the degree of truth of the dominant term. The evaluation of the fuzzy rules must follow the sequence proposed on the decision tree. Thus, the system inputs are fuzzified and successively used to assess other fuzzy variables in cascade to the point where the final priority is obtained. It is not particularly straightforward to evaluate how the resulting target priority is modified as a function of the main fuzzy variables by only examining the fuzzy rules. Therefore, graphic representations are a valuable tool to assess the inferential rules. These representations may be obtained by fixing all the variables except two involved in the evaluation of the priority. An example of the obtained surfaces is presented in Fig. 10.20. The configuration presented in Fig. 10.20 assumes that the values of three variables ("Track quality," "Position," and "Weapon capabilities") are maintained at 0.5, and both "Threat" and "Hostile" are varied over their respective domains. This configuration may represent a situation in which the target is located at medium range and has "medium" importance in respect of the weapon systems of the radar platform.

Having defined and tuned the fuzzy if–then rules, the method for prioritizing the relative importance of tracked targets must be validated against test trajectories. For all test trajectories, scenarios consist of targets with different identities. The analysis

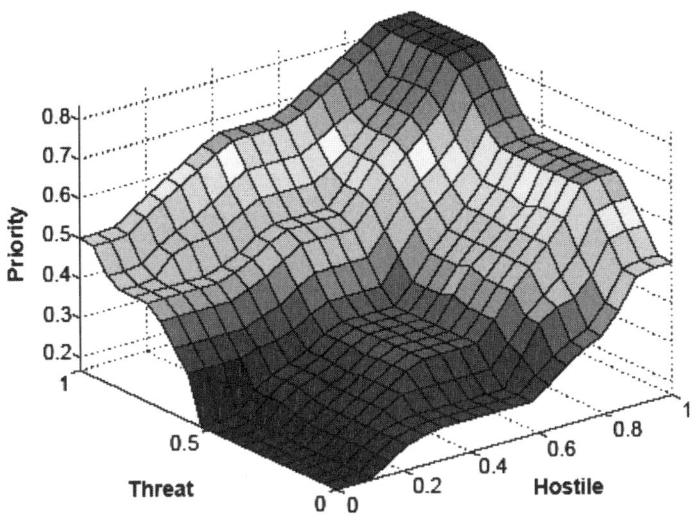

Figure 10.20 Graphic representations of the fuzzy rules, considering three fixed variables: "Position," "Track quality," and "Weapon systems."

shows that by knowing the identity of the targets their priorities may vary. This provides valuable information to be accounted for when deciding how to allocate radar resources in overload situations. Two cases are presented here for targets moving towards the radar platform on constant-velocity straight-line trajectories. These have been chosen as they represent situations of a high degree of threat where targets may be moving towards the radar platform in order to start an attack. In addition, they present the behavior of the method when a variable such as "Approach" is fixed. This helps to simplify the analysis and evaluation of the reasons for the prioritization. The system can also be examined in more complex scenarios where all variables involved in the prioritization are changing over the simulation.

The left-hand side of Fig. 10.21 shows the first test trajectory where a target moves towards the radar platform on a straight line, having a constant velocity of 300 m/s. The red circle indicates the origin of the trajectory. Three targets are assumed in the analysis. They have the same dynamics and flight height; however, their likelihoods of being hostile are different, as follows: 1 (hostile), 0.5 (unknown), and 0.1 (friendly), corresponding to the red, blue, and green curves respectively. The evolution of the resulting priorities is shown in the right-hand side of the figure and shows that, in general, all priorities increase as the targets move towards the radar platform; the greater the likelihood of being hostile, the greater the resulting priority. Figure 10.21 also suggests that priorities of targets that have unknown identity present a similar behavior to friendly targets in the early stages of the trajectory. This may be explained by the fact that during that period, the range of the targets is longer than the tactical range of the platform weapon systems. This happens until around 80 s. From that instant, as the target is moving close to the boundaries of this weapon systems tactical range, the degree of threat of the unknown target is likely to increase. Thus, its priority evolution has similar behavior to the priority evolution of the hostile target: the closer the unknown target, the higher will be its priority. At short ranges, if the identity of the target is still unknown, the target is assumed to be hostile, and its resulting priority is assessed as that.

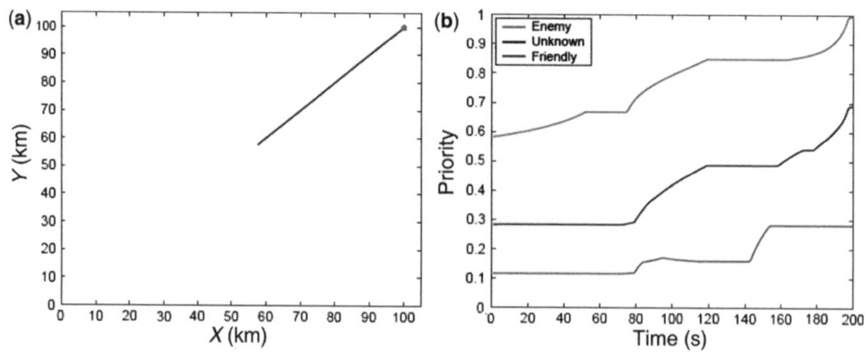

Figure 10.21 Resulting priorities for three targets with different threat levels of being hostile, moving on the same trajectory. (See color insert.)

Figure 10.22 presents the results of a simulation where targets are assumed to move on a straight-line trajectory with a velocity of 800 m/s. The same possibilities of being hostile as the previous analysis are considered. Owing to the high velocity and short ranges, the evolution of the priorities is rather different from the previous case. During the first few seconds of simulation, both unknown and hostile targets have slightly higher priorities than in the first example. This may be explained by their high velocities. All target priorities remain fixed until about 30 s, when the target position is getting close to the weapons system operational range. Before 30 s, all targets have the maximum priority possible for the set of characteristics of their dynamics, identity, and the capabilities of the weapon systems. Thereafter, the priorities are increased in order to allow the radar platform to face the threat. It is observed that, from around 30 s to 60 s of simulation time, the priority of the unknown target presents a high rate of increase. The analysis indicates that more importance is progressively given to this target, which is gradually assumed to be like a hostile target, because its velocity is very high, the target is approaching the radar quickly, and its identity is unknown over this period. From around 60 s to 85 s, the unknown target has the highest priority possible for the combination of input variables that determine its importance. From 85 s, its priority increases again, reaching its highest at around 100 s, when the target position is within operational range of the platform weapons systems, and as a consequence both hostile and unknown targets have the same priority. Such an unknown approaching target is considered to be of highest importance because of its potential degree of danger, represented by its velocity, and the way it is approaching the radar platform. Like the unknown target, the priorities of both hostile and friendly targets increase from around 30 s, as they are getting close to the weapons system operational range. These priorities continue to increase, reaching their maximum values not later than after 100 s of simulation.

We now consider using fuzzy logic to assign priorities to sectors of surveillance. Figure 10.23 shows the evolution of target numbers in a surveillance sector over a period of 60 s. This configuration may represent an environment where a large

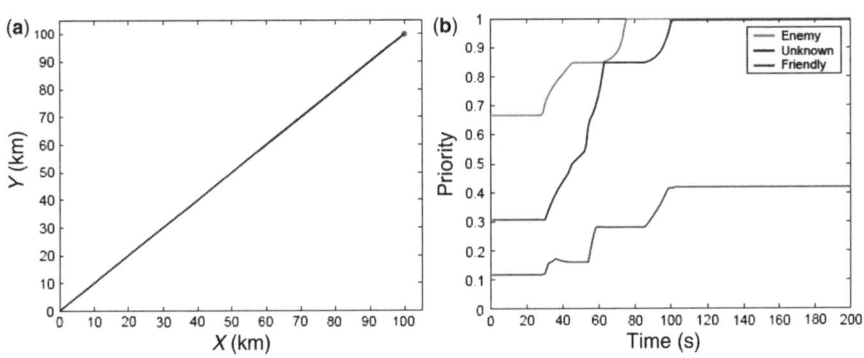

Figure 10.22 Resulting priorities for three targets with different threat levels of being hostile, moving on the same trajectory. Target velocity: 800 m/s. (See color insert.)

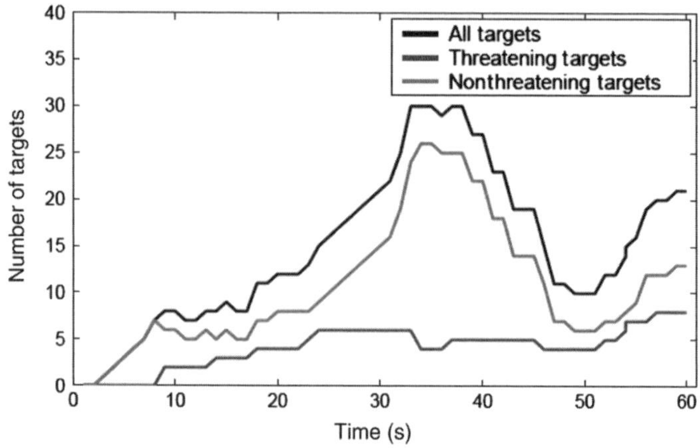

Figure 10.23 Evolution of target numbers detected in the surveillance sector.

number of small and unimportant targets are detected at long range in the same sector and a few targets are assessed as threatening. This may occur, for example, when a ship-borne multifunction radar operates in areas close to naval ports.

Figure 10.24 shows the evolution of the resulting priorities of the sector using the fuzzy approach, where "OP" stands for original priority. During the first few seconds of simulation the priority is maintained constant, being defined by the prior information about the sector. This happens because no targets detected during this period are assessed as threatening and the rate of detection of new targets is small. At around 9 s, the first targets evaluated as threatening are detected, and the resulting sector priority is increased. The comparison of the three curves shows that the lower the initial value of the variable "Original priority," the greater is the increase observed in the resulting priority. However, the absolute magnitude of the priority is determined by the original priority setting. This behavior occucs because threatening targets were not expected in the sector. This increase in the importance aims to

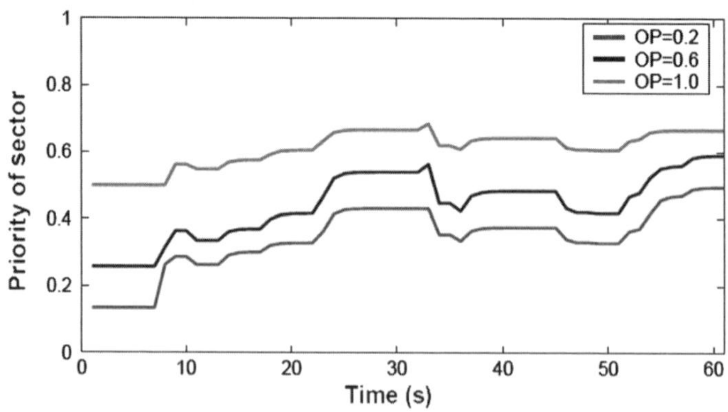

Figure 10.24 Evolution of priorities using the adaptive fuzzy method.

signal potential changes in the previous assumptions about the sector. In overload situations, such behavior may lead to the decision of preserving the surveillance performance in the sector. From 9 s, the numbers of both threatening and nonthreatening targets change over time, causing the priorities to be changed adaptively. Figure 10.24 suggests that the resulting priorities tend to follow the trends of the curve of evolution of threatening targets. This may be explained by examining the decision tree (Fig. 10.17) used for the prioritization. When a target is detected, it is assumed that during the observation period, its degree of threat is evaluated. As both the identity and the trajectory are used in this evaluation, a target does not have to be identified to be considered threatening. This explains why the total number of nonthreatening targets is not considered in assessing the priority of the sector, but the rate at which they are detected in the sector is.

Some points of interest are highlighted in the resulting priorities. Initially, after around 12 s of simulation, the priorities in all curves decrease as a result of the decrease in the number of new targets entering the sector, although the number of threatening targets remains unchanged. This is explained by the fact that the density of threatening targets is small, and hence the variation of the total number of targets has a greater effect on sector priority.

At around 32 s, an increase is observed in the sector priority when OP is either 1 or 0.6, due to the fact that the rate of new targets in the sector is 4 targets/s over the previous observation time. Finally, during the last few seconds of the simulation, regardless of the original importance of the sector, all priorities are increased as a result of the combination of positive rates of new nonthreatening and threatening targets in the sector. Once again, it is observed that these factors have greatest effects in situations where the original importance of the sector is small, signalling for changes in the original expectation in respect of threats.

The results of the situations examined here suggest that the fuzzy logic approach is an intelligent and valid means for evaluating the priority of targets and sectors. By imitating the human decision-making process, and by combining dynamic characteristics about radar tracking and military aspects, such as the ability of the weapon systems of the radar platform to face potential threats, the fuzzy approach may represent an effective and efficient and intelligent support for decisions regarding radar resource management. Both examples show a clear logical relationship between the apparent degree of threat and the ensuing priority. However, it is not clear whether this represents an optimal solution or even if it truly represents an improvement over alternative simpler schemes. Thus this approach is compared with other possible prioritizing methods in order to examine their performance in terms of the allocation of resources. This comparison is made in the next section.

10.8 COMPARISON OF THE DIFFERENT PRIORITIZATION METHODS

Three prioritization methods are compared here. The first is based on a fuzzy reasoning algorithm. The second is a fixed prioritization scheme commonly used when scheduling algorithms are examined. In this case, the original task ranking does

TABLE 10.8 Radar tasks ranking order.

Priority	Radar Task
1	Track maintenance
2	Plot confirmation
3	Track initiation
4	Track update
5	Surveillance

not evolve, even though the environment is changing. The priority order in Table 10.8 was used as a reference for the comparison. Finally, a third method called a "hard logic" approach can be described by a set of rules similar to the ones proposed for the fuzzy logic approach. However, for each operational condition, only one rule is fired to determine the priority of the radar task.

For the comparison, a region of coverage spanning from -45° to $+45^\circ$ away from the antenna's broadside was considered. This region was divided into three different sectors. In each sector, both frame times and dwell times are defined in order to determine the desired surveillance performance. Sector 1 spans from 0° to $+45^\circ$ in azimuth and from 0° to 20° in elevation. Sector 2 spans from 0° to -45° in azimuth and from 0° to 20° in elevation. These sectors require the highest search rate, as targets may be detected at short ranges in low elevations. Finally, sector 3 extends from the top of sectors 1 and 2 to an elevation of 50°, spanning from -45° to $+45^\circ$. To achieve the surveillance performance, all radar resources need to be allocated to perform this radar function. All sectors have equal initial priorities, assigned according to the expected possibility of detecting threatening targets. The tracking load is then represented by a number of hostile targets that, after being inserted in the system, are detected and tracked using Kalman filters.

An example regarding prioritization of targets is examined using one of the test trajectories presented in the previous section. The example again assumes that three targets are moving towards the radar platform on a straight-line trajectory with a velocity of 300 m/s, as shown in Fig. 10.11. Like the earlier example, the targets have different possibilities of being hostile, being considered hostile (red curves), unknown (blue curves) and friendly (green curves). This time the solid lines represent the priority evolution for the targets when the fuzzy logic approach is used; the dotted lines represent the results using the "hard" method. The main difference that can be identified is the soft transition between the different levels of priority when using the fuzzy logic approach. The "hard" method produces such transitions in steps in a way that is clearly very different.

The analysis of priority evolution when the target is considered to be hostile suggests that both fuzzy and "hard" results have similar trends in respect of priority assignment in spite of the different shapes. Similar behavior is observed when comparing the priority evolution of an unknown target. As was seen earlier, this is because when the target is far away from the radar platform, its importance is evaluated as if a friendly target. In the case of the "hard" method, in the early stages of the simulation, the unknown target is considered to be as important as the friendly target. However, as the target moves to positions close to the radar platform, its priority increases as a consequence of the target being assessed as dangerous. The priority evolution is, therefore, similar to the evolution of a hostile target; the unknown is not assessed to be as important as the hostile target because of the ranges and absolute velocities considered in this example. Finally, the results for the friendly targets show that, unlike the fuzzy approach, the "hard" method maintains the target priority constant throughout the period of the simulation. This behavior may be explained by the fact that both the range and velocity of the targets are not sufficient to provoke a change in the priority level at which the target is assessed by the "hard" method.

This comparison shows that both methods may be used in order to evaluate the relative priority of the radar targets. Depending on the situation, the priority resulting from the utilization of the fuzzy logic may be greater or not than the priority obtained by the use of the "hard" method. At first, there is a tendency to consider that if two systems execute the same set of tasks, the system that assesses these tasks with lower priority should be considered more effective, because fewer resources would be allocated to execute the tasks. However, this analysis is not always valid in radar scheduling, where the resources are demanded by the radar function in order to achieve the performance requirements. The task priority is, therefore, important for preparing the set of measurements to be executed by the radar.

In overload situations the prioritization of tasks also plays an important role. The radar resource manager will rely on the priorities to decide which functions will have their performance maintained or degraded. In this case, the smooth transitions provided by the fuzzy logic may represent an advantage in respect of resource allocation, for they prevent this allocation from varying significantly over time. The use of the fixed priority would have a very different effect in resource allocation as the targets would be tracked using a track-and-scan mode, regardless of the danger represented by the target. This method might therefore lead to unnecessary allocation or resources in the early stages of the track life of the unknown target.

Next we consider an example regarding the prioritization of sectors of surveillance. We use the same evolution of the number of targets, nonthreatening and threatening, detected in a sector of surveillance as was shown in Fig. 10.23. The resulting priorities using both "hard" and fuzzy logic approaches are presented in Figs. 10.25 and 10.26. The solid curves identify the evolution of the sector priorities using the fuzzy logic approach and the dotted lines identify the results from the "hard" prioritization. Different situations in respect of a priori knowledge about the surveillance sector are considered. The curves in red represent situations where the importance of the sector as a result of that prior knowledge is very high, meaning that the possibility of detecting threatening targets in the sector is high. The curves in blue represent situations where the radar operates in an area from which no significant knowledge is known, making the sector medium in importance.

The use of the "hard" method leads to step transitions between priority levels and shows less sensitivity to tactical variations in the environment than the use of the fuzzy logic approach. However, it is worth noting that the trends for the mean values of priorities achieved by both methods are alike in terms of adapting sector

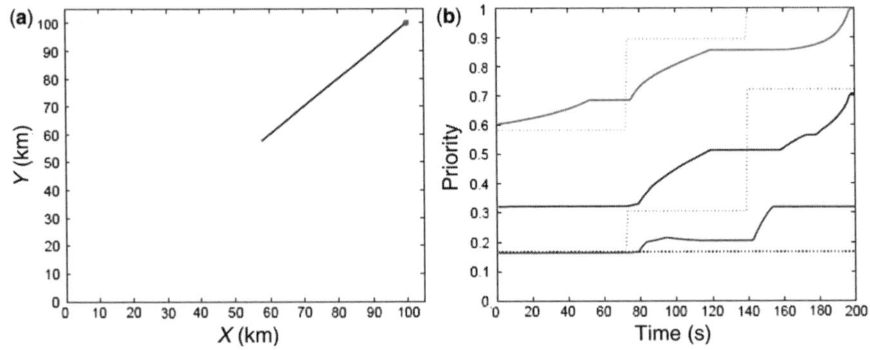

Figure 10.25 Comparison of fuzzy (solid curves) and hard logic (dotted lines) when evaluating target priorities. (See color insert.)

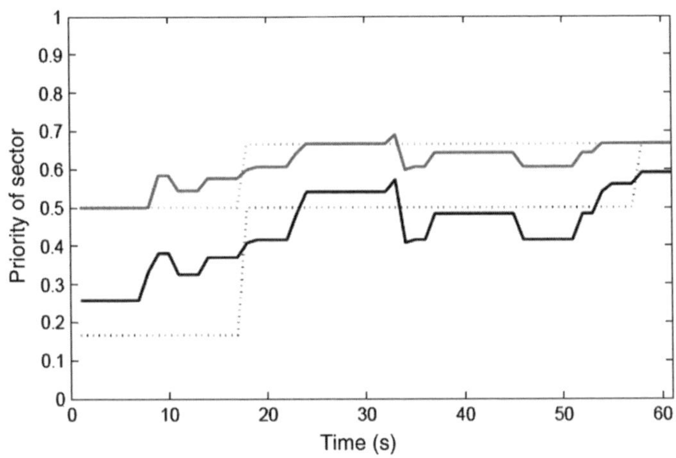

Figure 10.26 Results of the prioritization of sectors of surveillance. (Solidlines, fuzzy logic; dottedlines, hard logic.) (See color insert.)

priorities to major changes in the environment. Specifically, at around 19 s and 58 s, the rises observed in results of the "hard" method are also observed in the results of the fuzzy logic, although these changes are gradual. To assess the effects of using such prioritization methods for radar resource management, it is important to assume that, in a radar application, several surveillance sectors are likely to be used. Thus, a simple configuration may give some insights into those effects. Consider an overload situation where a number of hostile targets are detected within the range of the radar platform weapons systems and are unevenly distributed over the surveillance sectors. These targets are likely to be assessed as high priority targets. The radar resource manager will, therefore, allocate all required resources to perform accurate tracking of the targets. Thus, the remaining resources must be shared among the sectors of surveillance, affecting their surveillance performance.

Thus, step variations in sector priority as produced by the "hard" method may change significantly the surveillance performance in the sector. This might not represent a problem if surveillance is performed and, as a result of the assessment, the sector priority increases. The new priority may maintain surveillance performance to detect further tactical variations in the sector. Nevertheless, a different result would be achieved if the sector priority decreases, degrading the surveillance performance. The new level of surveillance may not be sufficient to perceive further changes in the tactical scenario, maintaining or even worsening the surveillance degradation; that is, the system may not be able to detect and track targets until they are much closer to the radar than the mission requirements for the radar.

We now consider an example of different tactical situations to further examine the resulting effects on radar resource management. The example was selected for the results of the comparison that is made in this section. The situation addressed is shown in Fig. 10.27, where seven targets are considered to be detected at different times in surveillance sectors 1 and 2. These targets move on trajectories within these sectors, as represented by the letters A to G. The red circles identify the positions where the targets are detected.

In addition, because it is assumed that little prior information is known about the area in which the radar platform is operating, the input variable "Original priority" for the adaptive prioritization systems associated with the a priori knowledge is assumed to be 0.5. The results of the radar load allocations using the prioritization methods are shown in Figs. 10.28 to 10.30. These figures also present the estimated track initiation ranges in relation to their desired values in the three sectors. Three time instants are relevant in the analysis: initially, at around 10 s, targets A and B are detected; then,

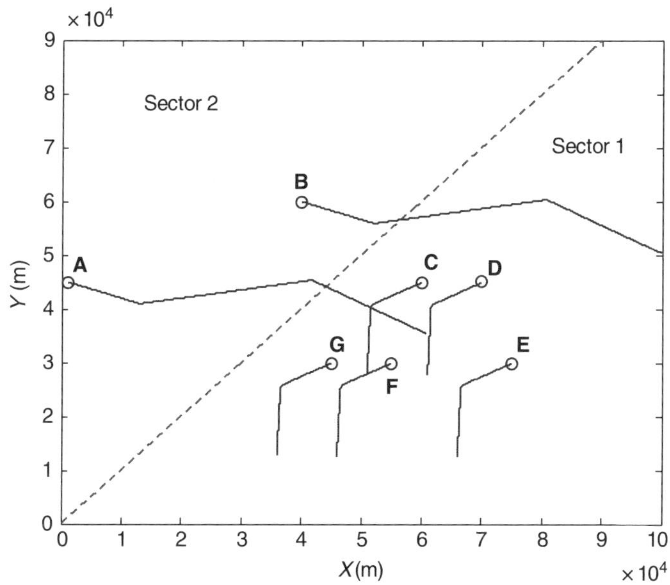

Figure 10.27 Trajectories of seven hostile targets. (See color insert.)

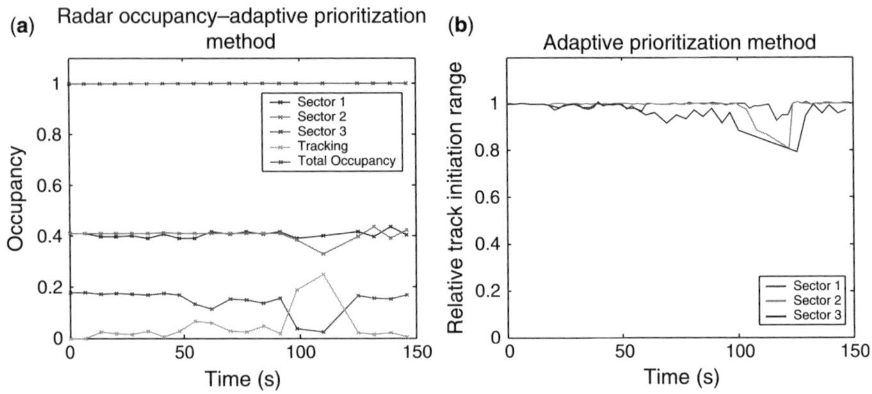

Figure 10.28 Evolution of (**a**) radar occupancy and (**b**) track initiation range using an adaptive fuzzy logic prioritization method.

from 45 s, the other targets (C to G) are detected; and, at around 100 s, all targets are maneuvering. Throughout the simulation period, the priorities of targets and sectors of surveillance are constantly reassessed.

The figures show that after the detection of targets A and B, the tracking load increases. Due to the identity and dynamics of the targets, all tracking tasks are assessed to be more important than surveillance. As a consequence, the occupancies in the surveillance sectors decrease to allow these tracking tasks to be carried out. At that time, the targets were in sector 2, making its priority higher than those of the other sectors, when using both hard and fuzzy logic methods. In this case, the outcome of the differences between sector priorities is that surveillance performance is maintained in sector 2, while it is degraded in the other sectors, as shown in Figs. 10.28 to 10.30. The use of the fixed priority, however, produces different results. From the start, the priority of the tracking tasks is assumed higher than the surveillance tasks, and the priorities in all surveillance sectors are assumed to be the

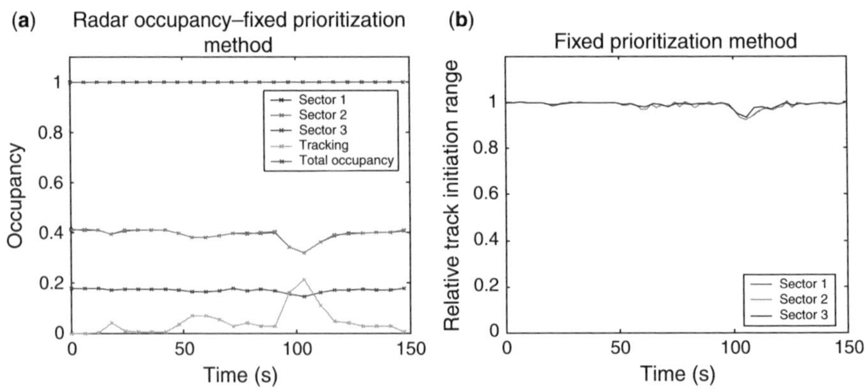

Figure 10.29 Evolution of (**a**) radar occupancy and (**b**) track initiation range using fixed prioritization method.

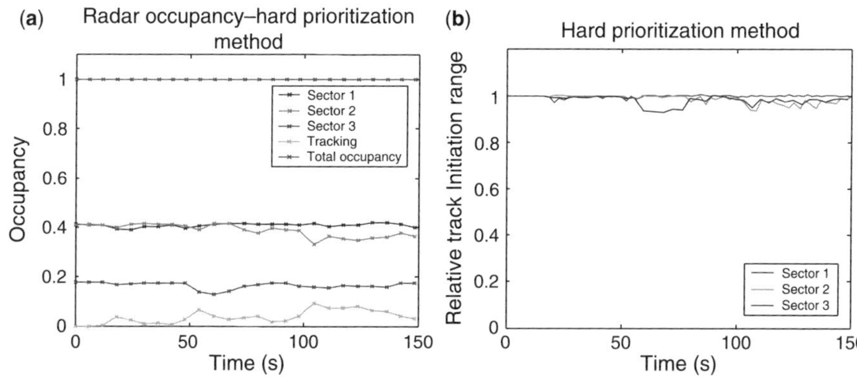

Figure 10.30 Evolution of (**a**) radar occupancy and (**b**) track initiation range using an adaptive "hard" logic prioritization method.

same. Thus, the resource allocation is predictable, regardless of the actual tactical situation of the environment. As the targets are detected, the surveillance is degraded in all surveillance sectors in proportion to their required occupancies. Because the priorities are fixed, resource allocation is not adapted to the new situation.

The detection of targets C to G, after 45 s of simulation, signals changes in the tactical situation in respect of the disposition of threatening targets over the surveillance volume. These targets are assessed as threatening by both the "hard" and fuzzy logic methods, making the priority of sector 1 rise. In this situation, the behavior of these priority assignment methods is alike, and the priority of sector 1 is assessed as having the highest priority among the surveillance sectors. The second highest is sector 2, where target A is still located. This explains the results of Figs. 10.28 and 10.30 from around 60 s to 100 s, when the occupancies in sectors 1 and 2 are maintained while the less important surveillance area in sector 3 has its performance degraded.

At around 100 s, when all targets are maneuvering, the tracking load increases rapidly. At that time, most of the targets are in sector 1, which, therefore, still has the highest priority for surveillance. Figure 10.28 a shows the resulting allocation using the adaptive fuzzy logic method, where the performance is maintained in sector 1, while being degraded in the other sectors. The hard prioritization method provides similar performance to that of the fuzzy logic method during the early stages of the simulation (Fig. 10.30a). However, when all targets are maneuvering, the rankings of targets A and B are reassessed using the "hard" method, being considered less important than surveillance tasks in sector 1. As a result, the system started tracking them using track-while-scan mode. This effect is not observed when the fuzzy logic method is applied, although the priorities of targets A and B are also reassessed. This is the reason the tracking load does not increase at the same rate as in Figs. 10.28a and 10.30a.

By using the fixed priority method, the radar system degraded surveillance performance proportionally to the initial occupancy in the three sectors (Fig. 10.29a). The evolution of the tactical situation is not considered, making the system behave in a similar way to the behavior observed in the first analysis in early stages of the simulation. Because the targets are mostly assessed as high priority targets throughout the period of the simulation, the required track accuracy is achieved in most cases. The only exception occurs when targets A and B are tracked in track-while-scan mode when the "hard" prioritization method is used. This is explained by the fact that targets with possibility of detection less that one are not detected at every scan, causing a reduced tracking performance mainly for low SNR targets.

A final analysis addresses the track initiation range in the sectors of surveillance. Given a track initiation range, the effect of increasing the search update is less restrictive than the effect of an equivalent reduction of radar power (Figs. 10.28 and 10.29). The increase of the search revisit interval is the method used in this paper to adapt this function to the available radar resources. Thus, using the radar equation, the impact of having this increase in the search revisit interval may be given by [11]

$$R_s = R_{s0}\sqrt[4]{\frac{T_s}{T_{s0}}} \tag{10.1}$$

where T_{s0} is the revisit time interval for which the track initiation range, R_{s0}, of the performance requirement is defined; T_s is the revisit time interval used in the surveillance sector; and R_s is the achieved track initiation range in the surveillance sector.

The results for the evolution of the relative tracking initiation range in all sectors of surveillance are presented in Figs. 10.28b to 10.30b. This relative track initiation is given by R_s/R_{s0}. The examination of these results corresponds to the analysis obtained from the evolution of radar occupancies in all sectors. During the first few seconds of simulation, as no targets are detected, the requirements in respect of track initiation range are achieved in all sectors. However, this performance is expected to change as a result of the changing tactical scenario and the way in which priorities are assessed to the sectors of surveillance. At around 10 s, when targets A and B are detected, the priority of sector 2 rises. It is important to note that the original surveillance requirements are very demanding, requesting 100% of the available radar resources just for surveillance. The detection of targets in any of the sectors may lead to degradation in the surveillance performance if targets are assessed as higher priority than the surveillance tasks. In the example examined here, this is exactly the case. Thus, the surveillance occupancy is maintained in sector 2, the highest priority surveillance sector, while being reduced in the other sectors. This decrease means that the revisit time intervals for surveillance in sectors 1 and 3 are increased, causing the reduction of the achieved track initiation range in these sectors, while maintaining it in sector 2. These results are repeated until the detection of targets C to G, after 45 s. Between 10 s and 45 s, the achieved track initiation ranges in sectors fluctuate as a consequence of the remaining resources after tracking allocation. After 45 s, the priorities of the surveillance sectors change as explained already. Because sector 3 has the lowest priority, it is the sector in which track initiation range is reduced from that time. It is only after 100 s, when all targets are maneuvering, that the behavior changes. First, the results of the system using the fuzzy logic prioritization method are examined. As a consequence of the high concentration of track update tasks over that period, the revisit times are increased and the track initiation ranges in sectors 1 and 2 are also degraded. After the maneuvers are finished, the resource allocation returns to the same levels prior to the maneuvers and the track initiation ranges are improved. When using the "hard" method, this behavior is not observed. This is explained by the fact that two targets are tracked using the track-while-scan mode, as a consequence of priority reassessment, reducing the effect of the increased tracking load over the maneuver period. The use of the fixed priority method produces different results in respect of track initiation range. Because all sectors are assumed to have the same priority, during increases in the tracking load, the track initiation ranges degrade in a similar way. At first, this analysis may lead one to conclude that this result is better. However, in situations where the tracking load is high, this behavior may lead to not tracking dangerous targets that are close to the required track initiation range, while avoiding degrading this performance measure in less important surveillance sectors where no targets are detected.

10.9 PRIORITIZATION ISSUES

By intelligently assessing the priorities of targets and sectors of surveillance according to a set of rules that imitates the human decision-making process in a similar tactical situation, it is expected that the resource manager can distribute the radar resources in a more effective way. The results suggest that the fuzzy approach is a valid means of evaluating the relative importance of the radar tasks; the resulting priorities have been adapted by the fuzzy logic prioritization method, according to how the radar system perceived the surrounding environment. The analysis of these fuzzy logic methods has been further assessed by comparing them with two other prioritization methods. The first is based on a fixed prioritization order and the second on a prioritization method that used the same inference rules developed for the fuzzy logic approach. In this method, however, only one rule is fired at a time. These methods are less computationally demanding than the fuzzy logic approach. For example, look-up tables may be used to evaluate the priority if the "hard" method is used. Despite using few computational resources, the hard prioritization method has some drawbacks.

From the test configurations examined, it is seen that the "hard" prioritization method leads in general to resource allocations that are similar to the ones provided by the fuzzy logic approaches. However, the nonsmooth transition characteristics associated with the "hard" prioritization may lead to very dissimilar and, sometimes, undesirable results. For example, the desired tracking accuracies for the two targets tracked using track-while-scan were not achieved. Assuming the hostile targets were close to the range of the weapons systems of the radar platform, it would be important to maintain accurate tracking of these targets. To avoid such an undesirable outcome, the radar operator would have to intervene in the prioritization to guarantee independent tracking for the targets in a real application.

Both fuzzy and fixed prioritization approaches have provided consistent results in low tracking load conditions, but the fuzzy approach may be more effective in high tracking load situations. Preserving the surveillance performance in high priority sectors may lead to detecting future changes in the tactical situation, such as target splitting. Such behavior may not be observed when using the fixed prioritization order. In this case, it is expected that surveillance is gradually degraded in all sectors in proportion to the remaining radar resources.

Overall, we conclude that the fuzzy logic approach has both intuitive appeal and a number of potential operational advantages, particularly under stressing conditions, which make it worthy of serious consideration in the design and development of radar resource managers.

10.10 SUMMARY AND CONCLUSIONS

Intelligent radar resource management is a necessary prerequisite if electronically scanned systems are ever to maximize their performance potential. Although we have considered two important and related subfunctions in scheduling and prioritization, they are not the sole components of an intelligent resource manager. The key underlying requirement for effective operation is for the radar system to be able to automatically and reliably interpret the scene being surveyed so that it is able to make intelligent decisions as to how best deploy resources to maximize performance and mission effectiveness. A simple example was introduced with the fuzzy logic priority assignment. To be at its most effective, it requires intelligent pre-attribution of the rules sets coupled with an ability to reliably interpret a detected target's degree of threat so that the correct set of radar parameters may be applied. Although very good progress is being made towards this end, there remains much research to be done, especially in the area of echo data interpretation.

REFERENCES

[1] S. Musick, R. Malhotra. Chasing the elusive sensor manager. *Proc. National Aerospace and Electronics Conference* 1994;1:606–613.

[2] E.R. Billam. Parameter optimisation in phased array. *Proc. RADAR 2002 Conference* 2002;490:34–37. 15–17 October 2002.

[3] G. Van Keuk, S.S. Blackman. On phased-array radar tracking and parameter control. *IEEE Trans. Aerospace and Electronic Systems* 1993;29(1):186–194.

[4] E.R. Billam. The problem of time in phased array radar. *Proc. RADAR'97 Conference* 1997;449:563–567. 14–16 October 1997.

[5] S.L. Coetzee, K. Woodbridge, C.J. Baker. Multifunction radar resource management using tracking optimisation. *Proc. International Conference RADAR* 2003, Adelaide, Australia, 3–5 September 2003. p. 578–583.

[6] W.K. Stafford. Real-time control of a Multifunction Electronically Scanned Adaptive Radar (MESAR). *IEE Colloquium on Real Time Management of Adaptive Radar Systems* 1990. p. 7/1–7/5.

[7] S. Sabatini, M. Tarantino. *Multifunction Array Radar — System Design and Analysis*. Artech House; Boston, Massachusetts, USA, 1994.

[8] A. Izquierdo-Fuente, J.R. Casar-Corredera. Optimal radar pulse scheduling using a neural network. *Proc. IEEE International Conference on Neural Networks* 1994;7;4558–4591.

[9] D. Strömberg, P. Grahn. Scheduling of tasks in phased array radar. *Proc. IEEE International Symposium on Phased Array Systems and Technology*, Boston, USA, 1996. p. 318–321.

[10] A.J. Orman, C.N. Potts, A.K. Shahani, A.R. Moore. Scheduling for a multifunction array radar system. *European Journal of Operational Research* 1996;90:13–25.

[11] J.M. Butler. Multi-function radar tracking and control. PhD thesis, UCL, University of London, 1998.

[12] M.T. Vine. Fuzzy logic in radar resource management. *Proc. IEE conference on Multifunction Radar and Sonar Sensor Management Techniques* 2001;173:5/1–4.

[13] S.L.C. Miranda, C.J. Baker, K. Woodbridge, H.D. Griffiths. Phased array radar resource management: a comparison of scheduling algorithms. *Proc. IEEE Radar Conference 2004*, Philadelphia, USA, 26–29 April 2004. p. 79–84.

[14] A. Izquierdo-Fuente, J.R. Casar-Corredera. Optimal radar pulse scheduling using a neural network. *IEEE International Conference on Neural Networks*, Orlando, USA, 1994.

[15] A.J. Orman, A.K. Shahani, A.R. Moore. Modelling for the control of a complex radar system. *Computers and Operations Research* 1998;25(3):239–249.

[16] A.P. Stoffel. Heuristic energy management for active array multifunction radars. *1994 IEEE National Telesystems Conference*, San Diego, USA, 1994.

[17] W. Komorniczak, T. Kuczerski, J.F. Pietrasinski. The priority assignment for detected targets in Multi-function Radar. *13th International Conference on Microwaves, Radar and Wireless Communications*, Poland, 2000.

[18] W. Komorniczak, J. Pietrasinski, B. Solaiman. The data fusion approach to the priority assignment in the Multifunction Radar. *14th International Conference on Microwaves, Radar and Wireless Communications*, Poland, 2002.

[19] J.M. Molina Lopez, F.J. Jimenez Rodriguez, J.R. Casar Corredera. Fuzzy management of surveillance sensors. *37th IEEE Conference on Decision and Control*, Tampa, USA, 1998.

[20] M.T Vine. Fuzzy logic in radar resource management. *IEE Colloquium on Multifunction Radar and Sonar Sensor Management Techniques* 2001, ref. no. 2001/173.

[21] S.L.C. Miranda, C.J. Baker, K. Woodbridge, H.D. Griffiths. Phased array radar resource management: a fuzzy logic approach for prioritisation of radar tasks and sectors of surveillance. *Proc. International Conference on Radar Systems*, Toulouse, France, 2004.

[22] G. Bojadziev, M. Bojadziev. *Fuzzy Sets, Fuzzy Logic, Applications*. World Scientific Publishing Co. Pte. Ltd; River Edge, NJ, USA, 1995.

INDEX

Knowledge Based Radar Detection, Tracking and Classification. Edited by Fulvio Gini and Muralidhar Rangaswamy